Lecture Notes in Mathematics

C.I.M.E. Foundation Subseries

Volume 2296

More information about this series at http://www.springer.com/series/3114

Fondazione C.I.M.E., Firenze

C.I.M.E. stands for *Centro Internazionale Matematico Estivo*, that is, International Mathematical Summer Centre. Conceived in the early fifties, it was born in 1954 in Florence, Italy, and welcomed by the world mathematical community: it continues successfully, year for year, to this day.

Many mathematicians from all over the world have been involved in a way or another in C.I.M.E.'s activities over the years. The main purpose and mode of functioning of the Centre may be summarised as follows: every year, during the summer, sessions on different themes from pure and applied mathematics are offered by application to mathematicians from all countries. A Session is generally based on three or four main courses given by specialists of international renown, plus a certain number of seminars, and is held in an attractive rural location in Italy.

The aim of a C.I.M.E. session is to bring to the attention of younger researchers the origins, development, and perspectives of some very active branch of mathematical research. The topics of the courses are generally of international resonance. The full immersion atmosphere of the courses and the daily exchange among participants are thus an initiation to international collaboration in mathematical research.

C.I.M.E. Director (2002 – 2014)
Pietro Zecca
Dipartimento di Energetica "S. Stecco"
Università di Firenze
Via S. Marta, 3
50139 Florence
Italy
e-mail: zecca@unifi.it

C.I.M.E. Director (2015 –)
Elvira Mascolo
Dipartimento di Matematica "U. Dini"
Università di Firenze
viale G.B. Morgagni 67/A
50134 Florence
Italy
e-mail: mascolo@math.unifi.it

C.I.M.E. Secretary
Paolo Salani
Dipartimento di Matematica "U. Dini"
Università di Firenze
viale G.B. Morgagni 67/A
50134 Florence
Italy
e-mail: salani@math.unifi.it

CIME activity is carried out with the collaboration and financial support of INdAM (Istituto Nazionale di Alta Matematica)

For more information see CIME's homepage: **http://www.cime.unifi.it**

Fabrice Baudoin • Séverine Rigot •
Giuseppe Savaré • Nageswari Shanmugalingam

New Trends on Analysis and Geometry in Metric Spaces

Levico Terme, Italy 2017

Luigi Ambrosio • Bruno Franchi • Irina Markina •
Francesco Serra Cassano
Editors

In collaboration with

CIRM
CENTRO INTERNAZIONALE
PER LA RICERCA MATEMATICA

Springer

FONDAZIONE
CIME
ROBERTO CONTI
CENTRO INTERNAZIONALE MATEMATICO ESTIVO
INTERNATIONAL MATHEMATICAL SUMMER CENTER

Authors

Fabrice Baudoin
Department of Mathematics
University of Connecticut
Storrs, Connecticut, USA

Séverine Rigot
Laboratoire de Mathématiques
J.A. Dieudonné
Université Côte d'Azur
Nice, France

Giuseppe Savaré
Department of Decision Sciences
Bocconi University
Milan, Italy

Nageswari Shanmugalingam
Department of Mathematical Sciences
University of Cincinnati
Cincinnati, Ohio, USA

Editors

Luigi Ambrosio
Scuola Normale Superiore
Pisa, Italy

Bruno Franchi
Department of Mathematics
University of Bologna
Bologna, Italy

Irina Markina
Department of Mathematics
University of Bergen
Bergen, Norway

Francesco Serra Cassano
Department of Mathematics
University of Trento
Trento, Italy

ISSN 0075-8434 ISSN 1617-9692 (electronic)
Lecture Notes in Mathematics
C.I.M.E. Foundation Subseries
ISBN 978-3-030-84140-9 ISBN 978-3-030-84141-6 (eBook)
https://doi.org/10.1007/978-3-030-84141-6

Mathematics Subject Classification: M26016, M12120, M12066, M21022, M11132

This Springer imprint is published by the registered company Springer Nature Switzerland AG.
The registered company address is: Gewerbestrasse 11, 6330 Cham, Switzerland

Contents

Introduction to the Notes of the School on Analysis and Geometry in Metric Spaces

Luigi Ambrosio, Bruno Franchi, Irina Markina, and Francesco Serra Cassano

The *Schools on Analysis and Geometry in Metric Spaces* have been meant to present different approaches to research topics in Geometric Measure Theory, Calculus of Variations, Partial Differential Equations, Differential Geometry, Control Theory with the common background of a relevant underlying geometric structure, usually non Riemannian.

Each School is one week long and provides four or five short courses taught by internationally known scholars. The courses are addressed to a young audience: young researchers, post-doc and PhD students from Italian and foreign Universities. Because one of the aims of the Schools is to establish valuable relationships between young researchers and internationally reputed scholars, fellowships for young researchers are always provided.

Since their beginning, all the Schools have been financed by

- CIRM (Centro Internazionale di Ricerca Matematica) of FBK (Fondazione Bruno Kessler—Trento);
- Dipartimento di Matematica of Università degli Studi di Trento.

L. Ambrosio
Scuola Normale Superiore, Pisa, Italy
e-mail: luigi.ambrosio@sns.it

B. Franchi
Department of Mathematics, University of Bologna, Bologna, Italy
e-mail: bruno.franchi@unibo.it

I. Markina
Department of Mathematics, University of Bergen, Bergen, Norway
e-mail: irina.markina@uib.no

F. Serra Cassano (✉)
Department of Mathematics, University of Trento, Trento, Italy
e-mail: francesco.serracassano@unitn.it

L. Ambrosio et al. (eds.), *New Trends on Analysis and Geometry in Metric Spaces*, C.I.M.E. Foundation Subseries 2296, https://doi.org/10.1007/978-3-030-84141-6_1

The Schools have been also financially supported, in different moments of their existence, by

- Analysis and Application, a Scandinavian nettwork, a project financed by NordForsk;
- GALA—Geometric Analysis in Lie groups and Applications, a project of the Sixth Framework Programme of the European Union;
- MANET—Metric ANalysis for Emergent Technologies, a Marie-Curie starting project. Funding Scheme: FP7-People-2013-ITN;
- I.N.D.A.M.—Istituto Nazionale di Alta Matematica—MIUR;
- SMI—Scuola Matematica Interuniversitaria—MIUR, CNR;

The organizers thank all the sponsors that, in different years, contributed to the Schools. Their help was particularly invaluable to provide fellowships for young researchers.

The Schools take place every second year, in late spring. Usually they are in Levico Terme (Trento) after the first School held in Povo (Trento) in May 1999.

All Short Courses Given at the Schools on Analysis and Geometry in Metric Spaces

(1) Andrei A. AGRACHEV (SISSA Trieste): *Abnormal Geodesics and Related Topics*. [4th School]
(2) Luigi AMBROSIO (SNS Pisa): *Geometric Measure Theory in Metric Spaces: Sets of Finite Perimeter and Currents in non Euclidean Spaces*. [2nd School]
(3) Zoltán BALOGH (Universität Bern): *Definition and Regularity of Quasiconformal Mappings in Metric Spaces*. [4th School]
(4) Fabrice BAUDOIN (University of Connecticut): *Geometric Inequalities in sub-Riemannian Manifold*. [10th School]
(5) Mario BONK (University of Michigan): *Quasiconformal Geometry of Fractals*. [6th School]
(6) Ugo BOSCAIN (CMAP - Ecole Polytechnique): *Introduction to Geodesics in sub-Riemannian Geometry*. [7th School]
(7) Sagun CHANILLO (Rutgers University): *CR Geometry in 3 Dimensions*. [11th School]
(8) Giovanna CITTI (Università di Bologna): *Riemannian Approximation of sub-Riemannian Structures*. [9th School]
(9) Thierry COULHON (Université de Cergy-Pontoise): *Analysis on Graphs*. [1st School]
(10) Michael COWLING (University of New South Wales): *1-Quasiconformal Maps on Carnot Groups*. [5th School]
(11) Guy DAVID (Université de Paris Sud (Orsay)): *Uniform Rectifiability for Sets with Nontrivial Topology, and Almgren Restricted (quasiminimal) Sets*. [1st School]
(12) Katrin FÄSSLER (University of Jyväskylä): *Quantitative Rectifiability in Heisenberg Groups*. [11th School]

(13) Nicola GAROFALO (Purdue University and Università di Padova): *Generalized Curvature-Dimension Inequalities, Li-Yau Inequalities and their Consequences*. [7th School]
(14) Piotr HAJŁASZ (Pittsburgh University): *Geometric Properties of the Heisenberg Groups*. [8th School]
(15) Juha HEINONEN (University of Michigan (Ann Arbor)): *Whitney Flat Forms and Applications*. [5th School]
(16) Ilkka HOLOPAINEN (University of Helsinki): *Minimal Submanifolds*. [9th School]
(17) Tadeusz IWANIEC (Syracuse University and University of Helsinki): *An Invitation to Sobolev Mappings, Extremal Harmonic Deformations, Minimal Surfaces and Hyperelasticity of Thin Plates*. [7th School]
(18) Bruce KLEINER (University of Michigan (Ann Arbor)): *Quasiconformal Geometry of Metric Spaces: Uniformization and Rigidity*. [4th School]
(19) Pekka KOSKELA (University of Jyväskylä): *Upper Gradients and Poincarè Inequalities*. [1st School]
(20) Enrico LE DONNE (University of Jyväskylä): *Carnot Groups*. [9th School]
(21) Ermanno LANCONELLI (Università di Bologna): *Potential Theory and Carnot Groups*. [3rd School]
(22) Valentino MAGNANI (Università di Pisa): *On the Area of Submanifolds in Homogeneous Groups*. [11th School]
(23) Jan MALY (Univerzita Karlova, Prague): *Nonabsolutely Convergent Integrals in Metric Spaces*. [7th School]
(24) Juan MANFREDI (University of Pittsburgh): *Nonlinear Subelliptic Equations on Carnot Groups*. [3rd School]
(25) Pertti MATTILA (University of Jyväskylä): *Rectifiability, Menger Curvature and Singular Integrals*. [3rd School]
(26) Andrea MONDINO (University of Oxford): *Metric Measure Spaces Satisfying Ricci Curvature Lower Bounds*. [11th School]
(27) Roberto MONTI (Università di Padova): *The Regularity Problem for Carnot-Carathodory Geodesics*. [8th School]
(28) Pierre PANSU (Université Paris-Sud): *Submanifolds and Differential Forms in Carnot Spaces*. [4th School]
(29) Hans-Martin REIMANN (Universität Bern): *Quasiconformal Mappings on Heisenberg Groups and Related Topics*. [2nd School]
(30) Fulvio RICCI (SNS Pisa): *Functional Calculus for Sublaplacians on Nilpotent Lie Groups and Singular Integrals*. [6th School]
(31) Séverine RIGOT (Université Côte d'Azure, Nice): *Differentiation of Measures in Metric Spaces*. [10th School]
(32) Manuel RITORÉ (Universidad de Granada): *Area-Stationary Surfaces and Isoperimetric Regions in the Heisenberg Group*. [5th School]
(33) Giuseppe SAVARÈ (Università di Pavia): *Sobolev Spaces, Optimal Transport and Heat Flow in Metric Measure Spaces*. [10th School]

(34) Stephen SEMMES (Rice University): *Lectures on Fractal Geometry, Decent Calculus and Structure among Geometry*. [1st School]
(35) Nageswari SHANMUGALINGAM (University of Cincinnati): *First Order Analysis in Metric Measure Spaces Using Upper Gradients: Functions of Bounded Variation*. [10th School]
(36) Karl-Theodor STURM (Universität Bonn): *Optimal Transports, Gradients Flows and Wasserstein Diffusion*. [6th School]
(37) Hector J. SUSSMANN (Rutgers University): *Geometric Aspects of the Theory of Necessary Conditions for Optimal Control*. [3rd School]
(38) Anton THALMAIER (Université du Luxembourg): *Subelliptic Brownian Motion and Curvature in sub-Riemannian Geometry*. [11th School]
(39) Tatiana TORO (University of Washington): *Structure of Measures*. [6th School]
(40) Emmanuel TRÉLAT (Laboratoire Jacques-Louis Lions. Sorbonne. Paris VI): *Sub-Riemannian Geometry: Singular Curves, Applications (Motion Planning, Shape Analysis, Semi-Classical Analysis)*. [8th School]
(41) Marc TROYANOV (EPFL Lausanne): *Axiomatic Sobolev Spaces on Measure Metric Spaces*. [2nd School]
(42) Jeremy TYSON (University of Illinois at Urbana-Champaign): *Distortion of Dimension by Sobolev and Quasiconformal Mappings*. [8th School]
(43) Alexander VASILIEV (University of Bergen): *Quadratic Differentials on Riemannian Surfaces*. [9th School]
(44) Stefan WENGER (Université de Fribourg): *Area Minimizing Discs in Metric Spaces and Geometric Applications*. [9th School]
(45) Richard WHEEDEN (Rutgers University): *Singular Integrals on Metric Spaces*. [1st School]

In particular we want to remember the speakers and friends that passed away

- Juha HEINONEN (1960–2007)
- Alexander VASILIEV (1962–2016)
- Richard WHEEDEN (1940–2020).

The program of extending geometric and analytic results, classical in the setting of Euclidean spaces or Riemannian manifolds, to sub-Riemannian manifolds, nilpotent groups or, even more generally, to metric measure spaces started at least 50 years ago. Clearly such a program is wide and it is articulated in very different streams. The subjects of the courses of the School, over more than 20 years, provide a picture, albeit an incomplete one, of the different research streams.

The **10th School** held in 2017 is an example of this multifaceted research. It has been articulated in four courses:

- Fabrice BAUDOIN (University of Connecticut): *Geometric Inequalities in sub-Riemannian Manifold*.

  The speaker presents some topics of the theory of diffusion operators and applications of heat semigroups methods in Riemannian geometry. In particular he shows how Ricci lower bounds and Bochner's formula lead to the notion of curvature dimension inequality for the Laplace-Beltrami operator and how this inequality only can be used to prove geometric and functional inequalities such as Li-Yau, Sobolev or isoperimetric inequalities. Generalizations to sub-Riemannian geometry are given at the end.

The subject is close, among others, to the courses by Karl-Theodor Sturm (6th School), Nicola Garofalo (7th School), Sagun Chanillo (11th School) and Anton Thalmaier (11th School).

- Séverine RIGOT (Université Côte d'Azure, Nice): *Differentiation of Measures in Metric Spaces.*

  The aim of this course is to recall classical results about differentiation of measures in the metric setting together with the covering properties on which they are based. The course focus on one of these covering properties, the so called weak Besicovitch covering property. This property plays a central role in the characterization of (complete separable) metric spaces where the differentiation theorem holds for every (locally finite Borel regular) measure.

  The subject is close, among others, to the courses by Luigi Ambrosio (2nd School), Zoltan Balogh (4th School) and Tatiana Toro (6th School).

- Giuseppe SAVARÈ (Università di Pavia): *Sobolev Spaces, Optimal Transport and Heat Flow in Metric Measure Spaces.*

  The notes are divided in four main parts. The *first one* is devoted to a preliminary study of the topological, metric, and measure-theoretic aspects of the underlying metric measure space. The *second part* is devoted to the construction of the Cheeger energy for Lipschitz functions in metric measure spaces. The *third part* deals with the dual characterization of Sobolev spaces. The *final part* is devoted to the dual/weak formulation of the Sobolev spaces $W^{1,p}(X)$.

  The subject is close, among others, to the courses by Pekka Koskela (1st School), Hans-Martin Reimann (2nd School), Marc Troyanov (2nd School), Zoltan Balogh (4th School) and Jeremy Tyson (8th School).

- Nageswari SHANMUGALINGAM (University of Cincinnati): *First Order Analysis in Metric Measure Spaces Using Upper Gradients: Functions of Bounded Variation.*

  The speaker gives a description of the theory of functions of bounded variation (BV) in the nonsmooth setting of metric measure spaces. The construction presented here is based on the one first proposed by Michele Miranda Jr. In particular the course focus on geometric properties of BV functions when the measure on the underlying metric measure space is doubling and supports a 1-Poincaré inequality with respect to function-upper gradient pairs.

  The subject is close, among others, to the courses by Pekka Koskela (1st School), Luigi Ambrosio (2nd School) and Nicola Garofalo (7th School).

Finally, the Scientific Committee

- Luigi AMBROSIO
- Bruno FRANCHI
- Irina MARKINA

- Raul Paolo SERAPIONI
- Francesco SERRA CASSANO

is pleased to remember that the organization of each School has always been strongly dependent on the work, the experience and the patience of Mr. Augusto MICHELETTI (of CIRM).

Geometric Inequalities on Riemannian and Sub-Riemannian Manifolds by Heat Semigroups Techniques

Fabrice Baudoin

Abstract In those lecture notes, we review some of the theory of diffusion operators and applications of heat semigroups methods in Riemannian geometry. In particular we will show how Ricci lower bounds and Bochner's formula lead to the notion of curvature dimension-equality for the Laplace-Beltrami operator and how this inequality only can be used to prove geometric and functional inequalities such as Li-Yau, Sobolev or isoperimetric inequalities. Some generalizations to sub-Riemannian geometry are given at the end.

1 Introduction

Those lecture notes correspond to a series of lectures given at the summer school: New Trends on Analysis and Geometry in Metric Spaces (10th School), June 26–30, 2017, Levico. The main goal of the lectures were to show how the theory of diffusion operators and associated semigroups could be used to study geometric and functional inequalities in the setting of Riemannian or even sub-Riemannian manifolds. While literature about the functional inequalities that can be studied using diffusion methods is extensive, see [2] and the references therein, it appeared useful to focus on concrete settings like Riemannian or sub-Riemannian manifolds to show the scope and applications of the theory. Essentially, our objective is to investigate how far one can go in those settings using Bochner's formula or its generalization and heat kernel techniques while using as little geometry of the ambient space as possible. The lecture notes are structured as follows:

Section 2 is introductory. It presents the basic definitions and results concerning the theory of subelliptic diffusion operators and their associated heat semigroups

Research was supported in part by NSF Grant DMS 1660031.

F. Baudoin (✉)
Department of Mathematics, University of Connecticut, Storrs, CT, USA
e-mail: fabrice.baudoin@uconn.edu

L. Ambrosio et al. (eds.), *New Trends on Analysis and Geometry in Metric Spaces*, C.I.M.E. Foundation Subseries 2296, https://doi.org/10.1007/978-3-030-84141-6_2

and heat kernels. Section 3 presents some applications of this theory to Riemannian manifolds. The Riemannian heat kernel is the main character of the section and we will show how it can used to prove geometric or functional inequalities such as volume doubling, Poincaré inequality, Sobolev inequality and isoperimetric inequality. Section 4 presents recent developments and applications of this theory in the context of sub-Riemannian manifolds, mostly after [5, 6].

2 Subelliptic Diffusion Operators

2.1 Diffusion Operators

Definition 2.1 A differential operator L on $\mathbb{R}^n$, is called a diffusion operator if it can be written

$$L = \sum_{i,j=1}^{n} \sigma_{ij}(x) \frac{\partial^2}{\partial x_i \partial x_j} + \sum_{i=1}^{n} b_i(x) \frac{\partial}{\partial x_i},$$

where b_i and σ_{ij} are continuous functions on $\mathbb{R}^n$, and if for every $x \in \mathbb{R}^n$ the matrix $\sigma(x) \overset{def}{=} (\sigma_{ij}(x))_{1 \le i,j \le n}$ is symmetric and nonnegative.

If for every $x \in \mathbb{R}^n$ the matrix $(\sigma_{ij}(x))_{1 \le i,j \le n}$ is positive definite, then the operator L is said to be elliptic. The canonical example of an elliptic diffusion operator is the Laplace operator in $\mathbb{R}^n$:

$$\Delta = \sum_{i=1}^{n} \frac{\partial^2}{\partial x_i^2}.$$

However, Definition 2.1 includes also non-elliptic operators such as for instance the heat operator

$$H = \Delta - \frac{\partial}{\partial t},$$

or very degenerate operators such as the so-called Kolmogorov operator in $\mathbb{R}^3$,

$$\mathcal{K} = \frac{\partial^2}{\partial x^2} + x \frac{\partial}{\partial y} - \frac{\partial}{\partial t}.$$

One of the most basic properties of diffusion operators is that they satisfy a maximum principle. Before we state it let us recall a simple result from linear algebra.

Lemma 2.2 *Let A and B be two symmetric and nonnegative matrices, then*

$$\operatorname{tr}(AB) \geq 0.$$

Proof Since A is symmetric and non negative, there exists a symmetric and nonnegative matrix S such that $S^2 = A$. We have then

$$\operatorname{tr}(AB) = \operatorname{tr}(S^2 B) = \operatorname{tr}(SBS) = \operatorname{tr}({}^t SBS).$$

The matrix ${}^t SBS$ is seen to be symmetric and nonnegative and therefore $\operatorname{tr}({}^t SBS) \geq 0$. □

Proposition 2.3 (Maximum Principle for Diffusion Operators) *Let $f : \mathbb{R}^n \to \mathbb{R}$ be a C^2 function that attains a local minimum at x. If L is a diffusion operator, then $Lf(x) \geq 0$.*

Proof Let

$$L = \sum_{i,j=1}^{n} \sigma_{ij}(x) \frac{\partial^2}{\partial x_i \partial x_j} + \sum_{i=1}^{n} b_i(x) \frac{\partial}{\partial x_i},$$

and let $f : \mathbb{R}^n \to \mathbb{R}$ be a C^2 function that attains a local minimum at x. We have

$$\begin{aligned} Lf(x) &= \sum_{i,j=1}^{n} \sigma_{ij}(x) \frac{\partial^2 f}{\partial x_i \partial x_j}(x) \\ &= \operatorname{tr}\left(\sigma(x) \operatorname{Hess} f(x)\right), \end{aligned}$$

where $\operatorname{Hess} f(x) = (\frac{\partial^2 f}{\partial x_i \partial x_j}(x))$ is the Hessian matrix of f. Since, by the assumption, we have $\operatorname{Hess} f(x) \geq 0$, the desired conclusion immediately follows from Lemma 2.2. □

Combined with the linearity, Proposition 2.3 actually characterizes the diffusion operators.

Theorem 2.4 *Let $L : C^\infty(\mathbb{R}^n) \to C^0(\mathbb{M})$ be an operator such that:*

(1) L is linear;
(2) for any $f \in C^\infty(\mathbb{R}^n)$ such that f has a local minimum at x, one has $Lf(x) \geq 0$.

Then, L is a diffusion operator.

Proof Suppose L satisfy the above properties. It is readily seen from (2) that $L1 = 0$. Fix now $y \in \mathbb{R}^n$. Our first observation is that if $g \in C^\infty(U)$, where $U \subset \mathbb{R}^n$ is a neighborhood of y, then

$$L(\| \cdot - y\|^3 g)(y) = 0. \tag{2.1}$$

To see this, for $\varepsilon > 0$ consider the function

$$x \to \|x - y\|^3 g(x) + \varepsilon \|x - y\|^2.$$

Since such function admits a local minimum at y, by 2) we have

$$L(\| \cdot - y\|^3 g)(y) \geq -\varepsilon L(\| \cdot - y\|^2)(y).$$

Letting $\varepsilon \to 0$, we thus obtain

$$L(\| \cdot - y\|^3 g)(y) \geq 0.$$

By considering now the function

$$x \to \|x - y\|^3 g(x) - \varepsilon \|x - y\|^2,$$

which has a local maximum at y, we obtain similarly that

$$L(\|x - y\|^3 g)(y) \leq 0.$$

In conclusion, we have proved (2.1).

In order to show that L is a diffusion operator we now consider $f \in C^\infty(\mathbb{R}^n)$, and would like to show that

$$Lf(y) = \sum_{i,j=1}^{n} \sigma_{ij}(y) \frac{\partial^2 f}{\partial x_i \partial x_j}(y) + \sum_{i=1}^{n} b_i(y) \frac{\partial f}{\partial x_i}(y). \tag{2.2}$$

for continuous functions σ_{ij} and b_i, with $(\sigma_{ij})_{1 \leq i,j \leq n} \geq 0$. By the Taylor formula there exists a neighborhood U of y, and a function $g \in C^\infty(U)$, such that for any $x \in U$ one has

$$f(x) = f(y) + \sum_{i=1}^{n} (x_i - y_i) \frac{\partial f}{\partial x_i}(y) + \frac{1}{2} \sum_{i,j=1}^{n} (x_i - y_i)(x_j - y_j) \frac{\partial^2 f}{\partial x_i \partial x_j}(y) + \|x - y\|^3 g(x).$$

By applying the operator L to the previous identity, and by taking 1), the fact that $L1 = 0$, and (2.1) into account, we obtain

$$Lf(y) = \sum_{i=1}^{n} L(x_i - y_i)(y) \frac{\partial f}{\partial x_i}(y) + \frac{1}{2} \sum_{i,j=1}^{n} L((x_i - y_i)(x_j - y_j))(y) \frac{\partial^2 f}{\partial x_i \partial x_j}(y).$$

By denoting

$$b_i(y) = L(x_i - y_i)(y), \quad \text{and } \sigma_{ij}(y) = \frac{1}{2} L((x_i - y_i)(x_j - y_j))(y),$$

we conclude that (2.2) holds. Furthermore, since L transforms smooth into continuous functions, the functions b_i's and σ_{ij}'s are continuous. To complete the proof it would thus suffice to show that the matrix $(\sigma_{ij}(y))_{1\le i,j\le n}$ be nonnegative. i.e., that for every $\xi \in \mathbb{R}^n$ one has $\sum_{i,j=1}^n \sigma_{ij}(y)\xi_i\xi_j \ge 0$. Indeed, by (1) again we have

$$\sum_{i,j=1}^n \sigma_{ij}(y)\xi_i\xi_j = \frac{1}{2}L(\langle \xi, x-y\rangle^2)(y).$$

Since the function $x \to \langle \xi, x-y\rangle^2$ attains a local minimum at y, we obtain from (2)

$$L(\langle \xi, x-y\rangle^2)(y) \ge 0.$$

From the arbitrariness of $y \in \mathbb{R}^n$, the proof is completed. □

The previous characterization of diffusion operators is intrinsic and has the advantage of being coordinate free. This suggests to adopt the following natural definition on manifolds.

Definition 2.5 Let $\mathbb{M}$ be a smooth manifold. A diffusion operator L on $\mathbb{M}$ is an operator $L : C^\infty(\mathbb{M}) \to C^0(\mathbb{M})$ such that:

1. L is linear;
2. for any $f \in C^\infty(\mathbb{M})$ which has a local minimum at x, one has $Lf(x) \ge 0$.

Of course, in any local chart, a diffusion operator L reads as

$$L = \sum_{i,j=1}^n \sigma_{ij}(x)\frac{\partial^2}{\partial x_i \partial x_j} + \sum_{i=1}^n b_i(x)\frac{\partial}{\partial x_i}.$$

where σ is symmetric and nonnegative.

2.2 *Subelliptic Diffusion Operators*

In this section, we give without proofs some results of regularity theory for subelliptic diffusion operators. For an introduction to those topics in the case of elliptic operators, we refer to Folland [13] in the Euclidean case and [14] in the manifold case. Regularity theory for hypoelliptic diffusion operators was developed by L. Hörmander [17] and we refer to [10] for a recent presentation.

Definition 2.6 Let L be a diffusion operator with smooth coefficients which is defined on an open set $\Omega \subset \mathbb{R}^n$. We say that L is subelliptic on Ω, if for every compact set $K \subset \Omega$, there exist a constant C and $\varepsilon > 0$ such that for every $u \in C_0^\infty(K)$,

$$\|u\|_{(2\varepsilon)}^2 \leq C\left(\|Lu\|_2^2 + \|u\|_2^2\right). \tag{2.3}$$

In the above definition, we denoted for $s \in \mathbb{R}$, the Sobolev norm

$$\|f\|_{(s)}^2 = \int_{\mathbb{R}^n} |\hat{f}(\xi)|^2(1+\|\xi\|^2)^s d\xi < +\infty,$$

where $\hat{f}(\xi)$ is the Fourier transform of f, and $\|\cdot\|_2$ is the classical L^2 norm. It is well-known that elliptic operators are subelliptic in the sense of the previous definition with $\varepsilon = 1$. There are many interesting examples of diffusion operators which are subelliptic but not elliptic. Let, for instance,

$$L = \sum_{i=1}^d V_i^2 + V_0$$

where $V_0, V_1, \cdots, V_d$ are smooth vector fields defined on an open set Ω. We denote by $\mathfrak{V}$ the Lie algebra generated by the V_i's, $1 \leq i \leq d$, and for $x \in \Omega$,

$$\mathfrak{V}(x) = \{V(x), V \in \mathfrak{V}\}.$$

The celebrated Hörmander's theorem states that if for every $x \in \Omega$, $\mathfrak{V}(x) = \mathbb{R}^n$, then L is a subelliptic operator. In that case ε is $1/d$, where d is the maximal length of the brackets that are needed to generate $\mathbb{R}^n$.

If L is a subelliptic diffusion operator, using the theory of pseudo-differential operators, it can be proved that the inequality (2.3) self-improves into a family of inequalities of the type

$$\|u\|_{(2\varepsilon+s)}^2 \leq C\left(\|Lu\|_{(s)}^2 + \|u\|_{(s)}^2\right), \quad u \in C_0^\infty(K),$$

where $s \in \mathbb{R}$ and the constant C only depends on K and s. This implies, in particular, by a usual bootstrap argument and Sobolev lemma that subelliptic operators are hypoelliptic. Iterating the latter inequality also leads to

$$\|u\|_{(2k\varepsilon)}^2 \leq C \sum_{j=0}^k \|L^j u\|_2^2, \quad u \in C_0^\infty(K),$$

where $k \geq 0$. This may be used to bound derivatives of u in terms of L^2 norms to iterated powers of u. Indeed, if α is a multi-index and k is such that $4k\varepsilon > 2|\alpha| + n$, then we get $\sup_{x\in K} |\partial^\alpha u(x)|^2 \le C\|u\|^2_{(2k\varepsilon)}$ and therefore

$$\sup_{x\in K} |\partial^\alpha u(x)|^2 \le C' \sum_{j=0}^{k} \|L^j u\|_2^2.$$

Along the same lines, we also get the following result.

Proposition 2.7 *Let L be a subelliptic diffusion operator with smooth coefficients on an open set $\Omega \subset \mathbb{R}^n$. Let $u \in L^2(\Omega)$ such that, in the sense of distributions,*

$$Lu, L^2u, \cdots, L^k u \in L^2(\Omega),$$

for some positive integer k. Let K be a compact subset of Ω and denote by ε the same constant as in (2.3). If $k > \frac{n}{4\varepsilon}$, then u is a continuous function on the interior of K and there exists a positive constant C such that

$$\sup_{x\in K} |u(x)|^2 \le C \sum_{j=0}^{k} \|L^j u\|^2_{L^2(\Omega)}.$$

More generally, if $k > \frac{m}{2\varepsilon} + \frac{n}{4\varepsilon}$ for some non negative integer m, then u is m-times continuously differentiable in the interior of K and there exists a positive constant C such that

$$\sup_{|\alpha|\le m} \sup_{x\in K} |\partial^\alpha u(x)|^2 \le C \sum_{j=0}^{k} \|L^j u\|^2_{L^2(\Omega)}.$$

As a consequence of the previous result, we see in particular that

$$\bigcap_{k\ge 0} \mathcal{D}(L^k) \subset C^\infty(\mathbb{R}^n).$$

We can define subelliptic operators on a manifold by using charts:

Definition 2.8 Let L be a diffusion operator on a manifold $\mathbb{M}$. We say that L is subelliptic on $\mathbb{M}$ if it is in any local chart.

Proposition 2.7 can then be extended to the manifold case:

Proposition 2.9 *Let $\mathbb{M}$ be a manifold endowed with a smooth positive measure μ, and let L be a subelliptic diffusion operator with smooth coefficients on an open set $\Omega \subset \mathbb{M}$. Let $u \in L^2(\Omega, \mu)$ such that, in the sense of distributions,*

$$Lu, L^2u, \cdots, L^k u \in L^2(\Omega, \mu),$$

for some positive integer k. Let K be a compact subset of Ω. *There exists a constant* $\varepsilon > 0$ *such that If* $k > \frac{n}{4\varepsilon}$, *then u is a continuous function on the interior of K and there exists a positive constant C such that*

$$\sup_{x \in K} |u(x)|^2 \le C \sum_{j=0}^{k} \|L^j u\|^2_{L^2(\Omega,\mu)}.$$

More generally, if $k > \frac{m}{2\varepsilon} + \frac{n}{4\varepsilon}$ *for some non negative integer m, then u is m-times continuously differentiable in the interior of K and there exists a positive constant C such that*

$$\sup_{|\alpha| \le m} \sup_{x \in K} |\partial^\alpha u(x)|^2 \le C \sum_{j=0}^{k} \|L^j u\|^2_{L^2(\Omega,\mu)}.$$

2.3 The Distance Associated to Subelliptic Diffusion Operators

Let L be a subelliptic diffusion operator defined on a manifold $\mathbb{M}$. For every smooth functions $f, g : \mathbb{M} \to \mathbb{R}$, let us define the so-called *carré du champ*, which is the symmetric first-order differential form defined by:

$$\Gamma(f, g) = \frac{1}{2} \left(L(fg) - fLg - gLf \right).$$

A straightforward computation shows that if, in a local chart,

$$L = \sum_{i,j=1}^{n} \sigma_{ij}(x) \frac{\partial^2}{\partial x_i \partial x_j} + \sum_{i=1}^{n} b_i(x) \frac{\partial}{\partial x_i},$$

then, in the same chart

$$\Gamma(f, g) = \sum_{i,j=1}^{n} \sigma_{ij}(x) \frac{\partial f}{\partial x_i} \frac{\partial g}{\partial x_j}.$$

As a consequence, for every smooth function f,

$$\Gamma(f) \ge 0.$$

Definition 2.10 An absolutely continuous curve $\gamma : [0, T] \to \mathbb{M}$ is said to be subunit for the operator L if for every smooth function $f : \mathbb{M} \to \mathbb{R}$ we have $\left| \frac{d}{dt} f(\gamma(t)) \right| \le \sqrt{(\Gamma f)(\gamma(t))}$. We then define the subunit length of γ as $\ell_s(\gamma) = T$.

Given $x, y \in \mathbb{M}$, we indicate with

$$S(x, y) = \{\gamma : [0, T] \to \mathbb{M} \mid \gamma \text{ is subunit for } \Gamma, \gamma(0) = x, \ \gamma(T) = y\}.$$

In these lecture notes we always assume that

$$S(x, y) \neq \emptyset, \quad \text{for every } x, y \in \mathbb{M}.$$

If L is an elliptic operator or if L is a sum of squares operator that satisfies Hörmander's condition, then this assumption is satisfied (this is the so-called Chow theorem).

Under such assumption it is easy to verify that

$$d(x, y) = \inf\{\ell_s(\gamma) \mid \gamma \in S(x, y)\}, \tag{2.4}$$

defines a true distance on $\mathbb{M}$. This is the intrinsic distance associated to the subelliptic operator L. A beautiful result by Fefferman and Phong [12] relates the subellipticity of L to the size of the balls for this metric:

Theorem 2.11 (Fefferman-Phong) *Let Ω be a relatively compact open subset of $\mathbb{M}$. For some $\varepsilon > 0$, there are constants $r_0 > 0$ and C such that for all x and r,*

$$B(x, r) \cap \Omega \subset B_d(x, Cr^{\varepsilon}) \cap \Omega$$

whenever $0 \leq r < r_0$. B_d denotes here the ball for the metric d and B the ball for an arbitrary Riemannian metric on Ω.

A corollary of this result is that the topology induced by d coincides with the manifold topology of $\mathbb{M}$. The distance d can also be computed using the following definition:

Proposition 2.12 *For every $x, y \in \mathbb{M}$,*

$$d(x, y) = \sup\left\{|f(x) - f(y)|, f \in C^{\infty}(\mathbb{M}), \|\Gamma(f)\|_{\infty} \leq 1\right\}, \quad x, y \in \mathbb{M}. \tag{2.5}$$

Proof Let $x, y \in \mathbb{M}$. We denote

$$\delta(x, y) = \sup\{|f(x) - f(y)|, f \in C_0^{\infty}(\mathbb{M}), \|\Gamma(f)\|_{\infty} \leq 1\}.$$

Let $\gamma : [0, T] \to \mathbb{M}$ be a sub-unit curve such that

$$\gamma(0) = x, \gamma(T) = x.$$

We have $\left|\frac{d}{dt} f(\gamma(t))\right| \le \sqrt{(\Gamma f)(\gamma(t))}$, therefore, if $\Gamma(f) \le 1$,

$$|f(y) - f(x)| \le T.$$

As a consequence

$$\delta(x, y) \le d(x, y).$$

We now prove the converse inequality which is trickier. The idea would be to use the function $f(y) = d(x, y)$ that satisfies $|f(x) - f(y)| = d(x, y)$ and "$\Gamma(f, f) = 1$". However, giving a precise meaning to $\Gamma(f, f) = 1$, is not so easy, because it turns out that f is not everywhere differentiable. If the operator L is elliptic, then the distance d is Riemannian and one can use an approximate identity to regularize f, that is to find C^∞ functions η_ε such that $|f(y) - \eta_\varepsilon(y)| \le \varepsilon$ and $\|\Gamma(\eta_\varepsilon)\|_\infty \le 1 + C\varepsilon$ for some constant $C > 0$ (see [18] for more details). We have then $|\eta_\varepsilon(y) - \eta_\varepsilon(x)| \le (1 + C\varepsilon)\delta(x, y)$, which implies $d(x, y) \le \delta(x, y)$ by letting $\varepsilon \to 0$. Thus, if L is elliptic then $d(x, y) = \delta(x, y)$. If L is only subelliptic, we consider an elliptic diffusion operator Δ on $\mathbb{M}$ and consider the sequence of operators $L_k = L + \frac{1}{k}\Delta$. We denote by d_k the distance associated to L_k. It is easy to see that d_k increases with k and that $d_k(x, y) \le d(x, y)$. We can find a curve $\gamma_k : [0, 1] \to \mathbb{M}$, such that $\gamma(0) = x, \gamma(1) = y$ and for every $f \in C^\infty(M)$,

$$\left|\frac{d}{dt} f(\gamma_k(t))\right|^2 \le \left(d_k^2(x, y) + \frac{1}{k}\right)\left(\Gamma(f)(\gamma_k(t)) + \frac{1}{k}\Gamma_\Delta(f)(\gamma_k(t))\right),$$

where Γ_Δ is the carré du champ operator of Δ. Since $d_k \le d$, we see that the sequence γ_k is uniformly equicontinuous. As a consequence of the Arzelà-Ascoli theorem, we deduce that there exists a subsequence which we continue to denote γ_k that converges uniformly to a curve $\gamma : [0, 1] \to \mathbb{M}$, such that $\gamma(0) = x, \gamma(1) = y$ and for every $f \in C^\infty(M)$,

$$\left|\frac{d}{dt} f(\gamma(t))\right|^2 \le \sup_k d_k^2(x, y)\Gamma(f)(\gamma_k(t)).$$

By definition of d, we deduce $d(x, y) \le \sup_k d_k(x, y)$. As a consequence, we proved that $d(x, y) = \lim_{k\to\infty} d_k(x, y)$. Since it is clear that

$$d_k(x, y) = \sup\left\{|f(x) - f(y)|, f \in C^\infty(\mathbb{M}), \left\|\Gamma(f) + \frac{1}{k}\Gamma_\Delta(f)\right\|_\infty \le 1\right\} \le \delta(x, y),$$

we finally conclude that $d(x, y) \le \delta(x, y)$, hence $d(x, y) = \delta(x, y)$. □

A straightforward corollary of the previous proposition is the following useful result:

Corollary 2.13 *If $f \in C^\infty(\mathbb{M})$ satisfies $\Gamma(f) = 0$, then f is constant.*

The following theorem is known as the Hopf-Rinow theorem, it provides a necessary and sufficient condition for the completeness of the metric space $(\mathbb{M}, d)$.

Theorem 2.14 (Hopf-Rinow Theorem) *The metric space $(\mathbb{M}, d)$ is complete (i.e. Cauchy sequences are convergent) if and only the compact sets are the closed and bounded sets.*

Proof It is clear that if closed and bounded sets are compact then the metric space $(\mathbb{M}, d)$ is complete; It comes from the fact that Cauchy sequence are convergent if and only if they have at least one cluster value. So, we need to prove that closed and bounded sets for the distance d are compact provided that $(\mathbb{M}, d)$ is complete. To check this, it is enough to prove that closed balls are compact.

Let $x \in \mathbb{M}$. Observe that if the closed ball $\bar{B}(x, r)$ is compact for some $r > 0$, then $\bar{B}(x, \rho)$ is closed for any $\rho < r$. Define

$$R = \sup\{r > 0, \bar{B}(x, r) \text{ is compact }\}.$$

Since d induces the manifold topology of $\mathbb{M}$, $R > 0$. Let us assume that $R < \infty$ and let us show that it leads to a contradiction.

We first show that $\bar{B}(x, R)$ is compact. Since $(\mathbb{M}, d)$ is assumed to be complete, it suffices to prove that $\bar{B}(x, R)$ is totally bounded: That is, for every $\varepsilon > 0$ there is a finite set S_ε such that every point of $\bar{B}(x, R)$ lies in a ε-neighborhood of S_ε.

So, let $\varepsilon > 0$ small enough. By definition of R, the ball $\bar{B}(x, R-\varepsilon/4)$ is compact; It is therefore totally bounded. We can find a finite set $S = \{y_1, \cdots, y_N\}$ such that every point of $\bar{B}(x, R - \varepsilon/4)$ lies in a $\varepsilon/2$-neighborhood of S. Let now $y \in \bar{B}(x, R)$. We claim that there exists $y' \in \bar{B}(x, R - \varepsilon/4)$ such that $d(y, y') \leq \varepsilon/2$. If $y \in \bar{B}(x, R - \varepsilon/4)$, there is nothing to prove, we may therefore assume that $y \notin \bar{B}(x, R - \varepsilon/4)$. Consider then a sub-unit curve $\gamma : [0, R + \varepsilon/4] \to \mathbb{M}$ such that $\gamma(0) = x, \gamma(R + \varepsilon/2) = y$. Let

$$\tau = \inf\{t, \gamma(t) \notin \bar{B}(x, R - \varepsilon/4)\}.$$

We have $\tau \geq R - \varepsilon/4$. On the other hand,

$$d(\gamma(\tau), \gamma(R + \varepsilon/2)) \leq R + \varepsilon/4 - \tau.$$

As a consequence,

$$d(\gamma(\tau), y) \leq \varepsilon/2.$$

In all cases, there exists therefore $y' \in \bar{B}(x, R - \varepsilon/4)$ such that $d(y, y') \leq \varepsilon/2$. We may then pick y_k in S such that $d(y_k, y') \leq \varepsilon/2$. From the triangle inequality, we have $d(y, y_k) \leq \varepsilon$.

So, at the end, it turns out that every point of $\bar{B}(x, R)$ lies in a ε-neighborhood of S. This shows that $\bar{B}(x, R)$ is totally bounded and therefore compact because $(\mathbb{R}^n, d)$ is assumed to be complete. Actually, the previous argument shows more, it shows that if every point of $\bar{B}(x, R)$ lies in a $\varepsilon/2$-neighborhood of a finite S, then every point of $\bar{B}(x, R + \varepsilon/4)$ will lie ε-neighborhood of S, so that the ball $\bar{B}(x, R+\varepsilon/4)$ is also compact. This contradicts the definition of R. Therefore every closed ball is a compact set, due to the arbitrariness of x. □

2.4 Essentially Self-Adjoint Subelliptic Operators

We consider on a smooth manifold $\mathbb{M}$ a subelliptic diffusion operator L. We assume that L is symmetric with respect to some measure μ, that is: For every smooth and compactly supported functions $f, g : \mathbb{M} \to \mathbb{R}$,

$$\int_{\mathbb{M}} gLf d\mu = \int_{\mathbb{M}} fLg d\mu.$$

We observe that if μ is a symmetric measure for L as above, then in the sense of distributions

$$L'\mu = 0,$$

where L' is the adjoint of L in distribution sense. As a consequence, μ needs to be a smooth measure.

The *carré du champ* introduced above allows to integrate by parts: For every $f, g \in C_0^\infty(\mathbb{M})$,

$$\int_{\mathbb{M}} fLg d\mu = -\int_{\mathbb{M}} \Gamma(f, g) d\mu = \int_{\mathbb{M}} gLf d\mu.$$

The operator L on its domain $\mathcal{D}(L) = C_0^\infty(\mathbb{M})$ is a densely defined non positive symmetric operator on the Hilbert space $L^2(\mathbb{M}, \mu)$. However, it is not self-adjoint, indeed it is easily checked that

$$\left\{ f \in C^\infty(\mathbb{M}), \|f\|_{L^2(\mathbb{M},\mu)} + \|Lf\|_{L^2(\mathbb{M},\mu)} < \infty \right\} \subset \mathcal{D}(L^*).$$

A famous theorem of Von Neumann asserts that any non negative and symmetric operator may be extended into a self-adjoint operator. The following construction, due to Friedrich, provides a canonical non negative self-adjoint extension.

Theorem 2.15 (Friedrichs Extension) *On the Hilbert space $L^2(\mathbb{M}, \mu)$, there exists a densely defined non positive self-adjoint extension of L.*

Proof On $C_0^\infty(\mathbb{M})$, let us consider the following norm

$$\|f\|_{\mathcal{E}}^2 = \|f\|_{L^2(\mathbb{M},\mu)}^2 + \int_{\mathbb{M}} \Gamma(f, f) d\mu.$$

By completing $C_0^\infty(\mathbb{M})$ with respect to this norm, we get a Hilbert space $(\mathcal{H}, \langle \cdot, \cdot \rangle_{\mathcal{E}})$. Since for $f \in C_0^\infty(\mathbb{M})$, $\|f\|_{L^2(\mathbb{M},\mu)} \le \|f\|_{\mathcal{E}}$, the injection map

$$\iota : (C_0^\infty(\mathbb{M}), \|\cdot\|_{\mathcal{E}}) \to (L^2(\mathbb{M}, \mu), \|\cdot\|_{L^2(\mathbb{M},\mu)})$$

is continuous and it may therefore be extended into a continuous map

$$\bar{\iota} : (\mathcal{H}, \|\cdot\|_{\mathcal{E}}) \to (L^2(\mathbb{M}, \mu), \|\cdot\|_{L^2(\mathbb{M},\mu)}).$$

Let us show that $\bar{\iota}$ is injective so that $\mathcal{H}$ may be identified with a subspace of $L^2(\mathbb{M}, \mu)$. So, let $f \in \mathcal{H}$ such that $\bar{\iota}(f) = 0$. We can find a sequence $f_n \in C_0^\infty(\mathbb{M})$, such that $\|f_n - f\|_{\mathcal{E}} \to 0$ and $\|f_n\|_{\mathbf{L}^2_\mu(\mathbb{M})} \to 0$. We have

$$\begin{aligned}
\|f\|_{\mathcal{E}} &= \lim_{m,n\to+\infty} \langle f_n, f_m \rangle_{\mathcal{E}} \\
&= \lim_{m\to+\infty} \lim_{n\to+\infty} \langle f_n, f_m \rangle_{\mathbf{L}^2_\mu(\mathbb{R}^n,\mathbb{R})} - \langle f_n, Lf_m \rangle_{\mathbf{L}^2_\mu(\mathbb{R}^n,\mathbb{R})} \\
&= 0,
\end{aligned}$$

thus $f = 0$ and $\bar{\iota}$ is injective. Let us now consider the map

$$B = \bar{\iota} \cdot \bar{\iota}^* : L^2(\mathbb{M}, \mu) \to L^2(\mathbb{M}, \mu).$$

It is well defined due to the fact that since $\bar{\iota}$ is bounded, it is easily checked that

$$\mathcal{D}(\bar{\iota}^*) = L^2(\mathbb{M}, \mu).$$

Moreover, B is easily seen to be symmetric, and thus self-adjoint because its domain is equal to $L^2(\mathbb{M}, \mu)$. Also, it is readily checked that the injectivity of $\bar{\iota}$ implies the injectivity of B. Therefore, we deduce that the inverse

$$A = B^{-1} : \mathcal{R}(\bar{\iota} \cdot \bar{\iota}^*) \subset L^2(\mathbb{M}, \mu) \to L^2(\mathbb{M}, \mu)$$

is a densely defined self-adjoint operator on $L^2(\mathbb{M}, \mu)$. Now, we observe that for $f, g \in C_0^\infty(\mathbb{M})$,

$$\begin{aligned}
&\langle f, g \rangle_{L^2(\mathbb{M},\mu)} - \langle Lf, g \rangle_{L^2(\mathbb{M},\mu)} \\
=&\langle \bar{i}^{-1}(f), \bar{i}^{-1}(g) \rangle_{\mathcal{E}}
\end{aligned}$$

$$=\langle (\bar{i}^{-1})^* \bar{i}^{-1} f, g \rangle_{L^2(\mathbb{M},\mu)}$$
$$=\langle (\bar{i}\bar{i}^*)^{-1} f, g \rangle_{L^2(\mathbb{M},\mu)}$$

Thus A and $\mathbf{Id} - L$ coincide on $C_0^\infty(\mathbb{M})$. By defining,

$$-\bar{L} = A - \mathbf{Id},$$

we get the required self-adjoint extension of $-L$. □

Remark 2.16 The operator $\bar{L}$, as constructed above, is called the Friedrichs extension of L.

Definition 2.17 If $\bar{L}$ is the unique non positive self-adjoint extension of L, then the operator L is said to be essentially self-adjoint on $C_0^\infty(\mathbb{M})$. In that case, there is no ambiguity and we shall denote $\bar{L} = L$.

We have the following first criterion for essential self-adjointness.

Lemma 2.18 *If for some* $\lambda > 0$,

$$\mathbf{Ker}(-L^* + \lambda \mathbf{Id}) = \{0\},$$

then the operator L is essentially self-adjoint on $C_0^\infty(\mathbb{M})$.

Proof We make the proof for $\lambda = 1$ and let the reader adapt it for $\lambda \neq 0$.

Let $-\tilde{L}$ be a non negative self-adjoint extension of $-L$. We want to prove that actually, $-\tilde{L} = -\bar{L}$. The assumption

$$\mathbf{Ker}(-L^* + \mathbf{Id}) = \{0\}$$

implies that $C_0^\infty(\mathbb{M})$ is dense in $\mathcal{D}(-L^*)$ for the norm

$$\|f\|_{\mathcal{E}}^2 = \|f\|_{L^2(\mathbb{M},\mu)}^2 - \langle f, L^* f \rangle_{L^2(\mathbb{M},\mu)}.$$

Since, $-\tilde{L}$ is a non negative self-adjoint extension of $-L$, we have

$$\mathcal{D}(-\tilde{L}) \subset \mathcal{D}(-L^*).$$

The space $C_0^\infty(\mathbb{M})$ is therefore dense in $\mathcal{D}(-\tilde{L})$ for the norm $\| \cdot \|_{\mathcal{E}}$. At that point, we use some notations introduced in the proof of the Friedrichs extension (Theorem 2.15). Since $C_0^\infty(\mathbb{M})$ is dense in $\mathcal{D}(-\tilde{L})$ for the norm $\| \cdot \|_{\mathcal{E}}$, we deduce that the equality

$$\langle f, g \rangle_{L^2(\mathbb{M},\mu)} - \langle \tilde{L} f, g \rangle_{L^2(\mathbb{M},\mu)} = \langle \bar{i}^{-1}(f), \bar{i}^{-1}(g) \rangle_{\mathcal{E}},$$

which is obviously satisfied for $f, g \in C_0^\infty(\mathbb{M})$ actually also holds for $f, g \in \mathcal{D}(\tilde{L})$. From the definition of the Friedrichs extension, we deduce that $\bar{L}$ and $\tilde{L}$ coincide on $\mathcal{D}(\tilde{L})$. Finally, since these two operators are self adjoint we conclude $\bar{L} = \tilde{L}$. □

Remark 2.19 Given the fact that $-L$ is given here with the domain $C_0^\infty(\mathbb{M})$, the condition

$$\mathbf{Ker}(-L^* + \lambda \mathbf{Id}) = \{0\},$$

is equivalent to the fact that if $f \in L^2(\mathbb{M}, \mu)$ is a function that satisfies in the sense of distributions

$$-Lf + \lambda f = 0,$$

then $f = 0$.

As a corollary of the previous lemma, the following proposition provides a useful sufficient condition for essential self-adjointness that is easy to check for several subelliptic diffusion operators (including Laplace-Beltrami operators on complete Riemannian manifolds).

Proposition 2.20 *If the diffusion operator L is a subelliptic diffusion operator and if there exists an increasing sequence $h_n \in C_0^\infty(\mathbb{M})$), $0 \le h_n \le 1$, such that $h_n \nearrow 1$ on $\mathbb{M}$, and $||\Gamma(h_n)||_\infty \to 0$, as $n \to \infty$, then the operator L is essentially self-adjoint on $C_0^\infty(\mathbb{M})$.*

Proof Let $\lambda > 0$. According to the previous lemma, it is enough to check that if $L^* f = \lambda f$ with $\lambda > 0$, then $f = 0$. As it was observed above, $L^* f = \lambda f$ is equivalent to the fact that, in sense of distributions, $Lf = \lambda f$. From the hypoellipticity of L, we deduce therefore that f is a smooth function. Now, for $h \in C_0^\infty(\mathbb{M})$,

$$\begin{aligned}
\int_{\mathbb{M}} \Gamma(f, h^2 f) d\mu &= -\langle f, L(h^2 f)\rangle_{L^2(\mathbb{M},\mu)} \\
&= -\langle L^* f, h^2 f\rangle_{L^2(\mathbb{M},\mu)} \\
&= -\lambda \langle f, h^2 f\rangle_{L^2(\mathbb{M},\mu)} \\
&= -\lambda \langle f^2, h^2\rangle_{L^2(\mathbb{M},\mu)} \\
&\le 0.
\end{aligned}$$

Since

$$\Gamma(f, h^2 f) = h^2 \Gamma(f) + 2 f h \Gamma(f, h),$$

we deduce that

$$\langle h^2, \Gamma(f) \rangle_{L^2(\mathbb{M},\mu)} + 2\langle fh, \Gamma(f,h) \rangle_{L^2(\mathbb{M},\mu)} \leq 0.$$

Therefore, by Cauchy-Schwarz inequality

$$\langle h^2, \Gamma(f) \rangle_{L^2(\mathbb{M},\mu)} \leq 4\|f\|^2_{L^2(\mathbb{M},\mu)} \|\Gamma(h)\|_\infty.$$

If we now use the sequence h_n and let $n \to \infty$, we obtain $\Gamma(f, f) = 0$ and therefore $f = 0$, as desired. □

Interestingly, the existence of such a sequence h_n is closely related to the completeness of the metric space $(\mathbb{M}, d)$ where d is the intrinsic distance associated to the operator L.

Proposition 2.21 *There exists an increasing sequence $h_n \in C_0^\infty(\mathbb{M})$, $0 \leq h_n \leq 1$, such that $h_n \nearrow 1$ on $\mathbb{M}$, and $\|\Gamma(h_n)\|_\infty \to 0$, as $n \to \infty$ if and only if the metric space $(\mathbb{M}, d)$ is complete.*

Proof Let us assume that the metric space $(\mathbb{M}, d)$ is complete. Let Δ be an elliptic operator on $\mathbb{M}$ and denote by d_R the Riemannian distance associated to the operator $L + \Delta$. We have $d_R \leq d$. Together with the Fefferman-Phong result recalled in Proposition 2.11, we deduce that the metric space $(\mathbb{M}, d_R)$ is complete.

If we fix a base point $x_0 \in \mathbb{M}$, we can find an exhaustion function $\rho \in C^\infty(\mathbb{M})$ such that

$$|\rho - d_R(x_0, \cdot)| \leq L, \qquad \|\nabla_R \rho\| \leq L \text{ on } \mathbb{M}.$$

By the completeness of $(\mathbb{M}, d_R)$ and the Hopf-Rinow theorem, the level sets $\Omega_s = \{x \in \mathbb{M} \mid \rho(x) < s\}$ are relatively compact and, furthermore, $\Omega_s \nearrow \mathbb{M}$ as $s \to \infty$. We now pick an increasing sequence of functions $\phi_n \in C^\infty([0, \infty))$ such that $\phi_n \equiv 1$ on $[0, n]$, $\phi_n \equiv 0$ outside $[0, 2n]$, and $|\phi_n'| \leq \frac{2}{n}$. If we set $h_n(x) = \phi_n(\rho(x))$, then we have $h_n \in C_0^\infty(\mathbb{M})$, $h_n \nearrow 1$ on $\mathbb{M}$ as $n \to \infty$, and

$$\|\sqrt{\Gamma(h_n)}\|_\infty \leq ||\nabla_R h_n||_\infty \leq \frac{2L}{n}.$$

Conversely, let us assume that here exists an increasing sequence $h_n \in C_0^\infty(\mathbb{M})$, $0 \leq h_n \leq 1$, such that $h_n \nearrow 1$ on $\mathbb{M}$, and $\|\Gamma(h_n)\|_\infty \to 0$, as $n \to \infty$. Let (x_k) be a Cauchy sequence in the metric space $(\mathbb{M}, d)$. We can find N such that $h_N(x_1) \geq 1/2$ and $d(x_1, x_k) \leq \frac{1}{4\|\Gamma(h_N)\|_\infty}$ for every k. For every k, we have then $h_N(x_k) \geq \frac{1}{4}$, so that x_k belongs to the support of h_N. By compactness, we deduce that x_k is convergent, hence $(\mathbb{M}, d)$ is complete. □

2.5 The Heat Semigroup Associated to a Subelliptic Diffusion Operator

We consider in this section a subelliptic diffusion operator L which is essentially self-adjoint on $C_0^\infty(\mathbb{M})$. As a consequence, L admits a unique self-adjoint extension (its Friedrichs extension). We shall continue to denote such extension by L. The domain of this extension shall be denoted by $\mathcal{D}(L)$.

If $L = -\int_0^{+\infty} \lambda dE_\lambda$ denotes the spectral decomposition of L in $L^2(\mathbb{M}, \mu)$, then by definition, the heat semigroup $(P_t)_{t\geq 0}$ is given by $P_t = \int_0^{+\infty} e^{-\lambda t} dE_\lambda$. It is a one-parameter family of bounded operators on $L^2(\mathbb{M}, \mu)$. Since the quadratic form $- < f, Lf >$ is a Dirichlet form, we deduce that $(P_t)_{t\geq 0}$ is a sub-Markov semigroup: it transforms non-negative functions into non-negative functions and satisfies

$$P_t 1 \leq 1. \tag{2.6}$$

The sub-Markov property and Riesz-Thorin interpolation classically allows to construct the semigroup $(P_t)_{t\geq 0}$ in $L^p(\mathbb{M}, \mu)$ and for $f \in L^p(\mathbb{M}, \mu)$ one has

$$||P_t f||_{L^p(\mathbb{M},\mu)} \leq ||f||_{L^p(\mathbb{M},\mu)}, \quad 1 \leq p \leq \infty. \tag{2.7}$$

For more details about the functional analytic construction of heat semigroups, we refer for instance to [4].

We now address the proof of the existence and regularity of the heat kernel associated to a locally subelliptic operator.

The key estimate to prove the existence of a heat kernel is the following:

Lemma 2.22 *Let K be a compact subset of $\mathbb{M}$. There exists a positive constant C and $k > 0$ such that for $f \in L^2(\mathbb{M}, \mu)$:*

$$\sup_{x \in K} |P_t f(x)| \leq C \left(1 + \frac{1}{t^k}\right) \|f\|_{L^2(\mathbb{M},\mu)}.$$

Proof Let us first observe that from the spectral theorem that if $f \in L^2(\mathbb{M}, \mu)$ then for every $k \geq 0$, $L^k P_t f \in L^2(\mathbb{M}, \mu)$ and

$$\|L^k P_t f\|_{L^2(\mathbb{M},\mu)} \leq \left(\sup_{\lambda \geq 0} \lambda^k e^{-\lambda t}\right) \|f\|_{L^2(\mathbb{M},\mu)}.$$

Now, let K be a compact set of $\mathbb{M}$. From Proposition 2.9 there exists a positive constant C such that

$$\sup_{x \in K} |P_t f(x)|^2 \leq C \sum_{k=0}^{k} \|L^k P_t f\|^2_{L^2(\mathbb{M},\mu)}.$$

Since it is immediately checked that

$$\sup_{\lambda \ge 0} \lambda^k e^{-\lambda t} = \left(\frac{k}{t}\right)^k e^{-k},$$

the bound

$$\sup_{x \in K} |P_t f(x)| \le C \left(1 + \frac{1}{t^k}\right) \|f\|_{L^2(\mathbb{M},\mu)}.$$

easily follows. □

We are now in position to prove the smoothing property of the semigroup.

Proposition 2.23 *For $f \in L^2(\mathbb{M},\mu)$, the function $(t,x) \to P_t f(x)$ is smooth on $(0,+\infty) \times \mathbb{M}$.*

Proof Let $f \in L^2(\mathbb{M},\mu)$. Let $t > 0$. As above, from the spectral theorem, $P_t f \in \cap_{k \ge 1} \mathcal{D}(L^k) \subset C^\infty(\mathbb{M})$. Hence $P_t f$ is a smooth function.

Next, we prove joint continuity in the variables $(t,x) \in (0,+\infty) \times \mathbb{R}^n$. It is enough to prove that if $t_0 > 0$ and if K is a compact set on $\mathbb{M}$,

$$\sup_{x \in K} |P_t f(x) - P_{t_0} f(x)| \to_{t \to t_0} 0.$$

From Proposition 2.9, there exists a positive constant C such that

$$\sup_{x \in K} |P_t f(x) - P_{t_0} f(x)| \le C \sum_{k=0}^{\kappa} \|L^k P_t f - L^k P_{t_0} f\|^2_{L^2(\mathbb{M},\mu)}.$$

Now, again from the spectral theorem, it is checked that

$$\lim_{t \to t_0} \sum_{k=0}^{\kappa} \|L^k P_t f - L^k P_{t_0} f\|^2_{L^2(\mathbb{M},\mu)} = 0.$$

This gives the expected joint continuity in (t,x). The joint smoothness in (t,x) is a consequence of the second part of Proposition 2.9 and the details are let to the reader. □

We now prove the following fundamental theorem about the existence of a heat kernel for the semigroup generated by subelliptic diffusion operators.

Theorem 2.24 *There is a smooth function $p(x,y,t)$, $t \in (0,+\infty)$, $x, y \in \mathbb{M}$, such that for every $f \in L^2(\mathbb{M},\mu)$ and $x \in \mathbb{M}$,*

$$P_t f(x) = \int_{\mathbb{M}} p(x,y,t) f(y) d\mu(y).$$

The function $p(x, y, t)$ *is called the heat kernel associated to* $(P_t)_{t\geq 0}$. *It satisfies furthermore:*

- *(Symmetry)* $p(x, y, t) = p(y, x, t)$;
- *(Chapman-Kolmogorov relation)* $p(x, y, t + s) = \int_{\mathbb{M}} p(x, z, t)p(z, y, s)d\mu(z)$.

Proof Let $x \in \mathbb{M}$ and $t > 0$. From the previous proposition, the linear form $f \to P_t f(x)$ is continuous on $L^2(\mathbb{M}, \mu)$, therefore from the Riesz representation theorem, there is a function $p(x, \cdot, t) \in L^2(\mathbb{M}, \mu)$, such that for $f \in L^2(\mathbb{M}, \mu)$,

$$P_t f(x) = \int_{\mathbb{M}} p(x, y, t) f(y) d\mu(y).$$

From the fact that P_t is self-adjoint on $L^2(\mathbb{M}, \mu)$:

$$\int_{\mathbb{M}} (P_t f) g d\mu = \int_{\mathbb{M}} f(P_t g) d\mu,$$

we easily deduce the symmetry property:

$$p(x, y, t) = p(y, x, t).$$

And the Chapman-Kolmogorov relation $p(t+s, x, y) = \int_{\mathbb{M}} p(x, z, t)p(z, y, s)d\mu(z)$ stems from the semigroup property $P_{t+s} = P_t P_s$. Finally, from the previous proposition the map $(t, x) \to p(x, \cdot, t) \in L^2(\mathbb{M}, \mu)$ is smooth on $\mathbb{M} \times (0, +\infty)$ for the weak topology on $L^2(\mathbb{M}, \mu)$. This implies that it is also smooth on $(0, +\infty) \times \mathbb{R}^n$ for the norm topology. Since, from the Chapman-Kolmogorov relation

$$p(x, y, t) = \langle p(x, \cdot, t/2), p(y, \cdot, t/2) \rangle_{L^2_\mu(\mathbb{M})},$$

we conclude that $(x, y, t) \to p(x, y, t)$ is smooth on $\mathbb{M} \times \mathbb{M} \times (0, +\infty)$. □

The semigroup $(P_t)_{t\geq 0}$ actually solves a parabolic Cauchy problem.

Lemma 2.25 *Let* $f \in L^p(\mathbb{M}, \mu)$, $1 \leq p \leq \infty$, *and let*

$$u(t, x) = P_t f(x), \quad t \geq 0, x \in \mathbb{M}.$$

Then u *is smooth on* $(0, +\infty) \times \mathbb{M}$ *and is a strong solution of the Cauchy problem*

$$\frac{\partial u}{\partial t} = Lu, \quad u(0, x) = f(x).$$

Proof For $\phi \in C_0^\infty((0,+\infty)\times \mathbb{M})$, we have

$$\begin{aligned}
&\int_{\mathbb{M}\times\mathbb{R}} \left(\left(-\frac{\partial}{\partial t}-L\right)\phi(t,x)\right) u(t,x)d\mu(x)dt \\
&\quad= \int_{\mathbb{R}}\int_{\mathbb{M}} \left(\left(-\frac{\partial}{\partial t}-L\right)\phi(t,x)\right) P_t f(x)dxdt \\
&\quad= \int_{\mathbb{R}}\int_{\mathbb{M}} P_t\left(\left(-\frac{\partial}{\partial t}-L\right)\phi(t,x)\right) f(x)dxdt \\
&\quad= \int_{\mathbb{R}}\int_{\mathbb{M}} -\frac{\partial}{\partial t}\left(P_t\phi(t,x)f(x)\right)dxdt \\
&= 0.
\end{aligned}$$

Therefore u is a weak solution of the equation $\frac{\partial u}{\partial t}=Lu$. Since u is smooth it is also a strong solution. □

We now address the uniqueness of solutions.

Lemma 2.26 *Let $v(x,t)$ be a non negative function such that*

$$\frac{\partial v}{\partial t}\le Lv, \quad v(x,0)=0,$$

and such that for every $t>0$,

$$\|v(\cdot,t)\|_{L^p(\mathbb{M},\mu)} < +\infty,$$

where $1<p<+\infty$. Then $v(x,t)=0$.

Proof Let $h\in C_0^\infty(\mathbb{M})$. Since u is a subsolution with the zero initial condition, for any $\tau\in(0,T)$,

$$\begin{aligned}
&\int_0^\tau\int_{\mathbb{M}} h^2(x)v^{p-1}(x,t)Lv(x,t)d\mu(x)dt \\
&\ge \int_0^\tau\int_{\mathbb{M}} h^2(x)v^{p-1}\frac{\partial v}{\partial t}d\mu(x)dt \\
&=\frac{1}{p}\int_0^\tau \frac{\partial}{\partial t}\left(\int_{\mathbb{M}} h^2(x)v^p d\mu(x)\right)dt \\
&=\frac{1}{p}\int_{\mathbb{M}} h^2(x)v^p(x,\tau)d\mu(x).
\end{aligned}$$

On the other hand, integrating by parts yields

$$\begin{aligned}&\int_0^\tau\int_{\mathbb{M}} h^2(x)v^{p-1}(x,t)Lv(x,t)d\mu(x)dt\\&=-\int_0^\tau\int_{\mathbb{M}} 2hv^{p-1}\Gamma(h,v)d\mu dt-\int_0^\tau\int_{\mathbb{M}} h^2(p-1)v^{p-2}\Gamma(v)d\mu dt.\end{aligned}$$

Observing that

$$\begin{aligned}0&\le\left(\sqrt{\frac{2}{p-1}\Gamma(h)v}-\sqrt{\frac{p-1}{2}\Gamma(v)h}\right)^2\\&\le\frac{2}{p-1}\Gamma(h)v^2+2\Gamma(h,v)hv+\frac{p-1}{2}\Gamma(v)h^2,\end{aligned}$$

we obtain the following estimate.

$$\begin{aligned}&\int_0^\tau\int_{\mathbb{M}} h^2(x)v^{p-1}(x,t)Lv(x,t)d\mu(x)dt\\&\le\int_0^\tau\int_{\mathbb{M}}\frac{2}{p-1}\Gamma(h)v^p d\mu dt-\int_0^\tau\int_{\mathbb{M}}\frac{p-1}{2}h^2v^{p-2}\Gamma(v)d\mu dt\\&=\int_0^\tau\int_{\mathbb{M}}\frac{2}{p-1}\Gamma(h)v^p d\mu dt-\frac{2(p-1)}{p^2}\int_0^\tau\int_{\mathbb{M}} h^2\Gamma(v^{p/2})d\mu dt.\end{aligned}$$

Combining with the previous conclusion we obtain,

$$\begin{aligned}\int_{\mathbb{M}} h^2(x)v^p(x,\tau)d\mu(x)+\frac{2(p-1)}{p}\int_0^\tau\int_{\mathbb{M}} h^2\Gamma(v^{p/2})d\mu dt\\\le\frac{2p}{(p-1)}\|\Gamma(h)\|_\infty^2\int_0^\tau\int_{\mathbb{M}} v^p d\mu dt.\end{aligned}$$

By using (H.1) and the previous inequality with an increasing sequence $h_n \in C_0^\infty(\mathbb{M})$, $0\le h_n\le 1$, such that $h_n \nearrow 1$ on $\mathbb{M}$, and $||\Gamma(h_n,h_n)||_\infty \to 0$, as $n\to\infty$, and letting $n\to+\infty$, we obtain $\int_{\mathbb{M}} v^p(x,\tau)d\mu(x)=0$ thus $v=0$. □

As a consequence of this result, any solution in $L^p(\mathbb{M},\mu)$, $1<p<+\infty$ of the heat equation $\frac{\partial u}{\partial t}=Lu$ is uniquely determined by its initial condition, and is therefore of the form $u(t,x)=P_t f(x)$. We stress that without further conditions, this result may fail when $p=1$ or $p=+\infty$.

3 The Heat Semigroup on a Complete Riemannian Manifold and Its Geometric Applications

In this section we shall consider a smooth and complete Riemannian manifold $(\mathbb{M}, g)$ with dimension n. The Riemannian measure will be denoted by μ and we will often use the notation $\langle \cdot, \cdot \rangle$ for $g(\cdot, \cdot)$. A canonical elliptic diffusion operator on $\mathbb{M}$ is the Laplace-Beltrami operator. In this section, we will show some applications to geometry of the Laplace-Beltrami operator and the semigroup it generates. We assume some basic knowledge about Riemannian geometry (see for instance [11, 20]). Some of the covered topics are inspired by M. Ledoux [19].

3.1 The Laplace-Beltrami Operator

The Laplace-Beltrami of $\mathbb{M}$ will be denoted by L. We recall that it is the generator of the pre-Dirichlet form

$$\mathcal{E}(f_1, f_2) = \int_{\mathbb{M}} g(\nabla f_1, \nabla f_2) d\mu, \quad f_1, f_2 \in C_0^\infty(\mathbb{M}),$$

where ∇ is the Riemannian gradient on $\mathbb{M}$. Since $\mathbb{M}$ is assumed to be complete, as we have seen in the previous section, the operator L is essentially self-adjoint on the space $C_0^\infty(\mathbb{M})$. More precisely, there exists an increasing sequence $h_n \in C_0^\infty(\mathbb{M})$ such that $h_n \nearrow 1$ on $\mathbb{M}$, and $||\Gamma(h_n, h_n)||_\infty \to 0$, as $n \to \infty$. Observe that for the Laplace-Beltrami operator L, the operator Γ is simply given by $\Gamma(f_1, f_2) = \langle \nabla f_1, \nabla f_2 \rangle$.

The Friedrichs extension of L, which is therefore the unique self-adjoint extension of L in $L^2(\mathbb{M}, \mu)$ will still be denoted by L and the domain of this extension is denoted by $\mathcal{D}(L)$.

Using the results of the previous section, we then have:

- By using the spectral theorem for L in the Hilbert space $L^2(\mathbb{M}, \mu)$, we may construct a strongly continuous contraction semigroup $(P_t)_{t \ge 0}$ in $L^2(\mathbb{M}, \mu)$ whose infinitesimal generator is L;
- By using the ellipticity of L, we may prove that $(P_t)_{t \ge 0}$ admits a heat kernel, that is: There is a smooth function $p(t, x, y)$, $t \in (0, +\infty)$, $x, y \in \mathbb{M}$, such that for every $f \in L^2(\mathbb{M}, \mu)$ and $x \in \mathbb{M}$,

$$P_t f(x) = \int_{\mathbb{M}} p(t, x, y) f(y) d\mu(y).$$

Moreover, the heat kernel satisfies the two following conditions:

– (Symmetry) $p(t, x, y) = p(t, y, x)$;
– (Chapman-Kolmogorov relation) $p(t+s, x, y) = \int_{\mathbb{M}} p(t, x, z)p(s, z, y)d\mu(z)$.

- The semigroup $(P_t)_{t\geq 0}$ is a sub-Markov semigroup: If $0 \leq f \leq 1$ is a function in $L^2(\mathbb{M}, \mu)$, then $0 \leq P_t f \leq 1$.
- By using the Riesz-Thorin interpolation theorem, $(P_t)_{t\geq 0}$ defines a contraction semigroup on $L^p(\mathbb{M}, \mu)$, $1 \leq p \leq \infty$.

3.2 The Heat Semigroup on a Compact Riemannian Manifold

In this section, we study some spectral properties of the Laplace-Beltrami operator and of the heat semigroup on a compact Riemannian manifold. So, let $(\mathbb{M}, g)$ be a compact Riemannian manifold. As usual, we denote by $(P_t)_{t\geq 0}$ the heat semigroup and by $p(t, x, y)$ the corresponding heat kernel. As a preliminary result, we have the following Liouville's type theorem.

Lemma 3.1 *Let $f \in \mathcal{D}(L)$ such that $Lf = 0$, then f is a constant function.*

Proof From the ellipticity of L, we first deduce that f is smooth. Then, since $\mathbb{M}$ is compact, the following equality holds

$$-\int_{\mathbb{M}} fLf d\mu = \int_{\mathbb{M}} \Gamma(f, f) d\mu.$$

Therefore $\Gamma(f, f) = 0$, which implies that f is a constant function. □

In the compact case, the heat semigroup satisfies the so-called stochastic completeness (or Markov) property.

Proposition 3.2 *For $t \geq 0$,*

$$P_t 1 = 1.$$

Proof Since the constant function 1 is in $L^2(\mathbb{M}, \mu)$, by compactness of $\mathbb{M}$, we may apply Proposition 2.26. □

It turns out that the compactness of $\mathbb{M}$ implies the compactness of the semigroup.

Proposition 3.3 *For $t > 0$ the operator P_t is a compact operator on the Hilbert space $L^2(\mathbb{M}, \mu)$. It is moreover trace class and*

$$\mathbf{Tr}(P_t) = \int_{\mathbb{M}} p(t, x, x)\mu(dx).$$

Proof From the existence of the heat kernel we have

$$P_t f(x) = \int_{\mathbb{M}} p(t, x, y) f(y) d\mu(y).$$

But from the compactness of $\mathbb{M}$, we have

$$\int_{\mathbb{M}} \int_{\mathbb{M}} p(t, x, y)^2 d\mu(x) d\mu(y) < +\infty.$$

Therefore, the operator

$$P_t : L^2(\mathbb{M}, \mu) \to L^2(\mathbb{M}, \mu)$$

is a Hilbert-Schmidt operator. It is thus in particular a compact operator.

Since $P_t = P_{t/2} P_{t/2}$, P_t is a product of two Hilbert-Schmidt operators. It is therefore a class trace operator and,

$$\mathbf{Tr}(P_t) = \int_{\mathbb{M}} \int_{\mathbb{M}} p(t/2, x, y) p(t/2, y, x) d\mu(x) d\mu(y).$$

We conclude then by applying the Chapman-Kolmogorov relation. □

In this compact framework, we have the following theorem

Theorem 3.4 *There exists a complete orthonormal basis* $(\phi_n)_{n \in \mathbb{N}}$ *of* $L^2(\mathbb{M}, \mu)$, *consisting of eigenfunctions of* $-L$, *with* ϕ_n *having an eigenvalue* λ_n *with finite multiplicity satisfying*

$$0 = \lambda_0 < \lambda_1 \le \lambda_2 \le \cdots \nearrow +\infty.$$

Moreover, for $t > 0$, $x, y \in \mathbb{M}$,

$$p(t, x, y) = \sum_{n=0}^{+\infty} e^{-\lambda_n t} \phi_n(x) \phi_n(y),$$

with convergence absolute and uniform for each $t > 0$.

Proof Let $t > 0$. From the Hilbert-Schmidt theorem for the non negative self adjoint compact operator P_t, there exists a complete orthonormal basis $(\phi_n(t))_{n \in \mathbb{N}}$ of $L^2(\mathbb{M}, \mu)$ and a non increasing sequence $\alpha_n(t) \ge 0$, $\alpha_n(t) \searrow 0$ such that

$$P_t \phi_n(t) = \alpha_n(t) \phi_n(t).$$

The semigroup property $P_{t+s} = P_t P_s$ implies first that for $k \in \mathbb{N}$, $k \ge 1$,

$$\phi_n(k) = \phi_n(1), \alpha_n(k) = \alpha_n(1)^k.$$

The same result is then seen to hold for $k \in \mathbb{Q}$, $k > 0$ and finally for $k \in \mathbb{R}$, due to the strong continuity of the semigroup. Since the map $t \to \| P_t \|_2$ is decreasing, we deduce that $\alpha_n(1) \leq 1$. Thus, there is a $\lambda_n \geq 0$ such that

$$\alpha_n(1) = e^{-\lambda_n}.$$

As a conclusion, there exists a complete orthonormal basis $(\phi_n)_{n\in\mathbb{N}}$ of $L^2(\mathbb{M}, \mu)$, and a sequence λ_n satisfying

$$0 \leq \lambda_0 \leq \lambda_1 \leq \lambda_2 \leq \cdots \nearrow +\infty,$$

such that

$$P_t \phi_n = e^{-\lambda_n t} \phi_n.$$

Since $P_t 1 = 1$, we actually have $\lambda_0 = 0$. Also, if $f \in L^2(\mathbb{M}, \mu)$ is such that $P_t f = f$, it is straightforward that $f \in \mathcal{D}(L)$ and that $Lf = 0$, so that thanks to Liouville theorem, f is a constant function. Therefore $\lambda_1 > 0$.

Since $P_t \phi_n = e^{-\lambda_n t} \phi_n$, by differentiating as $t \to 0$ in $L^2(\mathbb{M}, \mu)$, we obtain furthermore that $\phi_n \in \mathcal{D}(L)$ and that $L\phi_n = -\lambda_n \phi_n$.

The family $(x, y) \to \phi_n(x)\phi_m(y)$ forms an orthonormal basis of $L^2(\mathbb{M}\times\mathbb{M}, \mu\otimes\mu)$. We therefore have a decomposition in $L^2(\mathbb{M} \times \mathbb{M}, \mu \otimes \mu)$,

$$p(t, x, y) = \sum_{m,n\in\mathbb{M}} c_{mn} \phi_m(x)\phi_n(y).$$

Since $p(t, \cdot, \cdot)$ is the kernel of P_t, it is then straightforward that for $m \neq n$, $c_{mn} = 0$ and that $c_{nn} = e^{-\lambda_n t}$. Therefore in $L^2(\mathbb{M}, \mu)$,

$$p(t, x, y) = \sum_{n=0}^{+\infty} e^{-\lambda_n t} \phi_n(x)\phi_n(y).$$

The continuity of p, together with the positivity of P_t imply, via Mercer's theorem that actually, the above series is absolutely and uniformly convergent for $t > 0$. □

As we stressed it in the statement of the theorem, in the decomposition

$$p(t, x, y) = \sum_{n=0}^{+\infty} e^{-\lambda_n t} \phi_n(x)\phi_n(y),$$

the eigenvalue λ_n is repeated according to its multiplicity. It is often useful to rewrite this decomposition under the form

$$p(t, x, y) = \sum_{n=0}^{+\infty} e^{-\alpha_n t} \sum_{k=1}^{d_n} \phi_k^n(x)\phi_k^n(y),$$

where the eigenvalue α_n is not repeated, that is

$$0 = \alpha_0 < \alpha_1 < \alpha_2 < \cdots$$

In this decomposition, d_n is the dimension of the eigenspace $\mathcal{V}_n$ corresponding to the eigenvalue α_n and $(\phi_k^n)_{1 \le k \le d_n}$ is an orthonormal basis of $\mathcal{V}_n$. If we denote,

$$\mathcal{K}_n(x, y) = \sum_{k=1}^{d_n} \phi_k^n(x)\phi_k^n(y),$$

then $\mathcal{K}_n$ is called the reproducing kernel of the eigenspace $\mathcal{V}_n$. It satisfies the following properties whose proofs are let to the reader:

Proposition 3.5

- *$\mathcal{K}_n$ does not depend on the choice of the basis $(\phi_k^n)_{1 \le k \le d_n}$;*
- *If $f \in \mathcal{V}_n$, then $\int_{\mathbb{M}} \mathcal{K}_n(x, y) f(y) d\mu(y) = f(x)$.*

From the very definition of the reproducing kernels, we have

$$p(t, x, y) = \sum_{n=0}^{+\infty} e^{-\alpha_n t} \mathcal{K}_n(x, y). \tag{3.1}$$

The compactness of $\mathbb{M}$ also implies the convergence to equilibrium for the semigroup.

Proposition 3.6 *Let $f \in L^2(\mathbb{M}, \mu)$, then uniformly on $\mathbb{M}$, when $t \to +\infty$,*

$$P_t f \to \frac{1}{\mu(\mathbb{M})} \int_{\mathbb{M}} f d\mu.$$

Proof It is obvious from the previous proposition and from spectral theory that in $L^2(\mathbb{M}, \mu)$, $P_t f$ converges to a constant function that we denote $P_\infty f$. The convergence is also uniform, because for $s, t, T > 0$,

$$\begin{aligned} \| P_{t+T} f - P_{s+T} f \|_\infty &= \sup_{x \in \mathbb{M}} | P_T (P_t f - P_s f)(x) | \\ &= \sup_{x \in \mathbb{M}} \left| \int_{\mathbb{M}} p(T, x, y)(P_t f - P_s f)(y) d\mu(y) \right| \\ &\le \left(\sup_{x \in \mathbb{M}} \sqrt{\int_{\mathbb{M}} p(T, x, y)^2 d\mu(y)} \right) \| P_t f - P_s f \|_2. \end{aligned}$$

Moreover, for every $t \geq 0$, $\int_{\mathbb{M}} P_t f d\mu = \int_{\mathbb{M}} f d\mu$. Therefore

$$\int_{\mathbb{M}} P_\infty f d\mu = \int_{\mathbb{M}} f d\mu.$$

Since $P_\infty f$ is constant, we finally deduce the expected result:

$$P_\infty f = \frac{1}{\mu(\mathbb{M})} \int_{\mathbb{M}} f d\mu.$$

□

3.3 Bochner's Identity

The Bochner's identity is a fundamental identity that connects the Ricci curvature of a Riemannian manifold $(\mathbb{M}, g)$ to the Laplace-Beltrami-operator (see [11, 20]).

Theorem 3.7 *If $f \in C^\infty(\mathbb{M})$, then*

$$\frac{1}{2} L(\|\nabla f\|^2) - \langle \nabla f, \nabla L f \rangle = \|\mathbf{Hess} f\|_{HS}^2 + \mathbf{Ric}(\nabla f, \nabla f),$$

where **Ric** *denotes the Ricci curvature tensor of $\mathbb{M}$ and $\|\mathbf{Hess} f\|_{HS}$ the Hilbert-Schmidt norm of the Hessian of f.*

3.4 The Curvature Dimension Inequality

We now introduce the Bakry's Γ_2 operator. For $f, g \in C^\infty(\mathbb{M})$, it is defined as

$$\Gamma_2(f, f) = \frac{1}{2} L(\|\nabla f\|^2) - \langle \nabla f, \nabla L f \rangle.$$

In the previous section, we have seen that on our Riemannian manifold $\mathbb{M}$,

$$\Gamma_2(f, f) = \|\mathbf{Hess} f\|_{HS}^2 + \mathbf{Ric}(\nabla f, \nabla f),$$

Therefore it should come as no surprise that a lower bound on **Ric** translates into a lower bound on Γ_2.

Theorem 3.8 *Let $\mathbb{M}$ be a Riemannian manifold. We have, in the sense of bilinear forms,* **Ric** $\geq \rho$ *if and only if for every $f \in C^\infty(\mathbb{M})$,*

$$\Gamma_2(f, f) \geq \frac{1}{n}(Lf)^2 + \rho \Gamma(f, f).$$

Proof Let us assume that **Ric** $\geq \rho$. In that case, from Bochner's formula we deduce that

$$\Gamma_2(f, f) \geq \|\mathbf{Hess} f\|_{HS}^2 + \rho\Gamma(f, f).$$

From Cauchy-Schwartz inequality, we have the bound

$$\|\mathbf{Hess} f\|_{HS}^2 \geq \frac{1}{n}\mathbf{Tr}\,(\mathbf{Hess} f)^2.$$

Since $\mathbf{Tr}\,(\mathbf{Hess} f) = Lf$, we conclude that

$$\Gamma_2(f, f) \geq \frac{1}{n}(Lf)^2 + \rho\Gamma(f, f).$$

Conversely, let us now assume that for every $f \in C^\infty(\mathbb{M})$,

$$\Gamma_2(f, f) \geq \frac{1}{n}(Lf)^2 + \rho\Gamma(f, f).$$

Let $x \in \mathbb{M}$ and $v \in \mathbf{T}_x\mathbb{M}$. It is possible to find a function $f \in C^\infty(\mathbb{M})$ such that, at x, $\mathbf{Hess} f = 0$ and $\nabla f = v$. We have then, by using Bochner's identity at x,

$$\mathbf{Ric}(v, v) \geq \rho\|v\|^2.$$

□

Remark 3.9 The inequality

$$\Gamma_2(f, f) \geq \frac{1}{n}(Lf)^2 + \rho\Gamma(f, f).$$

is called the curvature dimension inequality. It is satisfied for more general operators than Laplace-Beltrami operators on manifolds with Ricci curvature lower bounds. Many of the results covered in those notes extend to those operators (see [2]).

We finally mention another consequence of Bochner's identity which shall be later used.

Lemma 3.10 *Let* $\mathbb{M}$ *be a Riemannian manifold such that* **Ric** $\geq \rho$. *For every* $f \in C^\infty(\mathbb{M})$,

$$\Gamma(\Gamma(f)) \leq 4\Gamma(f)\left(\Gamma_2(f) - \rho\Gamma(f)\right).$$

Proof It follows from the fact that

$$\Gamma_2(f, f) \geq \|\mathbf{Hess} f\|_{HS}^2 + \rho\Gamma(f, f).$$

and Cauchy-Schwarz inequality implies $\Gamma(\Gamma(f)) \le 4\|\mathbf{Hess} f\|_{HS}^2 \Gamma(f)$. Details are let to the reader. □

3.5 Stochastic Completeness

In this section, we will prove a first interesting consequence of the Bochner's identity: We will prove that if, on a complete Riemannian manifold $\mathbb{M}$, the Ricci curvature is bounded from below, then the heat semigroup is stochastically complete, that is $P_t 1 = 1$. This result is due to S.T. Yau, and we will see this property is also equivalent to the uniqueness in L^∞ for solutions of the heat equation. The proof we give is due to D. Bakry.

Let $\mathbb{M}$ be a complete Riemannian manifold and denote by L its Laplace-Beltrami operator. As usual, we denote by P_t the heat semigroup generated by L. Throughout the section, we will assume that the Ricci curvature of $\mathbb{M}$ is bounded from below by $\rho \in \mathbb{R}$. As seen in the previous section, this is equivalent to the fact that for every $f \in C^\infty(\mathbb{M})$,

$$\Gamma_2(f, f) \ge \frac{1}{n}(Lf)^2 + \rho\Gamma(f, f).$$

We start with a technical lemma:

Lemma 3.11 *If $f \in L^2(\mathbb{M}, \mu)$, then for every $t > 0$, the functions $\Gamma(P_t f)$, $L\Gamma(P_t f)$, $\Gamma(P_t f, LP_t f)$ and $\Gamma_2(P_t f)$ are in $L^1(\mathbb{M}, \mu)$.*

Proof It is straightforward to see from the spectral theorem that $\Gamma(P_t f) \in L^1(\mathbb{M}, \mu)$. Similarly, $|\Gamma(P_t f, LP_t f)| \le \sqrt{\Gamma(P_t f)\Gamma(LP_t f)} \in L^1(\mathbb{M}, \mu)$. Since,

$$\Gamma_2(P_t f) = \frac{1}{2}\left(L\Gamma(P_t f) - 2\Gamma(P_t f, LP_t f)\right)$$

we are let with the problem of proving that $\Gamma_2(P_t f) \in L^1(\mathbb{M}, \mu)$. If $g \in C_0^\infty(\mathbb{M})$, then an integration by parts easily yields

$$\int_{\mathbb{M}} \Gamma_2(g) d\mu = \int_{\mathbb{M}} (Lg)^2 d\mu.$$

As a consequence,

$$\int_{\mathbb{M}} \Gamma_2(g) - \rho\Gamma(g) d\mu = \int_{\mathbb{M}} (Lg)^2 + \rho g Lg d\mu,$$

and we obtain

$$\int_{\mathbb{M}} |\Gamma_2(g) - \rho\Gamma(g)| d\mu \le \left(1 + \frac{1}{2}|\rho|\right) \int_{\mathbb{M}} (Lg)^2 d\mu + \frac{1}{2}|\rho| \int_{\mathbb{M}} g^2 d\mu.$$

Using a density argument, it is then easily proved that for $g \in \mathcal{D}(L) \cap C^\infty(\mathbb{M})$ we have

$$\int_{\mathbb{M}} |\Gamma_2(g) - \rho\Gamma(g)| d\mu \le \left(1 + \frac{1}{2}|\rho|\right) \int_{\mathbb{M}} (Lg)^2 d\mu + \frac{1}{2}|\rho| \int_{\mathbb{M}} g^2 d\mu.$$

In particular, we deduce that if $g \in \mathcal{D}(L) \cap C^\infty(\mathbb{M})$, then $\Gamma_2(g) \in L^1(\mathbb{M}, \mu)$. □

We will also need the following fundamental parabolic comparison theorem that shall be extensively used throughout these lecture notes.

Proposition 3.12 *Let $T > 0$. Let $u, v : \mathbb{M} \times [0, T] \to \mathbb{R}$ be smooth functions such that:*

(i) For every $t \in [0, T]$, $u(\cdot, t) \in L^2(\mathbb{M})$ and $\int_0^T \|u(\cdot, t)\|_2 dt < \infty$;
(ii) $\int_0^T \|\sqrt{\Gamma(u)(\cdot, t)}\|_p dt < \infty$ for some $1 \le p \le \infty$;
(iii) For every $t \in [0, T]$, $v(\cdot, t) \in L^q(\mathbb{M})$ and $\int_0^T \|v(\cdot, t)\|_q dt < \infty$ for some $1 \le q \le \infty$.

If the inequality

$$Lu + \frac{\partial u}{\partial t} \ge v,$$

holds on $\mathbb{M} \times [0, T]$, then we have

$$P_T u(\cdot, T)(x) \ge u(x, 0) + \int_0^T P_s v(\cdot, s)(x) ds.$$

Proof Let $f, g \in C_0^\infty(\mathbb{M})$, $f, g \ge 0$. We claim that we must have

$$\int_{\mathbb{M}} g P_T(f u(\cdot, T)) d\mu - \int_{\mathbb{M}} g f u(x, 0) d\mu \ge -\|\sqrt{\Gamma(f)}\|_\infty \int_0^T \int_{\mathbb{M}} (P_t g)\sqrt{\Gamma(u)} d\mu dt \tag{3.2}$$

$$- \|\sqrt{\Gamma(f)}\|_\infty \int_0^T \|\sqrt{\Gamma(P_t g)}\|_2 \|u(\cdot, t)\|_2 dt + \int_{\mathbb{M}} g \int_0^T P_t(f v(\cdot, t)) d\mu dt,$$

where for every $1 \le p \le \infty$ and a measurable F, we have let $||F||_p = ||F||_{L^p(\mathbb{M})}$. To establish (3.2) we consider the function

$$\phi(t) = \int_{\mathbb{M}} g P_t(f u(\cdot, t)) d\mu.$$

Differentiating ϕ we find

$$\begin{aligned}\phi'(t) &= \int_{\mathbb{M}} gP_t \left(L(fu) + f\frac{\partial u}{\partial t} \right) d\mu \\ &= \int_{\mathbb{M}} gP_t \left((Lf)u + 2\Gamma(f,u) + fLu + f\frac{\partial u}{\partial t} \right) d\mu \\ &\geq \int_{\mathbb{M}} gP_t \left((Lf)u + 2\Gamma(f,u) \right) d\mu + \int_{\mathbb{M}} gP_t(fv) d\mu.\end{aligned}$$

Since

$$\begin{aligned}\int_{\mathbb{M}} gP_t \left((Lf)u \right) d\mu &= \int_{\mathbb{M}} (P_t g)(Lf) u d\mu \\ &= -\int_{\mathbb{M}} \Gamma(f, u(P_t g)) d\mu \\ &= -\left(\int_{\mathbb{M}} P_t g \Gamma(f,u) + u\Gamma(f, P_t g) d\mu \right),\end{aligned}$$

we obtain

$$\phi'(t) \geq \int_{\mathbb{M}} P_t g \Gamma(f,u) d\mu - \int_{\mathbb{M}} u\Gamma(f, P_t g) d\mu + \int_{\mathbb{M}} gP_t(fv) d\mu.$$

Now, we can bound

$$\left| \int_{\mathbb{M}} (P_t g)\Gamma(f,u) d\mu \right| \leq \|\sqrt{\Gamma(f)}\|_\infty \int_{\mathbb{M}} (P_t g)\sqrt{\Gamma(u)} d\mu,$$

and for a.e. $t \in [0, T]$ the integral in the right-hand side is finite in view of the assumption (ii) above. We have thus obtained

$$\phi'(t) \geq -\|\sqrt{\Gamma(f)}\|_\infty \int_{\mathbb{M}} (P_t g)\sqrt{\Gamma(u)} d\mu - \int_{\mathbb{M}} u\Gamma(f, P_t g) d\mu + \int_{\mathbb{M}} gP_t(fv(\cdot,t)) d\mu.$$

As a consequence, we find

$$\begin{aligned}&\int_{\mathbb{M}} gP_T(fu(\cdot,T)) d\mu - \int_{\mathbb{M}} gfu(x,0) d\mu \\ &\geq -\|\sqrt{\Gamma(f)}\|_\infty \int_0^T \int_{\mathbb{M}} (P_t g)\sqrt{\Gamma(u)} d\mu dt - \int_0^T \int_{\mathbb{M}} u\Gamma(f, P_t g) d\mu dt \\ &\quad + \int_0^T \int_{\mathbb{M}} gP_t(fv(\cdot,t)) d\mu dt\end{aligned}$$

$$\geq -\|\sqrt{\Gamma(f)}\|_\infty \int_0^T \int_{\mathbb{M}} (P_t g)\sqrt{\Gamma(u)}d\mu dt - \int_0^T \|u(\cdot,t)\|_2 \|\Gamma(f, P_t g)\|_2 dt$$

$$+ \int_{\mathbb{M}} g \int_0^T P_t(f v(\cdot,t))dt d\mu$$

$$\geq -\|\sqrt{\Gamma(f)}\|_\infty \int_0^T \int_{\mathbb{M}} (P_t g)\sqrt{\Gamma(u)}d\mu dt - \|\sqrt{\Gamma(f)}\|_\infty \int_0^T \|u(\cdot,t)\|_2 \|\sqrt{\Gamma(P_t g)}\|_2 dt$$

$$+ \int_{\mathbb{M}} g \int_0^T P_t(f v(\cdot,t))dt d\mu,$$

which proves (3.2). Let now $h_k \in C_0^\infty(\mathbb{M})$ be a sequence such that $0 \leq h_k \leq 1$, $\|\Gamma(h_k)\|_\infty \to 0$ and h_k increases to 1. Using h_k in place of f in (3.2), and letting $k \to \infty$, gives

$$\int_{\mathbb{M}} g P_T(u(\cdot,T))d\mu - \int_{\mathbb{M}} g u(x,0)d\mu \geq \int_{\mathbb{M}} g \int_0^T P_t(v(\cdot,t))dt d\mu.$$

We observe that the assumption on v and Minkowski's integral inequality guarantee that the function $x \to \int_0^T P_t(v(\cdot,t))(x)dt$ belongs to $L^q(\mathbb{M})$. We have in fact

$$\begin{aligned}
\left(\int_{\mathbb{M}} \left|\int_0^T P_t(v(\cdot,t))dt\right|^q d\mu\right)^{\frac{1}{q}} &\leq \int_0^T \left|\int_{\mathbb{M}} |P_t(v(\cdot,t))|^q d\mu\right|^{\frac{1}{q}} dt \\
&\leq \int_0^T \left|\int_{\mathbb{M}} |v(\cdot,t)|^q d\mu\right|^{\frac{1}{q}} dt \\
&\leq T^{\frac{1}{q'}} \left(\int_0^T \int_{\mathbb{M}} |v(\cdot,t)|^q d\mu dt\right)^{\frac{1}{q}} < \infty.
\end{aligned}$$

Since this must hold for every non negative $g \in C_0^\infty(\mathbb{M})$, we conclude that

$$P_T(u(\cdot,T))(x) \geq u(x,0) + \int_0^T P_s(v(\cdot,s))(x)ds,$$

which completes the proof. □

We are in position to prove the first gradient bound for the semigroup P_t.

Proposition 3.13 *If f is a smooth function in $\mathcal{D}(L)$, then for every $t \geq 0$ and $x \in \mathbb{M}$,*

$$\sqrt{\Gamma(P_t f)}(x) \leq e^{-\rho t} P_t \sqrt{\Gamma(f)}(x).$$

Proof We fix $T > 0$ and consider the functional

$$\Phi(x, t) = e^{-\rho t} \sqrt{\Gamma(P_{T-t} f)}(x).$$

We first assume that $(x, t) \to \Gamma(P_t f)(x) > 0$ on $\mathbb{M} \times [0, T]$. From the previous lemma, we have $\Phi(t) \in L^2(\mathbb{M})$. Moreover $\Gamma(\Phi)(t) = e^{-2\rho t} \frac{\Gamma(\Gamma(P_{T-t} f))}{4\Gamma(P_{T-t} f)}$. So, we have $\Gamma(\Phi)(t) \le e^{-2\rho t}(\Gamma_2(P_{T-t} f) - \rho \Gamma(P_t f))$. Therefore, again from the previous proposition , we deduce that $\Gamma(\Phi)(t) \in L^1(\mathbb{M})$. Next, we easily compute that

$$\frac{\partial \Phi}{\partial t} + L\Phi = e^{-\rho t} \left(\frac{\Gamma_2(P_{T-t} f)}{\sqrt{\Gamma(P_{T-t} f)}} - \frac{\Gamma(\Gamma(P_{T-t} f))}{4\Gamma(P_{T-t} f)^{3/2}} - \rho \sqrt{\Gamma(P_{T-t} f)} \right).$$

Thus,

$$\frac{\partial \Phi}{\partial t} + L\Phi \ge 0.$$

We can then use the parabolic comparison theorem to infer that

$$\sqrt{\Gamma(P_T f)} \le e^{-\rho T} P_T \left(\sqrt{\Gamma(f)} \right).$$

If $(x, t) \to \Gamma(P_t f)(x)$ vanishes on $\mathbb{M} \times [0, T]$, we consider the functional

$$\Phi(t) = e^{-\rho t} g_\varepsilon(\Gamma(P_{T-t} f)),$$

where, for $0 < \varepsilon < 1$,

$$g_\varepsilon(y) = \sqrt{y + \varepsilon^2} - \varepsilon.$$

Since $\Phi(t) \in L^2(\mathbb{M})$, an argument similar to that above (details are let to the reader) shows that

$$g_\varepsilon(\Gamma(P_T f)) \le e^{-\rho T} P_T \left(g_\varepsilon(\Gamma(f)) \right).$$

Letting $\varepsilon \to 0$, we conclude that

$$\sqrt{\Gamma(P_T f)} \le e^{-\rho T} P_T \left(\sqrt{\Gamma(f)} \right).$$

□

We now prove the promised stochastic completeness result:

Theorem 3.14 *For $t \ge 0$, one has $P_t 1 = 1$.*

Proof Let $f, g \in C_0^\infty(\mathbb{M})$, we have

$$\int_{\mathbb{M}} (P_t f - f) g d\mu = \int_0^t \int_{\mathbb{M}} \left(\frac{\partial}{\partial s} P_s f \right) g d\mu ds$$
$$= \int_0^t \int_{\mathbb{M}} (L P_s f) \, g d\mu ds = - \int_0^t \int_{\mathbb{M}} \Gamma(P_s f, g) d\mu ds.$$

By means of the previous Proposition and Cauchy-Schwarz inequality, we find

$$\left| \int_{\mathbb{M}} (P_t f - f) g d\mu \right| \le \left(\int_0^t e^{-\rho s} ds \right) \sqrt{\|\Gamma(f)\|_\infty} \int_{\mathbb{M}} \Gamma(g)^{\frac{1}{2}} d\mu. \tag{3.3}$$

We now apply the previous inequality with $f = h_n$, and then let $n \to \infty$. Since by Beppo Levi's monotone convergence theorem we have $P_t h_n(x) \nearrow P_t 1(x)$ for every $x \in \mathbb{M}$, we see that the left-hand side converges to $\int_{\mathbb{M}} (P_t 1 - 1) g d\mu$. We thus reach the conclusion

$$\int_{\mathbb{M}} (P_t 1 - 1) g d\mu = 0, \quad g \in C_0^\infty(\mathbb{M}).$$

It follows that $P_t 1 = 1$. □

3.6 Convergence to Equilibrium, Poincaré and Log-Sobolev Inequalities

Let $\mathbb{M}$ be a complete n-dimensional Riemannian manifold and denote by L its Laplace-Beltrami operator. As usual, we denote by P_t the heat semigroup generated by L. Throughout the section, we will assume that the Ricci curvature of $\mathbb{M}$ is bounded from below by $\rho > 0$. We recall that this is equivalent to the fact that for every $f \in C^\infty(\mathbb{M})$,

$$\Gamma_2(f, f) \ge \frac{1}{n} (Lf)^2 + \rho \Gamma(f, f).$$

Readers knowing Riemannian geometry know that from Bonnet-Myers theorem, the manifold needs to be compact and we therefore expect the semigroup to converge to equilibrium. However, our goal will be to not use the Bonnet-Myers theorem, because eventually we shall provide a proof of this fact using semigroup theory. Thus the results in this section will not use the compactness of $\mathbb{M}$.

Lemma 3.15 *The Riemannian measure μ is finite, i.e. $\mu(\mathbb{M}) < +\infty$ and for every $f \in L^2_\mu(\mathbb{M})$, the following convergence holds pointwise and in $L^2(\mathbb{M}, \mu)$,*

$$P_t f \to_{t\to+\infty} \frac{1}{\mu(\mathbb{M})} \int_{\mathbb{M}} f d\mu.$$

Proof Let $f, g \in C_0^\infty(\mathbb{M})$, we have

$$\begin{aligned}\int_{\mathbb{M}} (P_t f - f) g d\mu &= \int_0^t \int_{\mathbb{M}} \left(\frac{\partial}{\partial s} P_s f\right) g d\mu ds \\ &= \int_0^t \int_{\mathbb{M}} (L P_s f)\, g d\mu ds = -\int_0^t \int_{\mathbb{M}} \Gamma(P_s f, g) d\mu ds.\end{aligned}$$

By means of Cauchy-Schwarz inequality, we find

$$\left| \int_{\mathbb{M}} (P_t f - f) g d\mu \right| \le \left(\int_0^t e^{-2\rho s} ds \right) \sqrt{\|\Gamma(f)\|_\infty} \int_{\mathbb{M}} \Gamma(g)^{\frac{1}{2}} d\mu. \qquad (3.4)$$

Now it is seen from spectral theorem that in $L^2(\mathbb{M})$ we have a convergence $P_t f \to P_\infty f$, where $P_\infty f$ belongs to the domain of L. Moreover $L P_\infty f = 0$. By ellipticity of L we deduce that $P_\infty f$ is a smooth function. Since $L P_\infty f = 0$, we have $\Gamma(P_\infty f) = 0$ and therefore $P_\infty f$ is constant.

Let us now assume that $\mu(\mathbb{M}) = +\infty$. This implies in particular that $P_\infty f = 0$ because no constant besides 0 is in $L^2(\mathbb{M})$. Using then (3.4) and letting $t \to +\infty$, we infer

$$\left| \int_{\mathbb{M}} f g d\mu \right| \le \left(\int_0^{+\infty} e^{-2\rho s} ds \right) \sqrt{\|\Gamma(f)\|_\infty} \int_{\mathbb{M}} \Gamma(g)^{\frac{1}{2}} d\mu.$$

Let us assume $g \ge 0$, $g \ne 0$ and take for f the usual localizing sequence h_n. Letting $n \to \infty$, we deduce

$$\int_{\mathbb{M}} g d\mu \le 0,$$

which is clearly absurd. As a consequence $\mu(\mathbb{M}) < +\infty$.

The invariance of μ implies then

$$\int_{\mathbb{M}} P_\infty f d\mu = \int_{\mathbb{M}} f d\mu,$$

and thus

$$P_\infty f = \frac{1}{\mu(\mathbb{M})} \int_{\mathbb{M}} f d\mu.$$

Finally, using the Cauchy-Schwarz inequality, we find that for $x \in \mathbb{M}$, $f \in L^2(\mathbb{M})$, $s, t, \tau \geq 0$,

$$\begin{aligned}
|P_{t+\tau} f(x) - P_{s+\tau} f(x)| &= |P_\tau (P_t f - P_s f)(x)| \\
&= \left| \int_{\mathbb{M}} p(\tau, x, y)(P_t f - P_s f)(y)\mu(dy) \right| \\
&\leq \int_{\mathbb{M}} p(\tau, x, y)^2 \mu(dy) \| P_t f - P_s f \|_2^2 \\
&\leq p(2\tau, x, x) \| P_t f - P_s f \|_2^2 .
\end{aligned}$$

Thus, we also have

$$P_t f(x) \to_{t \to +\infty} \frac{1}{\mu(\mathbb{M})} \int_{\mathbb{M}} f d\mu .$$

□

Proposition 3.16 *The following Poincaré inequality is satisfied: For $f \in \mathcal{D}(L)$,*

$$\frac{1}{\mu(\mathbb{M})} \int_{\mathbb{M}} f^2 d\mu \leq \left(\frac{1}{\mu(\mathbb{M})} \int_{\mathbb{M}} f d\mu \right)^2 + \frac{n-1}{n\rho} \frac{1}{\mu(\mathbb{M})} \int_{\mathbb{M}} \Gamma(f, f) d\mu .$$

Proof Let $f \in C_0^\infty(\mathbb{M})$. We have by assumption

$$\Gamma_2(f, f) \geq \frac{1}{n}(Lf)^2 + \rho \Gamma(f, f).$$

Therefore, by integrating the latter inequality we obtain

$$\int_{\mathbb{M}} \Gamma_2(f, f) d\mu \geq \frac{1}{n}(Lf)^2 + \rho \int_{\mathbb{M}} \Gamma(f, f) d\mu .$$

But we have

$$\int_{\mathbb{M}} \Gamma_2(f, f) d\mu = - \int_{\mathbb{M}} \Gamma(f, Lf) d\mu = \int_{\mathbb{M}} (Lf)^2 d\mu .$$

Therefore we obtain

$$\int_{\mathbb{M}} (Lf)^2 d\mu \geq \rho \int_{\mathbb{M}} \Gamma(f, f) d\mu = - \frac{n\rho}{n-1} \int_{\mathbb{M}} f L f d\mu .$$

By density, this last inequality is seen to hold for every function $f \in \mathcal{D}(L)$. It means that the L^2 spectrum of $-L$ lies in $\{0\} \cup \left[\frac{n\rho}{n-1}, +\infty \right)$. Since from the previous proof the projection of f onto the 0-eigenspace is given by $\frac{1}{\mu(\mathbb{M})}\int_{\mathbb{M}} f d\mu$, we deduce that

$$\int_{\mathbb{M}}\left(f-\frac{1}{\mu(\mathbb{M})}\int_{\mathbb{M}} f d\mu\right)^2 d\mu \le \frac{n-1}{n\rho}\int_{\mathbb{M}}\Gamma(f,f)d\mu,$$

which is exactly Poincaré inequality □

As observed in the proof, the Poincaré inequality

$$\int_{\mathbb{M}} f^2 d\mu \le \left(\int_{\mathbb{M}} f d\mu\right)^2 + \frac{n-1}{n\rho}\int_{\mathbb{M}}\Gamma(f,f)d\mu.$$

is equivalent to the fact that the L^2 spectrum of $-L$ lies in $\{0\}\cup\left[\frac{n\rho}{n-1},+\infty\right)$, or in other words that $-L$ has a spectral gap of size at least $\frac{n\rho}{n-1}$. This is Lichnerowicz estimate. It is sharp, because on the n-dimensional sphere it is known that $\rho = n-1$ and that the first non zero eigenvalue is exactly equal to n.

As a basic consequence of the spectral theorem and of the above spectral gap estimate, we also get the rate convergence to equilibrium in $L^2(\mathbb{M},\mu)$ for P_t.

Proposition 3.17 *Let $f \in L^2(\mathbb{M},\mu)$, then for $t \ge 0$,*

$$\left\|P_t f - \int_{\mathbb{M}} f d\mu\right\|_2^2 \le e^{-\frac{2n\rho}{n-1}t}\left\|f-\int_{\mathbb{M}} f d\mu\right\|_2^2.$$

As we have just seen, the convergence in L^2 of P_t is connected and actually equivalent to the Poincaré inequality.

We now turn to the so-called log-Sobolev inequality which is connected to the convergence in entropy for P_t. This inequality is much stronger (and more useful) than the Poincaré inequality. To simplify a little the expressions, we assume in the sequel that $\mu(\mathbb{M}) = 1$ (Otherwise, just replace μ by $\frac{\mu}{\mu(\mathbb{M})}$ in the following results).

Theorem 3.18 *For $f \in \mathcal{D}(L)$, $f \ge 0$,*

$$\int_{\mathbb{M}} f^2 \ln f^2 d\mu \le \int_{\mathbb{M}} f^2 d\mu \ln\left(\int_{\mathbb{M}} f^2 d\mu\right) + \frac{2}{\rho}\int_{\mathbb{M}}\Gamma(f,f)d\mu.$$

Proof By considering $\sqrt{f}$ instead of f, it is enough to show that if f is positive,

$$\int_{\mathbb{M}} f\ln f d\mu \le \int_{\mathbb{M}} f d\mu \ln\left(\int_{\mathbb{M}} f d\mu\right) + \frac{1}{2\rho}\int_{\mathbb{M}}\frac{\Gamma(f,f)}{f}d\mu.$$

We now have

$$\begin{aligned}\int_{\mathbb{M}} f \ln f d\mu - \int_{\mathbb{M}} f d\mu \ln\left(\int_{\mathbb{M}} f d\mu\right) &= -\int_0^{+\infty} \frac{d}{dt}\int_{\mathbb{M}} P_t f \ln P_t f d\mu dt \\ &= -\int_0^{+\infty}\int_{\mathbb{M}} L P_t f \ln P_t f d\mu dt \\ &= \int_0^{+\infty}\int_{\mathbb{M}} \Gamma(P_t f, \ln P_t f) d\mu dt \\ &= \int_0^{+\infty}\int_{\mathbb{M}} \frac{\Gamma(P_t f, P_t f)}{P_t f} d\mu dt\end{aligned}$$

Now, we know that

$$\Gamma(P_t f, P_t f) \le e^{-2\rho t}\left(P_t\sqrt{\Gamma(f,f)}\right)^2.$$

And, from Cauchy-Schwarz inequality,

$$(P_t\sqrt{\Gamma(f,f)})^2 \le P_t \frac{\Gamma(f,f)}{f} P_t f.$$

Therefore,

$$\int_{\mathbb{M}} f \ln f d\mu - \int_{\mathbb{M}} f d\mu \ln\left(\int_{\mathbb{M}} f d\mu\right) \le \int_0^{+\infty} e^{-2\rho t} dt \int_{\mathbb{M}} \frac{\Gamma(f,f)}{f} d\mu,$$

which is the required inequality. □

We finally prove the entropic convergence of P_t.

Theorem 3.19 *Let $f \in L^2(\mathbb{M}, \mu)$, $f \ge 0$. For $t \ge 0$,*

$$\begin{aligned}&\int_{\mathbb{M}} P_t f \ln P_t f d\mu - \int_{\mathbb{M}} P_t f d\mu \ln\left(\int_{\mathbb{M}} P_t f d\mu\right) \\ &\le e^{-2\rho t}\left(\int_{\mathbb{M}} f \ln f d\mu - \int_{\mathbb{M}} f d\mu \ln\left(\int_{\mathbb{M}} f d\mu\right)\right).\end{aligned}$$

Proof Let us assume $\int_{\mathbb{M}} f d\mu = 1$, otherwise we use the following argument with $\frac{f}{\int_{\mathbb{R}^n} f d\mu}$ and consider the functional

$$\Phi(t) = \int_{\mathbb{M}} P_t f \ln P_t f d\mu,$$

which by differentiation gives

$$\Phi'(t) = \int_{\mathbb{M}} LP_t f \ln P_t f d\mu = -\int_{\mathbb{M}} \frac{\Gamma(P_t f)}{P_t f} d\mu.$$

Using now the log-Sobolev inequality, we obtain

$$\Phi'(t) \le -2\rho\Phi(t).$$

The Gronwall's differential inequality implies then:

$$\Phi(t) \le e^{-2\rho t}\Phi(0),$$

that is

$$\int_{\mathbb{M}} P_t f \ln P_t f d\mu \le e^{-2\rho t} \int_{\mathbb{M}} f \ln f d\mu.$$

□

3.7 The Li-Yau Inequality

Let $\mathbb{M}$ be a complete n-dimensional Riemannian manifold and, as usual, denote by L its Laplace-Beltrami operator. Throughout the section, we will assume again that the Ricci curvature of $\mathbb{M}$ is bounded from below by $\rho \in \mathbb{R}$. The section is devoted to the proof of a beautiful inequality due to P. Li and S.T. Yau.

Henceforth, we will indicate $C_b^\infty(\mathbb{M}) = C^\infty(\mathbb{M}) \cap L^\infty(\mathbb{M})$.

Lemma 3.20 *Let $f \in C_b^\infty(\mathbb{M})$, $f > 0$ and $T > 0$, and consider the function*

$$\phi(x,t) = (P_{T-t} f)(x)\Gamma(\ln P_{T-t} f)(x),$$

which is defined on $\mathbb{M} \times [0,T]$. We have

$$L\phi + \frac{\partial \phi}{\partial t} = 2(P_{T-t} f)\Gamma_2(\ln P_{T-t} f).$$

Proof Let for simplicity $g(x,t) = P_{T-t} f(x)$. A simple computation gives

$$\frac{\partial \phi}{\partial t} = g_t \Gamma(\ln g) + 2g\Gamma\left(\ln g, \frac{g_t}{g}\right).$$

On the other hand,

$$L\phi = Lg\Gamma(\ln g) + gL\Gamma(\ln g) + 2\Gamma(g, \Gamma(\ln g)).$$

Combining these equations we obtain

$$L\phi + \frac{\partial \phi}{\partial t} = gL\Gamma(\ln g) + 2\Gamma(g, \Gamma(\ln g)) + 2g\Gamma\left(\ln g, \frac{g_t}{g}\right).$$

From (4.4) we see that

$$\begin{aligned} 2g\Gamma_2(\ln g) &= g(L\Gamma(\ln g) - 2\Gamma(\ln g, L(\ln g))) \\ &= gL\Gamma(\ln g) - 2g\Gamma(\ln g, L(\ln g)). \end{aligned}$$

Observing that

$$L(\ln g) = -\frac{\Gamma(g)}{g^2} - \frac{g_t}{g},$$

we conclude that

$$L\phi + \frac{\partial \phi}{\partial t} = 2(P_{T-t}f)\Gamma_2(\ln P_{T-t}f).$$

□

We now turn to an important variational inequality that shall extensively be used throughout these sections. Given a function $f \in C_b^\infty(\mathbb{M})$ and $\varepsilon > 0$, we let $f_\varepsilon = f + \varepsilon$.

Suppose that $T > 0$, and $x \in \mathbb{M}$ be given. For a function $f \in C_b^\infty(\mathbb{M})$ with $f \geq 0$ we define for $t \in [0, T]$,

$$\Phi(t) = P_t\left((P_{T-t}f_\varepsilon)\Gamma(\ln P_{T-t}f_\varepsilon)\right).$$

Theorem 3.21 *Let $a \in C^1([0, T], [0, \infty))$ and $\gamma \in C((0, T), \mathbb{R})$. Given $f \in C_0^\infty(\mathbb{M})$, with $f \geq 0$, we have*

$$\begin{aligned} &a(T)P_T\left(f_\varepsilon\Gamma(\ln f_\varepsilon)\right) - a(0)(P_Tf_\varepsilon)\Gamma(\ln P_Tf_\varepsilon) \\ &\geq \int_0^T \left(a' + 2\rho a - \frac{4a\gamma}{n}\right)\Phi(s)ds + \left(\frac{4}{n}\int_0^T a\gamma ds\right)LP_Tf_\varepsilon - \left(\frac{2}{n}\int_0^T a\gamma^2 ds\right)P_Tf_\varepsilon. \end{aligned}$$

Proof Let $f \in C^\infty(\mathbb{M})$, $f \geq 0$. Consider the function

$$\phi(x, t) = a(t)(P_{T-t}f)(x)\Gamma(\ln P_{T-t}f)(x) + b(t)(P_{T-t}f)(x)\Gamma^Z(\ln P_{T-t}f)(x).$$

Applying the previous lemma and the curvature-dimension inequality, we obtain

$$\begin{aligned} L\phi + \frac{\partial \phi}{\partial t} &= a'(P_{T-t}f)\Gamma(\ln P_{T-t}f) + 2a(P_{T-t}f)\Gamma_2(\ln P_{T-t}f) \\ &\geq \left(a' + 2\rho a\right)(P_{T-t}f)\Gamma(\ln P_{T-t}f) + \frac{2a}{n}(P_{T-t}f)(L(\ln P_{T-t}f))^2. \end{aligned}$$

But, we have

$$(L(\ln P_{T-t}f))^2 \geq 2\gamma L(\ln P_{T-t}f) - \gamma^2,$$

and

$$L(\ln P_{T-t}f) = \frac{LP_{T-t}f}{P_{T-t}f} - \Gamma(\ln P_{T-t}f).$$

Therefore we obtain,

$$L\phi + \frac{\partial \phi}{\partial t} \geq \left(a' + 2\rho a - \frac{4a\gamma}{n}\right)(P_{T-t}f)\Gamma(\ln P_{T-t}f) + \frac{4a\gamma}{n}LP_{T-t}f - \frac{2a\gamma^2}{n}P_{T-t}f.$$

We then easily reach the conclusion by using the parabolic comparison theorem in L^∞. □

As a first application the previous result, we derive a family of Li-Yau type inequalities. We choose the function γ in a such a way that

$$a' - \frac{4a\gamma}{n} + 2\rho a = 0.$$

That is

$$\gamma = \frac{n}{4}\left(\frac{a'}{a} + 2\rho\right).$$

Integrating the inequality from 0 to T, and denoting $V = \sqrt{a}$, we obtain the following result.

Proposition 3.22 *Let* $V : [0, T] \to \mathbb{R}^+$ *be a smooth function such that*

$$V(0) = 1, V(T) = 0.$$

We have

$$\Gamma(\ln P_T f) \leq \left(1 - 2\rho\int_0^T V^2(s)ds\right)\frac{LP_T f}{P_T f} + \frac{n}{2}\left(\int_0^T V'(s)^2 ds + \rho^2\int_0^T V(s)^2 ds - \rho\right).$$

A first family of interesting inequalities may be obtained with the choice

$$V(t) = \left(1 - \frac{t}{T}\right)^\alpha, \alpha > \frac{1}{2}.$$

In this case we have

$$\int_0^T V(s)^2 ds = \frac{T}{2\alpha+1}$$

and

$$\int_0^T V'(s)^2 ds = \frac{\alpha^2}{(2\alpha-1)T},$$

In particular, we therefore proved the celebrated Li-Yau inequality:

Theorem 3.23 (Li-Yau Inequality) *If $f \in C_0^\infty(\mathbb{M})$, $f \geq 0$. For $\alpha > \frac{1}{2}$ and $T > 0$, we have*

$$\Gamma(\ln P_T f) \leq \left(1 - \frac{2\rho T}{2\alpha+1}\right)\frac{LP_T f}{P_T f} + \frac{n}{2}\left(\frac{\alpha^2}{(2\alpha-1)T} + \frac{\rho^2 T}{2\alpha+1} - \rho\right).$$

In the case, $\rho = 0$ and $\alpha = 1$, it reduces to the beautiful sharp inequality:

$$\Gamma(\ln P_t f) \leq \frac{LP_t f}{P_t f} + \frac{n}{2t}, \qquad t > 0. \tag{3.5}$$

Although in the sequel, we shall first focus on the case $\rho = 0$, let us presently briefly discuss the case $\rho > 0$.

Using the Li-Yau inequality with $\alpha = 3/2$ leads to the Bakry-Qian inequality:

$$\frac{LP_t f}{P_t f} \leq \frac{n\rho}{4}, \qquad t \geq \frac{2}{\rho}.$$

Also, by using

$$V(t) = \frac{e^{-\frac{\rho t}{3}}(e^{-\frac{2\rho t}{3}} - e^{-\frac{2\rho T}{3}})}{1 - e^{-\frac{2\rho T}{3}}},$$

we obtain the following inequality:

$$\Gamma(\ln P_t f) \leq e^{-\frac{2\rho t}{3}}\frac{LP_t f}{P_t f} + \frac{n\rho}{3}\frac{e^{-\frac{4\rho t}{3}}}{1 - e^{-\frac{2\rho t}{3}}}, \qquad t \geq 0.$$

3.8 The Parabolic Harnack Inequality

Let $\mathbb{M}$ be a complete n-dimensional Riemannian manifold and, as usual, denote by L its Laplace-Beltrami operator. Throughout the section, we will assume that the

Ricci curvature of $\mathbb{M}$ is bounded from below by $-K$ with $K \geq 0$. Our purpose is to prove a first important consequence of the Li-Yau inequality: The parabolic Harnack inequality.

Theorem 3.24 *Let $f \in L^\infty(\mathbb{M})$, $f \geq 0$. For every $s \leq t$ and $x, y \in \mathbb{M}$,*

$$P_s f(x) \leq P_t f(y) \left(\frac{t}{s}\right)^{\frac{n}{2}} \exp\left(\frac{d(x,y)^2}{4(t-s)} + \frac{Kd(x,y)^2}{6} + \frac{nK}{4}(t-s)\right). \tag{3.6}$$

Proof We first assume that $f \in C_0^\infty(\mathbb{M})$. Let $x, y \in \mathbb{M}$ and let $\gamma : [s,t] \to \mathbb{M}$, $s < t$ be an absolutely continuous path such that $\gamma(s) = x$, $\gamma(t) = y$. We write the Li-Yau inequality in the form

$$\Gamma(\ln P_u f(x)) \leq a(u) \frac{LP_u f(x)}{P_u f(x)} + b(u), \tag{3.7}$$

where

$$a(u) = 1 + \frac{2K}{3} u,$$

and

$$b(u) = \frac{n}{2}\left(\frac{1}{u} + \frac{K^2 u}{3} + K\right).$$

Let us now consider

$$\phi(u) = \ln P_u f(\gamma(u)).$$

We compute

$$\phi'(u) = (\partial_u \ln P_u f)(\gamma(u)) + \langle \nabla \ln P_u f(\gamma(u)), \gamma'(u) \rangle.$$

Now, for every $\lambda > 0$, we have

$$\langle \nabla \ln P_u f(\gamma(u)), \gamma'(u) \rangle \geq -\frac{1}{2\lambda^2} \| \nabla \ln P_u f(x) \|^2 - \frac{\lambda^2}{2} \| \gamma'(u) \|^2.$$

Choosing $\lambda = \sqrt{\frac{a(u)}{2}}$ and using then (3.7) yields

$$\phi'(u) \geq -\frac{b(u)}{a(u)} - \frac{1}{4} a(u) \| \gamma'(u) \|^2.$$

By integrating this inequality from s to t we get as a result.

$$\ln P_t f(y) - \ln P_s f(x) \geq -\int_s^t \frac{b(u)}{a(u)} du - \frac{1}{4}\int_s^t a(u)\|\gamma'(u)\|^2 du.$$

We now minimize the quantity $\int_s^t a(u)\|\gamma'(u)\|^2 du$ over the set of absolutely continuous paths such that $\gamma(s) = x, \gamma(t) = y$. By using reparametrization of paths, it is seen that

$$\int_s^t a(u)\|\gamma'(u)\|^2 du \geq \frac{d^2(x,y)}{\int_s^t \frac{dv}{a(v)}},$$

with equality achieved for $\gamma(u) = \sigma\left(\frac{\int_s^u \frac{dv}{a(v)}}{\int_s^t \frac{dv}{a(v)}}\right)$ where $\sigma : [0,1] \to \mathbb{M}$ is a unit geodesic joining x and y. As a conclusion,

$$P_s f(x) \leq \exp\left(\int_s^t \frac{b(u)}{a(u)} du + \frac{d^2(x,y)}{4\int_s^t \frac{dv}{a(v)}}\right) P_t f(y).$$

Now, from Cauchy-Schwarz inequality we have

$$\int_s^t \frac{dv}{a(v)} \geq \frac{(t-s)^2}{\int_s^t a(v)dv} = \frac{(t-s)^2}{(t-s) + \frac{2K}{3}(t-s)^2},$$

and also

$$\int_s^t \frac{b(u)}{a(u)} du = \frac{n}{2}\int_s^t \frac{1/u + K^2 u/3 + K}{1 + 2Ku/3} du \leq \frac{n}{2}\int_s^t \left(\frac{1}{u} + \frac{K}{2}\right) du$$

This proves (3.6) when $f \in C_0^\infty(\mathbb{M})$. We can then extend the result to $f \in L^\infty(\mathbb{M})$ by considering the approximations $h_n P_\tau f \in C_0^\infty(\mathbb{M})$, where $h_n \in C_0^\infty(\mathbb{M})$, $h_n \geq 0$, $h_n \to_{n\to\infty} 1$ and let $n \to \infty$ and $\tau \to 0$. □

The following result represents an important consequence of Theorem 3.24.

Corollary 3.25 *Let $p(x, y, t)$ be the heat kernel on $\mathbb{M}$. For every $x, y, z \in \mathbb{M}$ and every $0 < s < t < \infty$ one has*

$$p(x,y,s) \leq p(x,z,t)\left(\frac{t}{s}\right)^{\frac{n}{2}} \exp\left(\frac{d(y,z)^2}{4(t-s)} + \frac{Kd(y,z)^2}{6} + \frac{nK}{4}(t-s)\right).$$

Proof Let $\tau > 0$ and $x \in \mathbb{M}$ be fixed. By the hypoellipticity of $L - \partial_t$, we know that $p(x, \cdot, \cdot + \tau) \in C^\infty(\mathbb{M} \times (-\tau, \infty))$. From the semigroup property we have

$$p(x, y, s + \tau) = P_s(p(x, \cdot, \tau))(y)$$

and

$$p(x, z, t + \tau) = P_t(p(x, \cdot, \tau))(z)$$

Since we cannot apply Theorem 3.24 directly to $u(y, t) = P_t(p(x, \cdot, \tau))(y)$, we consider $u_n(y, t) = P_t(h_n p(x, \cdot, \tau))(y)$, where $h_n \in C_0^\infty(\mathbb{M})$, $0 \le h_n \le 1$, and $h_n \nearrow 1$. From (3.6) we find

$$\begin{aligned} P_s(h_n p(x, \cdot, \tau))(y) \le P_t(h_n p(x, \cdot, \tau))(z) \left(\frac{t}{s}\right)^{\frac{n}{2}} \\ \times \exp\left(\frac{d(y, z)^2}{4(t - s)} + \frac{K d(y, z)^2}{6} + \frac{nK}{4}(t - s)\right) \end{aligned}$$

Letting $n \to \infty$, by Beppo Levi's monotone convergence theorem we obtain

$$p(x, y, s + \tau) \le p(x, z, t + \tau) \left(\frac{t}{s}\right)^{\frac{n}{2}} \exp\left(\frac{d(y, z)^2}{4(t - s)} + \frac{K d(y, z)^2}{6} + \frac{nK}{4}(t - s)\right)$$

The desired conclusion follows by letting $\tau \to 0$. □

A nice consequence of the parabolic Harnack inequality for the heat kernel is the following lower bound for the heat kernel:

Proposition 3.26 (Cheeger-Yau Lower Bound) *For $x, z \in \mathbb{M}$ and $t > 0$,*

$$p(x, z, t) \ge \frac{1}{(4\pi t)^{n/2}} \exp\left(-\frac{d(x, z)^2}{4t} - \frac{K d(x, z)^2}{6} - \frac{nK}{4} t\right).$$

Proof We just need to use the above Harnack inequality with $y = x$ and let $s \to 0$ using the asymptotics $\lim_{s \to 0} s^{n/2} p_s(x, x) = \frac{1}{(4\pi)^{n/2}}$. □

Observe that when $K = 0$, the inequality is sharp, since it is actually an equality on the Euclidean space!

3.9 The Gaussian Upper Bound

Let $\mathbb{M}$ be a complete n-dimensional Riemannian manifold and, as usual, denote by L its Laplace-Beltrami operator. As in the previous section, we will assume that the

Ricci curvature of $\mathbb{M}$ is bounded from below by $-K$ with $K \geq 0$. Our purpose in this section is to prove a Gaussian upper bound for the heat kernel. Our main tools are the parabolic Harnack inequality proved in the previous section and the following integrated maximum principle:

Proposition 3.27 *Let $g : \mathbb{M} \times \mathbb{R}_{\geq 0} \to \mathbb{R}$ be a non positive continuous function such that, in the sense of distributions,*

$$\frac{\partial g}{\partial t} + \frac{1}{2}\Gamma(g) \leq 0,$$

then, for every $f \in L^2(\mathbb{M}, \mu)$, we have

$$\int_{\mathbb{M}} e^{g(y,t)}(P_t f)^2(y) d\mu(y) \leq \int_{\mathbb{M}} e^{g(y,0)} f^2(y) d\mu(y).$$

Proof Since

$$\left(L - \frac{\partial}{\partial t}\right)(P_t f)^2 = 2P_t f \left(L - \frac{\partial}{\partial t}\right)(P_t f) + 2\Gamma(P_t f) = 2\Gamma(P_t f),$$

multiplying this identity by $h_n^2(y)e^{g(y,t)}$, where h_n is the usual localizing sequence, and integrating by parts, we obtain

$$\begin{aligned}
0 &= 2\int_0^\tau \int_{\mathbb{M}} h_n^2 e^g \Gamma(P_t f) d\mu(y) dt - \int_0^\tau \int_{\mathbb{M}} h_n^2 e^g \left(L - \frac{\partial}{\partial t}\right)(P_t f)^2 d\mu(y) dt \\
&= 2\int_0^\tau \int_{\mathbb{M}} h_n^2 e^g \Gamma(P_t f) d\mu(y) dt \\
&+ 4\int_0^\tau \int_{\mathbb{M}} h_n e^g P_t f \Gamma(h_n, P_t f) d\mu(y) dt \\
&+ 2\int_0^\tau \int_{\mathbb{M}} h_n^2 e^g P_t f \Gamma(P_t f, g) d\mu(y) dt - \int_0^\tau \int_{\mathbb{M}} h_n e^g (P_t f)^2 \frac{\partial g}{\partial t} d\mu(y) dt \\
&- \int_{\mathbb{M}} h_n e^g (P_t f)^2 d\mu(y)\Big|_{t=0} + \int_{\mathbb{M}} h_n e^g (P_t f)^2 d\mu(y)\Big|_{t=\tau} \\
&\geq 2\int_0^\tau \int_{\mathbb{M}} h_n^2 e^g \left(\Gamma(P_t f) + P_t f \Gamma(P_t f, g) + \frac{P_t f^2}{4}\Gamma(g)\right) d\mu(y) dt \\
&+ 4\int_0^\tau \int_{\mathbb{M}} h_n e^g P_t f \Gamma(h_n, P_t f) d\mu(y) dt \\
&+ \int_{\mathbb{M}} h_n e^g (P_t f)^2 d\mu(y)\Big|_{t=\tau} - \int_{\mathbb{M}} h_n e^g (P_t f)^2 d\mu(y)\Big|_{t=0}.
\end{aligned}$$

From this we conclude

$$\int_{\mathbb{M}} h_n e^g (P_t f)^2 d\mu(y)\Big|_{t=\tau} \le \int_{\mathbb{M}} h_n e^g (P_t f)^2 d\mu(y)\Big|_{t=0} - 4\int_0^\tau \int_{\mathbb{M}} h_n e^g P_t f \Gamma(h_n, P_t f) d\mu(y)dt.$$

We now claim that

$$\lim_{n\to\infty} \int_0^\tau \int_{\mathbb{M}} h_n e^g P_t f \Gamma(h_n, P_t f) d\mu(y)dt = 0.$$

To see this we apply Cauchy-Schwarz inequality which gives

$$\begin{aligned}
&\left|\int_0^\tau \int_{\mathbb{M}} h_n e^g P_t f \Gamma(h_n, P_t f) d\mu(y)dt\right| \\
&\le \left(\int_0^\tau \int_{\mathbb{M}} h_n^2 e^g (P_t f)^2 \Gamma(h_n) d\mu(y)dt\right)^{\frac{1}{2}} \left(\int_0^\tau \int_{\mathbb{M}} e^g \Gamma(P_t f) d\mu(y)dt\right)^{\frac{1}{2}} \\
&\le \left(\int_0^\tau \int_{\mathbb{M}} e^g (P_t f)^2 \Gamma(h_n) d\mu(y)dt\right)^{\frac{1}{2}} \left(\int_0^\tau \int_{\mathbb{M}} e^g \Gamma(P_t f) d\mu(y)dt\right)^{\frac{1}{2}} \to 0,
\end{aligned}$$

as $n \to \infty$. With the claim in hands we now let $n \to \infty$ in the above inequality obtaining

$$\int_{\mathbb{M}} e^{g(y,t)})(P_t f)^2(y) d\mu(y) \le \int_{\mathbb{M}} e^{g(y,0)}) f^2(y) d\mu(y)$$

□

We are now ready for the main bound of this section:

Theorem 3.28 *For any* $0 < \epsilon < 1$ *there exist positive constants* $C_1 = C_1(\epsilon)$ *and* $C_2 = C_2(n, \epsilon) > 0$, *such that for every* $x, y \in \mathbb{M}$ *and* $t > 0$ *one has*

$$p(x, y, t) \le \frac{C_1}{\mu(B(x, \sqrt{t}))^{\frac{1}{2}} \mu(B(y, \sqrt{t}))^{\frac{1}{2}}} \exp\left(C_2 K t - \frac{d(x, y)^2}{(4+\epsilon)t}\right).$$

Proof Given $T > 0$, and $\alpha > 0$ we fix $0 < \tau \le (1+\alpha)T$. For a function $\psi \in C_0^\infty(\mathbb{M})$, with $\psi \ge 0$, in $\mathbb{M} \times (0, \tau)$ we consider the function

$$f(y, t) = \int_{\mathbb{M}} p(y, z, t) p(x, z, T) \psi(z) d\mu(z), \quad x \in \mathbb{M}.$$

Since $f = P_t(p(x,\cdot,T)\psi)$, it satisfies the Cauchy problem

$$\begin{cases} Lf - f_t = 0 & \text{in } \mathbb{M} \times (0,\tau), \\ f(z,0) = p(x,z,T)\psi(z), & z \in \mathbb{M}. \end{cases}$$

Let $g : \mathbb{M} \times [0,\tau] \to \mathbb{R}$ be a non positive continuous function such that, in the sense of distributions,

$$\frac{\partial g}{\partial t} + \frac{1}{2}\Gamma(g) \leq 0.$$

From the previous lemma, we know that:

$$\int_{\mathbb{M}} e^{g(y,\tau)} f^2(y,\tau) d\mu(y) \leq \int_{\mathbb{M}} e^{g(y,0)} f^2(y,0) d\mu(y). \tag{3.8}$$

At this point we fix $x \in \mathbb{M}$ and for $0 < t \leq \tau$ consider the indicator function $\mathbf{1}_{B(x,\sqrt{t})}$ of the ball $B(x,\sqrt{t})$. Let $\psi_k \in C_0^\infty(\mathbb{M})$, $\psi_k \geq 0$, be a sequence such that $\psi_k \to \mathbf{1}_{B(x,\sqrt{t})}$ in $L^2(\mathbb{M})$, with supp $\psi_k \subset B(x, 100\sqrt{t})$. Slightly abusing the notation we now set

$$f(y,s) = P_s(p(x,\cdot,T)\mathbf{1}_{B(x,\sqrt{t})})(y) = \int_{B(x,\sqrt{t})} p(y,z,s)p(x,z,T)d\mu(z).$$

Thanks to the symmetry of $p(x,y,s) = p(y,x,s)$, we have

$$f(x,T) = \int_{B(x,\sqrt{t})} p(x,z,T)^2 d\mu(z). \tag{3.9}$$

Applying (3.8) to $f_k(y,s) = P_s(p(x,\cdot,T)\psi_k)(y)$, we find

$$\int_{\mathbb{M}} e^{g(y,\tau)} f_k^2(y,\tau) d\mu(y) \leq \int_{\mathbb{M}} e^{g(y,0)} f_k^2(y,0) d\mu(y). \tag{3.10}$$

At this point we observe that as $k \to \infty$

$$\left| \int_{\mathbb{M}} e^{g(y,\tau)} f_k^2(y,\tau) d\mu(y) - \int_{\mathbb{M}} e^{g(y,\tau)} f^2(y,\tau) d\mu(y) \right|$$
$$\leq 2 \| e^{g(\cdot,\tau)} \|_{L^\infty(\mathbb{M})} \| p(x,\cdot,T) \|_{L^2(\mathbb{M})} \| p(x,\cdot,\tau) \|_{L^\infty(B(x,110\sqrt{t}))} \| \psi_k - \mathbf{1}_{B(x,\sqrt{t})} \|_{L^2(\mathbb{M})} \to 0.$$

By similar considerations we find

$$\left| \int_{\mathbb{M}} e^{g(y,0)} f_k^2(y,0) d\mu(y) - \int_{\mathbb{M}} e^{g(y,0)} f^2(y,0) d\mu(y) \right|$$
$$\leq 2 \| e^{g(\cdot,0)} \|_{L^\infty(\mathbb{M})} \| p(x,\cdot,T) \|_{L^\infty(B(x,110\sqrt{t}))} \| \psi_k - \mathbf{1}_{B(x,\sqrt{t})} \|_{L^2(\mathbb{M})} \to 0.$$

Letting $k \to \infty$ in (3.10) we thus conclude that the same inequality holds with f_k replaced by $f(y,s) = P_s(p(x,\cdot,T)1_{B(x,\sqrt{t})})(y)$. This implies in particular the basic estimate

$$\inf_{z\in B(x,\sqrt{t})} e^{g(z,\tau)} \int_{B(x,\sqrt{t})} f^2(z,\tau)d\mu(z) \tag{3.11}$$

$$\le \int_{B(x,\sqrt{t})} e^{g(z,\tau)} f^2(z,\tau)d\mu(z) \le \int_{\mathbb{M}} e^{g(z,\tau)} f^2(z,\tau)d\mu(z)$$

$$\le \int_{\mathbb{M}} e^{g(z,0)} f^2(z,0)d\mu(z) = \int_{B(y,\sqrt{t})} e^{g(z,0)} p(x,z,T)^2 d\mu(z)$$

$$\le \sup_{z\in B(y,\sqrt{t})} e^{g(z,0)} \int_{B(y,\sqrt{t})} p(x,z,T)^2 d\mu(z).$$

At this point we choose in (3.11)

$$g(y,t) = g_x(y,t) = -\frac{d(x,y)^2}{2((1+2\alpha)T - t)}.$$

Using the fact that $\Gamma(d) \le 1$, one can easily check that g satisfies

$$\frac{\partial g}{\partial t} + \frac{1}{2}\Gamma(g) \le 0.$$

Taking into account that

$$\inf_{z\in B(x,\sqrt{t})} e^{g_x(z,\tau)} = \inf_{z\in B(x,\sqrt{t})} e^{-\frac{d(x,z)^2}{2((1+2\alpha)T-\tau)}} \ge e^{\frac{-t}{2((1+2\alpha)T-\tau)}},$$

if we now choose $\tau = (1+\alpha)T$, then from the previous inequality and from (3.9) we conclude that

$$\int_{B(x,\sqrt{t})} f^2(z,(1+\alpha)T)d\mu(z) \le \left(\sup_{z\in B(y,\sqrt{t})} e^{-\frac{d(x,z)^2}{2(1+2\alpha)T} + \frac{t}{2\alpha T}}\right) \int_{B(y,\sqrt{t})} p(x,z,T)^2 d\mu(z). \tag{3.12}$$

We now apply Theorem 3.24 which gives for every $z \in B(x,\sqrt{t})$

$$f(x,T)^2 \le f(z,(1+\alpha)T)^2(1+\alpha)^n e^{\frac{t}{2\alpha T} + \frac{Kt}{3} + \frac{nK\alpha T}{2}}.$$

Integrating this inequality on $B(x, \sqrt{t})$ we find

$$\left(\int_{B(y,\sqrt{t})} p(x,z,T)^2 d\mu(z)\right)^2 = f(x,T)^2$$
$$\leq \frac{(1+\alpha)^n e^{\frac{t}{2\alpha T}+\frac{Kt}{3}+\frac{nK\alpha T}{2}}}{\mu(B(x,\sqrt{t}))} \int_{B(x,\sqrt{t})} f^2(z,(1+\alpha)T)d\mu(z).$$

If we now use (3.12) in the last inequality we obtain

$$\int_{B(y,\sqrt{t})} p(x,z,T)^2 d\mu(z) \leq \frac{(1+\alpha)^n e^{\frac{t}{2\alpha T}+\frac{Kt}{3}+\frac{nK\alpha T}{2}}}{\mu(B(x,\sqrt{t}))} \left(\sup_{z\in B(y,\sqrt{t})} e^{-\frac{d(x,z)^2}{2(1+2\alpha)T}+\frac{t}{2\alpha T}}\right).$$

Choosing $T = (1+\alpha)t$ in this inequality we find

$$\int_{B(y,\sqrt{t})} p(x,z,(1+\alpha)t)^2 d\mu(z)$$
$$\leq \frac{(1+\alpha)^n e^{\frac{Kt}{3}+\frac{nK}{2}\alpha(1+\alpha)t+\frac{1}{2\alpha(1+\alpha)}}}{\mu(B(x,\sqrt{t}))} \left(\sup_{z\in B(y,\sqrt{t})} e^{-\frac{d(x,z)^2}{2(1+2\alpha)(1+\alpha)t}+\frac{1}{2\alpha(1+\alpha)}}\right). \quad (3.13)$$

We now apply Corollary 3.32 obtaining for every $z \in B(y, \sqrt{t})$

$$p(x,y,t)^2 \leq p(x,z,(1+\alpha)t)^2(1+\alpha)^n \exp\left(\frac{1}{2\alpha}+\frac{Kt}{3}+\frac{nK\alpha t}{4}\right).$$

Integrating this inequality in $z \in B(y, \sqrt{t})$, we have

$$\mu(B(y,\sqrt{t}))p(x,y,t)^2 \leq (1+\alpha)^n \exp\left(\frac{1}{2\alpha}+\frac{Kt}{3}+\frac{nK\alpha t}{4}\right)$$
$$\times \int_{B(y,\sqrt{t})} p(x,z,(1+\alpha)t)^2 d\mu(z).$$

Combining this inequality with (3.13) we conclude

$$p(x,y,t) \leq \frac{(1+\alpha)^n e^{\frac{3+\alpha}{4\alpha(1+\alpha)}+\frac{Kt}{3}+\frac{nKt}{4}\left(\alpha^2+\frac{3}{2}\alpha\right)}}{\mu(B(x,\sqrt{t}))^{\frac{1}{2}}\mu(B(y,\sqrt{t}))^{\frac{1}{2}}} \left(\sup_{z\in B(y,\sqrt{t})} e^{-\frac{d(x,z)^2}{4(1+2\alpha)(1+\alpha)t}}\right).$$

If now $x \in B(y, \sqrt{t})$, then

$$d(x,z)^2 \geq (d(x,y)-\sqrt{t})^2 > d(x,y)^2 - t,$$

and therefore

$$\sup_{z\in B(y,\sqrt{t})} e^{-\frac{d(x,z)^2}{4(1+2\alpha)(1+\alpha)t}} \le e^{\frac{1}{4(1+2\alpha)(1+\alpha)}} e^{-\frac{d(x,y)^2}{4(1+2\alpha)(1+\alpha)t}}.$$

If instead $x \notin B(y, \sqrt{t})$, then for every $\delta > 0$ we have

$$d(x,z)^2 \ge (1-\delta)d(x,y)^2 - (1+\delta^{-1})t$$

Choosing $\delta = \alpha/(\alpha+1)$ we find

$$d(x,z)^2 \ge \frac{d(x,y)^2}{1+\alpha} - (2+\alpha^{-1})t,$$

and therefore

$$\sup_{z\in B(y,\sqrt{t})} e^{-\frac{d(x,z)^2}{4(1+2\alpha)(1+\alpha)t}} \le e^{-\frac{d(x,y)^2}{4(1+2\alpha)(1+\alpha)^2 t}+\frac{2+\alpha^{-1}}{4(1+2\alpha)(1+\alpha)}}$$

For any $\epsilon > 0$ we now choose $\alpha > 0$ such that $4(1+2\alpha)(1+\alpha)^2 = 4+\epsilon$ to reach the desired conclusion. □

3.10 Volume Doubling Property

In this section we consider a complete and n-dimensional Riemannian manifold $(\mathbb{M}, g)$ with non negative Ricci curvature. Our goal is to prove the following fundamental result, which is known as the volume doubling property. We follow an approach developed by N. Garofalo and the author.

Theorem 3.29 *There exists a constant $C = C(n) > 0$ such that for every $x \in \mathbb{M}$ and every $r > 0$ one has*

$$\mu(B(x,2r)) \le C\mu(B(x,r)).$$

Actually by suitably adapting the arguments given in the sequel, the previous result can be extended to the case of negative Ricci curvature as follows:

Theorem 3.30 *Assume* **Ric** $\ge -K$ *with* $K \ge 0$. *There exist positive constants $C_1 = C_1(n,K)$, $C_2 = C_2(n,K)$ such that for every $x \in \mathbb{M}$ and every $r > 0$ one has*

$$\mu(B(x,2r)) \le C_1 e^{Kr^2}\mu(B(x,r)).$$

For simplicity, we show the arguments in the case $K = 0$ and let the reader work out the arguments in the case $K \neq 0$.

This result can be obtained from geometric methods as a consequence of the Bishop-Gromov comparison theorem. The proof we give instead only relies on the previous methods and has the advantage to generalize to a much larger class of operators than Laplace-Beltrami on Riemannian manifolds.

The key heat kernel estimate that leads to the doubling property is the following uniform and scale invariant lower bound on the heat kernel measure of balls.

Theorem 3.31 *There exist an absolute constant $K > 0$, and $A > 0$, depending only on n, such that*

$$P_{Ar^2}(\mathbf{1}_{B(x,r)})(x) \geq K, \qquad x \in \mathbb{M}, r > 0.$$

Proof We first recall the following result that was proved in a previous section: Let $a \in C^1([0,T],[0,\infty))$ and $\gamma \in C((0,T),\mathbb{R})$. Given $f \geq 0$, which is bounded and such that $\sqrt{f}$ is Lipschitz, we have

$$a(T)P_T(f\Gamma(\ln f)) - a(0)(P_T f)\Gamma(\ln P_T f)$$
$$\geq \int_0^T \left(a' + 2\rho a - \frac{4a\gamma}{n}\right)\Phi(s)ds + \left(\frac{4}{n}\int_0^T a\gamma ds\right) LP_T f - \left(\frac{2}{n}\int_0^T a\gamma^2 ds\right) P_T f.$$

We choose

$$a(t) = \tau + T - t,$$
$$\gamma(t) = -\frac{n}{4(\tau + T - t)}$$

where $\tau > 0$ will later be optimized. Noting that we presently have

$$a' = 1, \quad a\gamma = -\frac{n}{4}, \quad a\gamma^2 = \frac{n^2}{16(\tau + T - t)^2},$$

we obtain the inequality

$$\tau P_T(f\Gamma(\ln f)) - (T+\tau)P_T f\Gamma(\ln P_T f) \geq -TLP_T f - \frac{n}{8}\ln\left(1 + \frac{T}{\tau}\right) P_T f$$

In what follows we consider a bounded function f on $\mathbb{M}$ such that $\Gamma(f) \leq 1$ almost everywhere on $\mathbb{M}$. For any $\lambda \in \mathbb{R}$ we consider the function ψ defined by

$$\psi(\lambda, t) = \frac{1}{\lambda}\log P_t(e^{\lambda f}), \quad \text{or alternatively} \quad P_t(e^{\lambda f}) = e^{\lambda\psi}.$$

Notice that Jensen's inequality gives $\lambda\psi \geq \lambda P_t f$, and so we have $P_t f \leq \psi$.

We now apply the previous inequality to the function $e^{\lambda f}$, obtaining

$$\lambda^2 \tau P_T\left(e^{\lambda f}\Gamma(f)\right) - \lambda^2 (T+\tau) e^{\lambda\psi}\Gamma(\psi) \geq -TLP_T(e^{\lambda f}) - \frac{n}{8} e^{\lambda\psi} \ln\left(1+\frac{T}{\tau}\right).$$

Keeping in mind that $\Gamma(f) \leq 1$, we see that $P_T(e^{\lambda f}\Gamma(f)) \leq e^{\lambda\psi}$. Using this observation in combination with the fact that

$$L\left(P_t(e^{\lambda f})\right) = \frac{\partial}{\partial t}\left(P_t(e^{\lambda f})\right) = \frac{\partial e^{\lambda\psi}}{\partial t} = \lambda e^{\lambda\psi} \frac{\partial \psi}{\partial t},$$

and switching notation from T to t, we infer

$$\lambda^2 \tau \geq \lambda^2 (t+\tau) e^{\lambda\psi}\Gamma(\psi) - \lambda t \frac{\partial \psi}{\partial t} - \frac{n}{8} \ln\left(1+\frac{t}{\tau}\right).$$

The latter inequality finally gives

$$\frac{\partial \psi}{\partial t} \geq -\frac{\lambda}{t}\left(\tau + \frac{n}{8\lambda^2}\ln\left(1+\frac{t}{\tau}\right)\right) \geq -\frac{\lambda}{t}\left(\tau + \frac{nt}{8\lambda^2\tau}\right). \tag{3.14}$$

We now optimize the right-hand side of the inequality with respect to τ. We notice explicitly that the maximum value of the right-hand side is attained at

$$\tau_0 = \sqrt{\frac{nt}{8\lambda^2}}.$$

We find therefore

$$\frac{\partial \psi}{\partial t} \geq -\sqrt{\frac{n}{2t}}$$

We now integrate the inequality between s and t, obtaining

$$\psi(\lambda, s) \leq \psi(\lambda, t) + \sqrt{\frac{n}{2}} \int_s^t \frac{d\tau}{\sqrt{\tau}}.$$

We infer then

$$P_s(\lambda f) \leq \lambda\psi(\lambda, t) + \lambda\sqrt{2nt}.$$

Letting $s \to 0^+$ we conclude

$$\lambda f \leq \lambda\psi(\lambda, t) + \lambda\sqrt{2nt}. \tag{3.15}$$

At this point we let $B = B(x,r) = \{x \in \mathbb{M} \mid d(y,x) < r\}$, and consider the function $f(y) = -d(y,x)$. Since we clearly have

$$e^{\lambda f} \le e^{-\lambda r} \mathbf{1}_{B^c} + \mathbf{1}_B,$$

it follows that for every $t > 0$ one has

$$e^{\lambda \psi(\lambda,t)(x)} = P_t(e^{\lambda f})(x) \le e^{-\lambda r} + P_t(\mathbf{1}_B)(x).$$

This gives the lower bound

$$P_t(\mathbf{1}_B)(x) \ge e^{\lambda \psi(\lambda,t)(x)} - e^{-\lambda r}.$$

To estimate the first term in the right-hand side of the latter inequality, we use the previous estimate which gives

$$P_t(\mathbf{1}_B)(x) \ge e^{-\lambda\sqrt{2nt}} - e^{-\lambda r}.$$

To make use of this estimate, we now choose $\lambda = \frac{1}{r}$, $t = Ar^2$, obtaining

$$P_{Ar^2}(\mathbf{1}_B)(x) \ge e^{-A\sqrt{2n}} - e^{-1}.$$

The conclusion follows then easily. □

We now turn to the proof of the volume doubling property. We first recall the following basic result which is a straightforward consequence of the Li Yau inequality.

Corollary 3.32 *Let $p(x,y,t)$ be the heat kernel on $\mathbb{M}$. For every $x, y, z \in \mathbb{M}$ and every $0 < s < t < \infty$ one has*

$$p(x,y,s) \le p(x,z,t) \left(\frac{t}{s}\right)^{\frac{n}{2}} \exp\left(\frac{d(y,z)^2}{4(t-s)}\right).$$

We are now in position to prove the volume doubling property.

From the semigroup property and the symmetry of the heat kernel we have for any $y \in \mathbb{M}$ and $t > 0$

$$p(y,y,2t) = \int_{\mathbb{M}} p(y,z,t)^2 d\mu(z).$$

Consider now a function $h \in C_0^\infty(\mathbb{M})$ such that $0 \le h \le 1$, $h \equiv 1$ on $B(x, \sqrt{t}/2)$ and $h \equiv 0$ outside $B(x, \sqrt{t})$. We thus have

$$\begin{aligned} P_t h(y) &= \int_{\mathbb{M}} p(y, z, t) h(z) d\mu(z) \\ &\le \left(\int_{B(x,\sqrt{t})} p(y, z, t)^2 d\mu(z) \right)^{\frac{1}{2}} \left(\int_{\mathbb{M}} h(z)^2 d\mu(z) \right)^{\frac{1}{2}} \\ &\le p(y, y, 2t)^{\frac{1}{2}} \mu(B(x, \sqrt{t}))^{\frac{1}{2}}. \end{aligned}$$

If we take $y = x$, and $t = r^2$, we obtain

$$P_{r^2}\left(\mathbf{1}_{B(x,r)}\right)(x)^2 \le P_{r^2} h(x)^2 \le p(x, x, 2r^2)\, \mu(B(x, r)). \tag{3.16}$$

At this point we use the crucial previous theorem, which gives for some $0 < A = A(n) < 1$

$$P_{Ar^2}(\mathbf{1}_{B(x,r)})(x) \ge K, \qquad x \in \mathbb{M}, r > 0.$$

Combining the latter inequality with the Harnack inequality and with (3.16), we obtain the following on-diagonal lower bound

$$p(x, x, 2r^2) \ge \frac{K^*}{\mu(B(x, r))}, \qquad x \in \mathbb{M},\ r > 0. \tag{3.17}$$

Applying Corollary 3.32 to $(y, t) \to p(x, y, t)$ for every $y \in B(x, \sqrt{t})$ we find

$$p(x, x, t) \le C(n) p(x, y, 2t).$$

Integration over $B(x, \sqrt{t})$ gives

$$p(x, x, t)\mu(B(x, \sqrt{t})) \le C(n) \int_{B(x,\sqrt{t})} p(x, y, 2t) d\mu(y) \le C(n),$$

where we have used $P_t 1 \le 1$. Letting $t = r^2$, we obtain from this the on-diagonal upper bound

$$p(x, x, r^2) \le \frac{C(n)}{\mu(B(x, r))}. \tag{3.18}$$

Combining (3.17) with (3.18) we finally obtain

$$\mu(B(x, 2r)) \le \frac{C}{p(x, x, 4r^2)} \le \frac{C^*}{p(x, x, 2r^2)} \le C^{**} \mu(B(x, r)),$$

where we have used once more Corollary 3.32. which gives

$$\frac{p(x,x,2r^2)}{p(x,x,4r^2)} \le C.$$

3.11 Upper and Lower Gaussian Bounds for the Heat Kernel

In this short section, as in the previous one, we consider a complete and n-dimensional Riemannian manifold $(\mathbb{M}, g)$ with non negative Ricci curvature. The volume doubling property that was proved is closely related to sharp lower and upper Gaussian bounds that are due to P. Li and S.T. Yau. We first record a basic consequence of the volume doubling property whose proof is let to the reader.

Theorem 3.33 *Let $C > 0$ be the constant such that for every $x \in \mathbb{M}$, $R > 0$,*

$$\mu(B(x, 2R)) \le C\mu(B(x, R)).$$

Let $Q = \log_2 C$. For any $x \in \mathbb{M}$ and $r > 0$ one has

$$\mu(B(x, tr)) \ge C^{-1}t^Q\mu(B(x, r)), \quad 0 \le t \le 1.$$

We are now in position to prove the main result of the section.

Theorem 3.34 *For any $0 < \varepsilon < 1$ there exists a constant $C = C(n, \varepsilon) > 0$, which tends to ∞ as $\varepsilon \to 0^+$, such that for every $x, y \in \mathbb{M}$ and $t > 0$ one has*

$$\frac{C^{-1}}{\mu(B(x,\sqrt{t}))}\exp\left(-\frac{d(x,y)^2}{(4-\varepsilon)t}\right) \le p(x,y,t) \le \frac{C}{\mu(B(x,\sqrt{t}))}\exp\left(-\frac{d(x,y)^2}{(4+\varepsilon)t}\right).$$

Proof We begin by establishing the lower bound. First, from the Harnack inequality we obtain for all $y \in \mathbb{M}$, $t > 0$, and every $0 < \varepsilon < 1$,

$$p(x,y,t) \ge p(x,x,\varepsilon t)\varepsilon^{\frac{n}{2}}\exp\left(-\frac{d(x,y)^2}{(4-\varepsilon)t}\right).$$

We thus need to estimate $p(x, x, \varepsilon t)$ from below. But this has already been done in the proof of the volume doubling property where we established:

$$p(x,x,\varepsilon t) \ge \frac{C^*}{\mu(B(x,\sqrt{\varepsilon/2}\sqrt{t}))}, \qquad x \in \mathbb{M},\ t > 0.$$

On the other hand, since $\sqrt{\varepsilon/2} < 1$, by the trivial inequality $\mu(B(x, \sqrt{\varepsilon/2}\sqrt{t})) \leq \mu(B(x, \sqrt{t}))$, we conclude

$$p(x, y, t) \geq \frac{C^*}{\mu(B(x, \sqrt{t}))} \varepsilon^{\frac{n}{2}} \exp\left(-\frac{d(x, y)^2}{(4-\varepsilon)t}\right).$$

This proves the Gaussian lower bound.

For the Gaussian upper bound, we first observe that the following upper bound was proved in a previous section:

$$p(x, y, t) \leq \frac{C}{\mu(B(x, \sqrt{t}))^{\frac{1}{2}} \mu(B(y, \sqrt{t}))^{\frac{1}{2}}} \exp\left(-\frac{d(x, y)^2}{(4+\varepsilon')t}\right).$$

At this point, by the triangle inequality and the volume doubling property we find.

$$\begin{aligned}\mu(B(x, \sqrt{t})) &\leq \mu(B(y, d(x, y) + \sqrt{t})) \\ &\leq C_1 \mu(B(y, \sqrt{t})) \left(\frac{d(x, y) + \sqrt{t}}{\sqrt{t}}\right)^Q.\end{aligned}$$

with $Q = \log_2 C$, where C is the doubling constant. This gives

$$\frac{1}{\mu(B(y, \sqrt{t}))} \leq \frac{C_1}{\mu(B(x, \sqrt{t}))} \left(\frac{d(x, y)}{\sqrt{t}} + 1\right)^Q.$$

Combining this with the above estimate we obtain

$$p(x, y, t) \leq \frac{C_1^{1/2} C}{\mu(B(x, \sqrt{t}))} \left(\frac{d(x, y)}{\sqrt{t}} + 1\right)^{\frac{Q}{2}} \exp\left(-\frac{d(x, y)^2}{(4+\varepsilon')t}\right).$$

If now $0 < \varepsilon < 1$, it is clear that we can choose $0 < \varepsilon' < \varepsilon$ such that

$$\frac{C_1^{1/2} C}{\mu(B(x, \sqrt{t}))} \left(\frac{d(x, y)}{\sqrt{t}} + 1\right)^{\frac{Q}{2}} \exp\left(-\frac{d(x, y)^2}{(4+\varepsilon')t}\right) \leq \frac{C^*}{\mu(B(x, \sqrt{t}))} \exp\left(-\frac{d(x, y)^2}{(4+\varepsilon)t}\right),$$

where C^* is a constant which tends to ∞ as $\varepsilon \to 0^+$. The desired conclusion follows by suitably adjusting the values of both ε' and of the constant in the right-hand side of the estimate. □

To conclude, we finally mention without proof, what the previous arguments give in the case where $\mathbf{Ric} \geq -K$ with $K \geq 0$. We encourage the reader to do the proof by herself/himself as an exercise.

Theorem 3.35 *Let us assume* **Ric** $\geq -K$ *with* $K \geq 0$. *For any* $0 < \varepsilon < 1$ *there exist constants* $C_1, C_2 = C(n, K, \varepsilon) > 0$, *such that for every* $x, y \in \mathbb{M}$ *and* $t > 0$ *one has*

$$p(x, y, t) \leq \frac{C_1}{\mu(B(x, \sqrt{t}))} \exp\left(-\frac{d(x, y)^2}{(4+\varepsilon)t} + KC_2(t + d(x, y)^2)\right).$$

$$p(x, y, t) \geq \frac{C_1^{-1}}{\mu(B(x, \sqrt{t}))} \exp\left(-\frac{d(x, y)^2}{(4-\varepsilon)t} - KC_2(t + d(x, y)^2)\right).$$

3.12 The Poincaré Inequality on Domains

Let $(\mathbb{M}, g)$ be a complete Riemannian manifold and $\Omega \subset \mathbb{M}$ be a non empty bounded set. Let $\mathcal{D}^\infty$ be the set of smooth functions $f \in C^\infty(\bar{\Omega})$ such that for every $g \in C^\infty(\bar{\Omega})$,

$$\int_\Omega gLf d\mu = -\int_\Omega \Gamma(f, g) d\mu.$$

It is easy to see that L is essentially self-adjoint on $\mathcal{D}^\infty$. Its Friedrichs extension, still denoted L, is called the Neumann Laplacian on Ω and the semigroup it generates, the Neumann semigroup. If the boundary $\partial\Omega$ is smooth, then it is known from the Green's formula that

$$\int_\Omega gLf d\mu = -\int_\Omega \Gamma(f, g) d\mu + \int_{\partial\Omega} gNf d\mu,$$

where N is the normal unit vector. As a consequence, $f \in \mathcal{D}^\infty$ if and only if $Nf = 0$. However, we stress that no regularity assumption on the boundary $\partial\Omega$ is needed to define the Neumann Laplacian and the Neumann semigroup.

Since $\bar{\Omega}$ is compact, the Neumann semigroup is a compact operator and $-L$ has a discrete spectrum $0 = \lambda_0 < \lambda_1 \leq \cdots$. We get then, the so-called Poincaré inequality on Ω: For every $f \in C^\infty(\bar{\Omega})$,

$$\int_\Omega (f - f_\Omega)^2 d\mu \leq \frac{1}{\lambda_1} \int_\Omega \Gamma(f) d\mu.$$

Our goal is to understand how the constant λ_1 depends on the size of the set Ω. A first step in that direction was made by Poincaré himself in the Euclidean case.

Theorem 3.36 *If $\Omega \subset \mathbb{R}^n$ is a bounded open convex set then for a smooth $f : \bar{\Omega} \to \mathbb{R}$ with $\int_\Omega f(x)dx = 0$,*

$$\frac{C_n}{\mathbf{diam}(\Omega)^2} \int_\Omega f^2(x)dx \le \int_\Omega \|\nabla f(x)\|^2 dx.$$

where C_n is a constant depending on n only.

Proof The argument of Poincaré is beautifully simple.

$$\begin{aligned}\frac{1}{\mu(\Omega)} \int_\Omega f^2(x)dx &= \frac{1}{2}\frac{1}{\mu(\Omega)} \int_\Omega f^2(x)dx + \frac{1}{2}\frac{1}{\mu(\Omega)} \int_\Omega f^2(y)dy \\ &= \frac{1}{2}\frac{1}{\mu(\Omega)^2} \int_\Omega \int_\Omega (f(x) - f(y))^2 dxdy.\end{aligned}$$

We now have

$$f(x) - f(y) = \int_0^1 (x-y) \cdot \nabla f(tx + (1-t)y)dt,$$

which implies

$$(f(x) - f(y))^2 \le \mathbf{diam}(\Omega)^2 \int_0^1 \|\nabla f\|^2(tx + (1-t)y)dt.$$

By a simple change of variables, we see that

$$\begin{aligned}\int_\Omega \int_\Omega \|\nabla f\|^2(tx + (1-t)y)dxdy &= \frac{1}{t^n} \int_\Omega \int_{t\Omega+(1-t)y} \|\nabla f\|^2(u)dudy \\ &= \frac{1}{t^n} \int_\Omega \int_\Omega \mathbf{1}_{t\Omega+(1-t)y}(u) \|\nabla f\|^2(u)dudy.\end{aligned}$$

Now, we compute

$$\int_\Omega \mathbf{1}_{t\Omega+(1-t)y}(u)dy = \mu\left(\Omega \cap \frac{1}{1-t}(u - t\Omega)\right) \le \min\left(1, \frac{t^n}{(1-t)^n}\right)\mu(\Omega).$$

As a consequence we obtain

$$\frac{1}{\mu(\Omega)} \int_\Omega f^2(x)dx \le \frac{\mathbf{diam}(\Omega)^2}{2\mu(\Omega)} \int_0^1 \min\left(1, \frac{t^n}{(1-t)^n}\right)\frac{dt}{t^n} \int_\Omega \|\nabla f(x)\|^2 dx$$

□

It is known (Payne-Weinberger) that the optimal C_n is π^2.

In this section, we extend the above inequality to the case of Riemannian manifolds with non negative Ricci curvature. The key point is a lower bound on the Neumann heat kernel of Ω. From now on we assume that $\mathbf{Ric} \geq 0$ and consider an open set in $\mathbb{M}$ that has a smooth and convex boundary in the sense the second fundamental form of $\partial\Omega$ is non negative. Due to the convexity of the boundary, all the results we obtained so far may be extended to the Neumann semigroup (see [9]). In particular, we have the following lower bound on the Neumann heat kernel:

Theorem 3.37 *Let $p^N(x, y, t)$ be the Neumann heat kernel of Ω. There exists a constant C depending only on the dimension of $\mathbb{M}$ such that for every $t > 0$, $x, y \in \mathbb{M}$,*

$$p^N(x, y, t) \geq \frac{C}{\mu(B(x, \sqrt{t}))} \exp\left(-\frac{d(x, y)^2}{3t}\right).$$

As we shall see, this implies the following Poincaré inequality:

Theorem 3.38 *For a smooth $f : \bar{\Omega} \to \mathbb{R}$ with $\int_\Omega f d\mu = 0$,*

$$\frac{C_n}{\mathbf{diam}(\Omega)^2} \int_\Omega f^2 d\mu \leq \int_\Omega \Gamma(f) d\mu.$$

where C is a constant depending on the dimension of $\mathbb{M}$ only.

Proof We denote by R the diameter of Ω. From the previous lower bound on the Neumann kernel of Ω, we have

$$p^N(x, y, R^2) \geq \frac{C}{\mu(\Omega)},$$

where C only depends on n. Denote now by P_t^N the Neumann semigroup. We have for $f \in \mathcal{D}^\infty$

$$P_{R^2}^N(f^2) - (P_{R^2}^N f)^2 = \int_0^{R^2} \frac{d}{dt} P_t^N((P_{R^2-t}^N f)^2) dt.$$

By integrating over Ω, we find then,

$$\begin{aligned}
\int_\Omega P_{R^2}^N(f^2) - (P_{R^2}^N f)^2 d\mu &= -\int_0^{R^2} \int_\Omega \frac{d}{dt}(P_t^N f)^2 d\mu dt \\
&= 2\int_0^{R^2} \int_\Omega \Gamma(P_t^N f, P_t^N f) d\mu dt \\
&\leq 2R^2 \int_\Omega \Gamma(f) d\mu.
\end{aligned}$$

But on the other hand, we have

$$P^N_{R^2}(f^2)(x) - (P^N_{R^2}f)^2(x) = P^N_{R^2}\left[\left(f - (P^N_{R^2}f)(x)\right)^2\right](x)$$
$$\geq \frac{C}{\mu(\Omega)}\int_\Omega (f(y) - (P^N_{R^2}f)(x))^2 d\mu(y)$$

which gives

$$\int_\Omega P^N_{R^2}(f^2) - (P^N_{R^2}f)^2 d\mu \geq C\int_\Omega \left(f(x) - \frac{1}{\mu(\Omega)}\int_\Omega f d\mu\right)^2 d\mu(x)$$

The proof is complete. □

In applications, it is often interesting to have a scale invariant Poincaré inequality on balls. If the manifold $\mathbb{M}$ has conjugate points, the geodesic spheres may not be convex and thus the previous argument does not work. However the following result still holds true:

Theorem 3.39 *There exists a constant $C_n > 0$ depending only on the dimension of $\mathbb{M}$ such that for every $r > 0$ and every smooth $f : B(x,r) \to \mathbb{R}$ with $\int_{B(x,r)} f d\mu = 0$,*

$$\frac{C_n}{r^2}\int_{B(x,r)} f^2 d\mu \leq \int_{B(x,r)} \Gamma(f) d\mu.$$

We only sketch the argument. By using the global lower bound

$$p(x,y,t) \geq \frac{C}{\mu(B(x,\sqrt{t}))}\exp\left(-\frac{d(x,y)^2}{3t}\right),$$

for the heat kernel, it is possible to prove a lower bound for the Neumman heat kernel on the ball $B(x_0, r)$: For $x, y \in B(x_0.r/2)$,

$$p^N(x,y,r^2) \geq \frac{C}{\mu(B(x_0,r))},$$

Arguing as before, we get

$$\frac{C_n}{r^2}\int_{B(x_0,r/2)} f^2 d\mu \leq \int_{B(x,r)} \Gamma(f) d\mu.$$

and show then that the integral on the left hand side can be taken on $B(x_0, r)$ by using a Whitney's type covering argument.

3.13 Sobolev Inequality and Volume Growth

In this section, we show how Sobolev inequalities on a Riemannian manifold are related to the volume growth of metric balls. The link between the Hardy-Littlewood-Sobolev theory and heat kernel upper bounds is due to Varopoulos,

Let $(\mathbb{M}, g)$ be a complete Riemannian manifold and let L be the Laplace-Beltrami operator of $\mathbb{M}$. As usual, we denote by P_t the semigroup generated by P_t and we assume $P_t 1 = 1$.

We have the following so-called maximal ergodic lemma, which was first proved by Stein. We give here the probabilistic proof since it comes with a nice constant.

Lemma 3.40 (Stein's Maximal Ergodic Theorem) *Let $p > 1$. For $f \in L^p(\mathbb{M}, \mu)$, denote $f^*(x) = \sup_{t \geq 0} |P_t f(x)|$. We have*

$$\| f^* \|_{L^p_\mu(\mathbb{M})} \leq \frac{p}{p-1} \| f \|_{L^p_\mu(\mathbb{M})}.$$

Proof For $x \in \mathbb{M}$, we denote by $(X_t^x)_{t \geq 0}$ the Markov process with generator L and started at x. We fix $T > 0$. By construction, for $t \leq T$, we have,

$$P_{T-t} f(X_T^x) = \mathbb{E}\left(f(X_{2T-t}^x) | X_T^x \right),$$

and thus

$$P_{2(T-t)} f(X_T^x) = \mathbb{E}\left((P_{T-t} f)(X_{2T-t}^x) | X_T^x \right).$$

As a consequence, we obtain

$$\sup_{0 \leq t \leq T} |P_{2(T-t)} f(X_T^x)| \leq \mathbb{E}\left(\sup_{0 \leq t \leq T} |(P_{T-t} f)(X_{2T-t}^x)| \mid X_T^x \right).$$

Jensen's inequality yields then

$$\sup_{0 \leq t \leq T} |P_{2(T-t)} f(X_T^x)|^p \leq \mathbb{E}\left(\sup_{0 \leq t \leq T} |(P_{T-t} f)(X_{2T-t}^x)|^p \mid X_T^x \right).$$

We deduce

$$\mathbb{E}\left(\sup_{0 \leq t \leq T} |P_{2(T-t)} f(X_T^x)|^p \right) \leq \mathbb{E}\left(\sup_{0 \leq t \leq T} |(P_{T-t} f)(X_{2T-t}^x)|^p \right).$$

Integrating the inequality with respect to the Riemannian measure μ, we obtain

$$\left\| \sup_{0\le t\le T} |P_{2(T-t)} f| \right\|_p \le \left(\int_{\mathbb{M}} \mathbb{E}\left(\sup_{0\le t\le T} |(P_{T-t} f)(X^x_{2T-t})|^p \right) d\mu(x) \right)^{1/p}.$$

By reversibility, we get then

$$\left\| \sup_{0\le t\le T} |P_{2(T-t)} f| \right\|_p \le \left(\int_{\mathbb{M}} \mathbb{E}\left(\sup_{0\le t\le T} |(P_{T-t} f)(X^x_t)|^p \right) d\mu(x) \right)^{1/p}.$$

We now observe that the process $(P_{T-t} f)(X^x_t)$ is martingale and thus Doob's maximal inequality gives

$$\mathbb{E}\left(\sup_{0\le t\le T} |(P_{T-t} f)(X^x_t)|^p \right)^{1/p} \le \frac{p}{p-1} \mathbb{E}\left(|f(X^x_T)|^p\right)^{1/p}.$$

The proof is complete. □

We now turn to the theorem by Varopoulos.

Theorem 3.41 *Let $n > 0$, $0 < \alpha < n$, and $1 < p < \frac{n}{\alpha}$. If there exists $C > 0$ such that for every $t > 0$, $x, y \in \mathbb{M}$,*

$$p(x, y, t) \le \frac{C}{t^{n/2}},$$

then for every $f \in L^p_\mu(\mathbb{M})$,

$$\|(-L)^{-\alpha/2} f\|_{\frac{np}{n-p\alpha}} \le \left(\frac{p}{p-1}\right)^{1-\alpha/n} \frac{2nC^{\alpha/n}}{\alpha(n-p\alpha)\Gamma(\alpha/2)} \|f\|_p$$

Proof We first observe that the bound

$$p(x, y, t) \le \frac{C}{t^{n/2}},$$

implies that $|P_t f(x)| \le \frac{C^{1/p}}{t^{n/2p}} \|f\|_p$. Denote $I_\alpha f(x) = (-L)^{-\alpha/2} f(x)$. We have

$$I_\alpha f(x) = \frac{1}{\Gamma(\alpha/2)} \int_0^{+\infty} t^{\alpha/2-1} P_t f(x) dt$$

Pick $\delta > 0$, to be later chosen, and split the integral in two parts:

$$I_\alpha f(x) = J_\alpha f(x) + K_\alpha f(x),$$

where $J_\alpha f(x) = \frac{1}{\Gamma(\alpha/2)} \int_0^\delta t^{\alpha/2-1} P_t f(x) dt$ and $K_\alpha f(x) = \frac{1}{\Gamma(\alpha/2)} \int_\delta^{+\infty} t^{\alpha/2-1} P_t f(x) dt$. We have

$$|J_\alpha f(x)| \le \frac{1}{\Gamma(\alpha/2)} \int_0^{+\infty} t^{\alpha/2-1} dt |f^*(x)| = \frac{2}{\alpha \Gamma(\alpha/2)} \delta^{\alpha/2} |f^*(x)|.$$

On the other hand,

$$\begin{aligned} |K_\alpha f(x)| &\le \frac{1}{\Gamma(\alpha/2)} \int_\delta^{+\infty} t^{\alpha/2-1} |P_t f(x)| dt \\ &\le \frac{C^{1/p}}{\Gamma(\alpha/2)} \int_\delta^{+\infty} t^{\frac{\alpha}{2} - \frac{n}{2p} - 1} dt \|f\|_p \\ &\le \frac{C^{1/p}}{\Gamma(\alpha/2)} \frac{1}{-\frac{\alpha}{2} + \frac{n}{2p}} \delta^{\frac{\alpha}{2} - \frac{n}{2p}} \|f\|_p. \end{aligned}$$

We deduce

$$|I_\alpha f(x)| \le \frac{2}{\alpha \Gamma(\alpha/2)} \delta^{\alpha/2} |f^*(x)| + \frac{C^{1/p}}{\Gamma(\alpha/2)} \frac{1}{-\frac{\alpha}{2} + \frac{n}{2p}} \delta^{\frac{\alpha}{2} - \frac{n}{2p}} \|f\|_p.$$

Optimizing the right hand side of the latter inequality with respect to δ yields

$$|I_\alpha f(x)| \le \frac{2nC^{\alpha/n}}{\alpha(n - p\alpha)\Gamma(\alpha/2)} \|f\|_p^{\alpha p/n} |f^*(x)|^{1 - p\alpha/n}.$$

The proof is then completed by using Stein's maximal ergodic theorem. □

A special case, of particular interest, is when $\alpha = 1$ and $p = 2$. We get in that case the following Sobolev inequality:

Theorem 3.42 *Let $n > 2$. If there exists $C > 0$ such that for every $t > 0$, $x, y \in \mathbb{M}$,*

$$p(x, y, t) \le \frac{C}{t^{n/2}},$$

then for every $f \in C_0^\infty(\mathbb{M})$,

$$\|f\|_{\frac{2n}{n-2}} \le 2^{1-1/n} \frac{2nC^{1/n}}{(n-2)\sqrt{\pi}} \|\sqrt{\Gamma(f)}\|_2$$

We mention that the constant in the above Sobolev inequality is not sharp.

Combining the above with the Li-Yau upper bound for the heat kernel, we deduce the following theorem:

Theorem 3.43 *Assume that* **Ric** ≥ 0 *and that there exists a constant* $C > 0$ *such that for every* $x \in \mathbb{M}$ *and* $r \geq 0$, $\mu(B(x, r)) \geq C r^n$, *then there exists a constant* $C' = C'(n) > 0$ *such that for every* $f \in C_0^\infty(\mathbb{M})$,

$$\| f \|_{\frac{2n}{n-2}} \leq C' \| \sqrt{\Gamma(f)} \|_2 .$$

In many situations, heat kernel upper bounds with a polynomial decay are only available in small times the following result is thus useful:

Theorem 3.44 *Let* $n > 0$, $0 < \alpha < n$, *and* $1 < p < \frac{n}{\alpha}$. *If there exists* $C > 0$ *such that for every* $0 < t \leq 1$, $x, y \in \mathbb{M}$,

$$p(x, y, t) \leq \frac{C}{t^{n/2}},$$

then, there is constant C' *such that for every* $f \in L^p_\mu(\mathbb{M})$,

$$\| (-L + 1)^{-\alpha/2} f \|_{\frac{np}{n - p\alpha}} \leq C' \| f \|_p .$$

Proof We apply the Varopoulos theorem to the semigroup $Q_t = e^{-t} P_t$. Details are let to the reader. □

The following corollary shall be later used:

Corollary 3.45 *Let* $n > 2$. *If there exists* $C > 0$ *such that for every* $0 < t \leq 1$, $x, y \in \mathbb{M}$,

$$p(x, y, t) \leq \frac{C}{t^{n/2}},$$

then there is constant C' *such that for every* $f \in C_0^\infty(\mathbb{M})$,

$$\| f \|_{\frac{2n}{n-2}} \leq C' \left(\| \sqrt{\Gamma(f)} \|_2 + \| f \|_2 \right).$$

3.14 Isoperimetric Inequality and Volume Growth

In this section, we study in further details the connection between volume growth of metric balls, heat kernel upper bounds and the L^1 Sobolev inequality. As we shall see, on a manifold with non negative Ricci curvature, all these properties are equivalent one to each other and equivalent to the isoperimetric inequality as well. We start with some preliminaries about geometric measure theory on Riemannian manifolds.

Let $(\mathbb{M}, g)$ be a complete and non compact Riemannian manifold.

In what follows, given an open set $\Omega \subset \mathbb{M}$ we will indicate with $\mathcal{F}(\Omega)$ the set of C^1 vector fields V's, on Ω such that $\|V\|_\infty \le 1$.

Given a function $f \in L^1_{loc}(\Omega)$ we define the total variation of f in Ω as

$$\mathrm{Var}(f;\Omega) = \sup_{\phi \in \mathcal{F}(\Omega)} \int_\Omega f \mathbf{div} \phi d\mu.$$

The space

$$BV(\Omega) = \{f \in L^1(\Omega) \mid \mathrm{Var}(f;\Omega) < \infty\},$$

endowed with the norm

$$||f||_{BV(\Omega)} = ||f||_{L^1(\mathbb{M})} + \mathrm{Var}(f;\Omega),$$

is a Banach space. It is well-known that $W^{1,1}(\Omega) = \{f \in L^1(\Omega) \mid \|\nabla f\| \in L^1(\Omega)\}$ is a strict subspace of $BV(\Omega)$. It is important to note that when $f \in W^{1,1}(\Omega)$, then $f \in BV(\Omega)$, and one has in fact

$$\mathrm{Var}(f;\Omega) = ||\sqrt{\Gamma(f)}||_{L^1(\Omega)}.$$

Given a measurable set $E \subset \mathbb{M}$ we say that it has finite perimeter in Ω if $\mathbf{1}_E \in BV(\Omega)$. In such case the horizontal perimeter of E relative to Ω is by definition

$$P(E;\Omega) = \mathrm{Var}(\mathbf{1}_E;\Omega).$$

We say that a measurable set $E \subset \mathbb{M}$ is a Caccioppoli set if $P(E;\Omega) < \infty$ for any $\Omega \subset \mathbb{M}$. For instance, O is an open relatively compact set in $\mathbb{M}$ whose boundary E is $n-1$ dimensional sub manifold of $\mathbb{M}$, then it is a Caccioppoli set and $P(E;\mathbb{M}) = \mu_{n-1}(E)$ where μ_{n-1} is the Riemannian measure on E. We will need the following approximation result.

Proposition 3.46 *Let $f \in BV(\Omega)$, then there exists a sequence $\{f_n\}_{n\in\mathbb{N}}$ of functions in $C^\infty(\Omega)$ such that:*

(i) $||f_n - f||_{L^1(\Omega)} \to 0$;
(ii) $\int_\Omega \sqrt{\Gamma(f_n)} d\mu \to \mathrm{Var}(f;\Omega)$.

If $\Omega = \mathbb{M}$, then the sequence $\{f_n\}_{n\in\mathbb{N}}$ can be taken in $C_0^\infty(\mathbb{M})$.

Our main result of the section is the following:

Theorem 3.47 *Let $n > 1$. Let us assume that* $\mathbf{Ric} \ge 0$. *then the following assertions are equivalent:*

(1) There exists a constant $C_1 > 0$ such that for every $x \in \mathbb{M}$, $r \ge 0$,

$$\mu(B(x,r)) \ge C_1 r^n.$$

(2) There exists a constant $C_2 > 0$ *such that for* $x \in \mathbb{M}$, $t > 0$,

$$p(x, x, t) \le \frac{C_2}{t^{\frac{n}{2}}}.$$

(3) There exists a constant $C_3 > 0$ *such that for every Caccioppoli set* $E \subset \mathbb{M}$ *one has*

$$\mu(E)^{\frac{n-1}{n}} \le C_3 P(E; \mathbb{M}).$$

(4) With the same constant $C_3 > 0$ *as in (3), for every* $f \in BV(\mathbb{M})$ *one has*

$$\left(\int_{\mathbb{M}} |f|^{\frac{n}{n-1}} d\mu \right)^{\frac{n-1}{n}} \le C_3 Var(f; \mathbb{M}).$$

Proof In the proof, we denote by d the dimension of $\mathbb{M}$. That (1) $\to$ (2) follows immediately from the Li-Yau upper Gaussian bound.

The proof that (2) $\to$ (3) is not straightforward, it relies on the Li-Yau inequality. Let $f \in C_0(\mathbb{M})$ with $f \ge 0$. By Li-Yau inequality, we obtain

$$\Gamma(P_t f) - P_t f \frac{\partial P_t f}{\partial t} \le \frac{d}{2t}(P_t f)^2. \tag{3.19}$$

This gives in particular,

$$\left(\frac{\partial P_t f}{\partial t} \right)^- \le \frac{d}{2t} P_t f, \tag{3.20}$$

where we have denoted $a^+ = \sup\{a, 0\}$, $a^- = \sup\{-a, 0\}$. Since $\int_{\mathbb{M}} \frac{\partial P_t f}{\partial t} d\mu = 0$, we deduce

$$|| \frac{\partial P_t f}{\partial t} ||_{L^1(\mathbb{M})} \le \frac{d}{t} ||f||_{L^1(\mathbb{M})}, \quad t > 0.$$

By duality, we deduce that for every $f \in C_0^\infty(\mathbb{M})$, $f \ge 0$,

$$\| \frac{\partial P_t f}{\partial t} \|_{L^\infty(\mathbb{M})} \le \frac{d}{t} \| f \|_{L^\infty(\mathbb{M})}.$$

Once we have this crucial information we can return to (3.19) and infer

$$\Gamma(P_t f) \le \frac{1}{t} \frac{3d}{2} \| f \|^2_{L^\infty(\mathbb{M})}, \quad t > 0.$$

Thus,

$$\|\sqrt{\Gamma(P_t f)}\|_{L^\infty(\mathbb{M})} \le \sqrt{\frac{3d}{2t}}\|f\|_{L^\infty(\mathbb{M})}.$$

Applying this inequality to $g \in C_0^\infty(\mathbb{M})$, with $g \ge 0$ and $||g||_{L^\infty(\mathbb{M})} \le 1$, if $f \in C_0^1(\mathbb{M})$ we have

$$\begin{aligned}
\int_{\mathbb{M}} g(f - P_t f)d\mu &= \int_0^t \int_{\mathbb{M}} g\frac{\partial P_s f}{\partial s} d\mu ds = \int_0^t \int_{\mathbb{M}} gLP_s f d\mu ds \\
&= \int_0^t \int_{\mathbb{M}} Lg P_s f d\mu ds \\
&= \int_0^t \int_{\mathbb{M}} P_s Lg f d\mu ds = \int_0^t \int_{\mathbb{M}} LP_s g f d\mu ds \\
&= -\int_0^t \int_{\mathbb{M}} \Gamma(P_s g, f) d\mu ds \\
&\le \int_0^t \|\sqrt{\Gamma(P_s g)}\|_{L^\infty(\mathbb{M})} \int_{\mathbb{M}} \sqrt{\Gamma(f)} d\mu ds \\
&\le \sqrt{6d}\,\sqrt{t} \int_{\mathbb{M}} \sqrt{\Gamma(f)} d\mu.
\end{aligned}$$

We thus obtain the following basic inequality: for $f \in C_0^1(\mathbb{M})$,

$$\|P_t f - f\|_{L^1(\mathbb{M})} \le \sqrt{6d}\ \sqrt{t}\ \|\sqrt{\Gamma(f)}\|_{L^1(\mathbb{M})}, \quad t > 0. \tag{3.21}$$

Suppose now that $E \subset \mathbb{M}$ is a bounded Caccioppoli set. But then, $\mathbf{1}_E \in BV(\Omega)$, for any bounded open set $\Omega \supset E$. It is easy to see that $\mathrm{Var}(\mathbf{1}_E; \Omega) = \mathrm{Var}_{\mathcal{H}}(\mathbf{1}_E; \mathbb{M})$, and therefore $\mathbf{1}_E \in BV(\mathbb{M})$. There exists a sequence $\{f_n\}_{n\in\mathbb{N}}$ in $C_0^\infty(\mathbb{M})$ satisfying (i) and (ii) above. Applying (3.21) to f_n we obtain

$$\|P_t f_n - f_n\|_{L^1(\mathbb{M})} \le \sqrt{6d}\,\sqrt{t}\ \|\sqrt{\Gamma(f_n)}\|_{L^1(\mathbb{M})} = \sqrt{6d}\,\sqrt{t}\ Var_{\mathcal{H}}(f_n, \mathbb{M}), \quad n \in \mathbb{N}.$$

Letting $n \to \infty$ in this inequality, we conclude

$$\|P_t\mathbf{1}_E - \mathbf{1}_E\|_{L^1(\mathbb{M})} \le \sqrt{6d}\,\sqrt{t}\ Var_{\mathcal{H}}(\mathbf{1}_E, \mathbb{M}) = \sqrt{6d}\,\sqrt{t}\ P(E; \mathbb{M}), \quad t > 0.$$

Observe now that, using $P_t 1 = 1$, we have

$$||P_t\mathbf{1}_E - \mathbf{1}_E||_{L^1(\mathbb{M})} = 2\left(\mu(E) - \int_E P_t\mathbf{1}_E d\mu\right).$$

On the other hand,

$$\int_E P_t \mathbf{1}_E d\mu = \int_{\mathbb{M}} \left(P_{t/2}\mathbf{1}_{\mathbb{M}}\right)^2 d\mu.$$

We thus obtain

$$||P_t \mathbf{1}_E - \mathbf{1}_E||_{L^1(\mathbb{M})} = 2\left(\mu(E) - \int_{\mathbb{M}} \left(P_{t/2}\mathbf{1}_E\right)^2 d\mu\right).$$

We now observe that the assumption (1) implies

$$p(x,x,t) \le \frac{C_4}{t^{n/2}}, \quad x \in \mathbb{M}, t > 0.$$

This gives

$$\begin{aligned}\int_{\mathbb{M}} (P_{t/2}\mathbf{1}_E)^2 d\mu &\le \left(\int_E \left(\int_{\mathbb{M}} p(x,y,t/2)^2 d\mu(y)\right)^{\frac{1}{2}} d\mu(x)\right)^2 \\ &= \left(\int_E p(x,x,t)^{\frac{1}{2}} d\mu(x)\right)^2 \le \frac{C_4}{t^{n/2}}\mu(E)^2.\end{aligned}$$

Combining these equations we reach the conclusion

$$\mu(E) \le \frac{\sqrt{6d}}{2}\sqrt{t}\, P(E;\mathbb{M}) + \frac{C_4}{t^{n/2}}\mu(E)^2, \quad t > 0.$$

Now the absolute minimum of the function $g(t) = At^{\alpha} + Bt^{-\beta}$, $t > 0$, where $A, B, \alpha, \beta > 0$, is given by

$$g_{\min} = \left[\left(\frac{\alpha}{\beta}\right)^{\frac{\beta}{\alpha+\beta}} + \left(\frac{\beta}{\alpha}\right)^{\frac{\alpha}{\alpha+\beta}}\right] A^{\frac{\beta}{\alpha+\beta}} B^{\frac{\alpha}{\alpha+\beta}}$$

Applying this observation with $\alpha = \frac{1}{2}$, $\beta = \frac{n}{2}$, we conclude

$$\mu(E)^{\frac{n-1}{n}} \le C_3 P(E,\mathbb{M}).$$

The fact that (3) implies (4) is classical geometric measure theory. It relies on the Federer co-area formula that we recall: For every $f, g \in C_0^{\infty}(\mathbb{M})$,

$$\int_{\mathbb{M}} g\|\nabla f\| d\mu = \int_{-\infty}^{+\infty} \left(\int_{f(x)=t} g(x) d\mu_{n-1}(x)\right) dt.$$

Let now $f \in C_0^\infty(\mathbb{M})$. We have

$$f(x) = \int_0^{+\infty} \mathbf{1}_{f(x)>t}(t)dt.$$

By using Minkowski inequality, we get then

$$\begin{aligned}
\|f\|_{\frac{n}{n-1}} &\le \int_0^\infty \|\mathbf{1}_{f(\cdot)>t}\|_{\frac{n}{n-1}} dt \\
&\le \int_0^\infty \mu(f > t)^{\frac{n}{n-1}} dt \\
&\le C_3 \int_0^\infty \mu_{n-1}(f = t)dt = C_3 \int_{\mathbb{M}} \sqrt{\Gamma(f)} d\mu
\end{aligned}$$

Finally, we show that (4) $\to$ (1). In what follows we let $\nu = n/(n-1)$. Let $p, q \in (0, \infty)$ and $0 < \theta \le 1$ be such that

$$\frac{1}{p} = \frac{\theta}{\nu} + \frac{1-\theta}{q}.$$

Hölder inequality, combined with assumption (4), gives for any $f \in Lip_d(\mathbb{M})$ with compact support

$$||f||_{L^p(\mathbb{M})} \le ||f||^\theta_{L^\nu(\mathbb{M})} ||f||^{1-\theta}_{L^q(\mathbb{M})} \le \left(C_3 ||\sqrt{\Gamma(f)}||_{L^1(\mathbb{M})} \right)^\theta ||f||^{1-\theta}_{L^q(\mathbb{M})}.$$

For any $x \in \mathbb{M}$ and $r > 0$ we now let $f(y) = (r - d(y,x))^+$. Clearly such $f \in Lip_d(\mathbb{M})$ and supp $f = \overline{B}(x,r)$. Since with this choice $||\sqrt{\Gamma(f)}||^\theta_{L^1(\mathbb{M})} \le \mu(B(x,r))^\theta$, the above inequality implies

$$\frac{r}{2} \mu(B(x, \frac{r}{2}))^{\frac{1}{p}} \le r^{1-\theta} \left(C_3 \mu(B(x,r))^\theta \, \mu(B(x,r)) \right)^{\frac{1-\theta}{q}},$$

which, noting that $\frac{1-\theta}{q} + \theta = \frac{n+\theta p}{pn}$, we can rewrite as follows

$$\mu(B(x,r)) \ge \left(\frac{1}{2C_3^\theta} \right)^{pa} \mu(B(x, \frac{r}{2}))^a r^{\theta p a},$$

where we have let $a = \frac{n}{n+\theta p}$. Notice that $0 < a < 1$. Iterating the latter inequality we find

$$\mu(B(x,r)) \ge \left(\frac{1}{2C_3^\theta} \right)^{p \sum_{j=1}^k a^j} r^{\theta p \sum_{j=1}^k a^j} 2^{-\theta p \sum_{j=1}^k (j-1) a^j} \mu(B(x, \frac{r}{2^k}))^{a^k}, \quad k \in \mathbb{N}.$$

From the doubling property for any $x \in \mathbb{M}$ there exist constants $C(x), R(x) > 0$ such that with $Q(x) = \log_2 C(x)$ one has

$$\mu(B(x,tr)) \geq C(x)^{-1} t^{Q(x)} \mu(B(x,r)), \quad 0 \leq t \leq 1, 0 < r \leq R(x).$$

This estimate implies that

$$\liminf_{k \to \infty} \mu(B(x, \frac{r}{2^k}))^{a^k} \geq 1, \quad x \in \mathbb{M}, r > 0.$$

Since on the other hand $\sum_{j=1}^{\infty} a^j = \frac{n}{\theta p}$, and $\sum_{j=1}^{\infty} (j-1) a^j = \frac{n^2}{\theta^2 p^2}$, we conclude that

$$\mu(B(x,r)) \geq \left(2^{-\frac{1}{\theta}(1+\frac{n}{p})} C_3^{-1}\right)^n r^n, \quad x \in \mathbb{M}, r > 0.$$

This establishes (1), thus completing the proof. □

3.15 *Sharp Sobolev Inequalities*

In this section we are interested in sharp Sobolev inequalities in positive curvature. Let $(\mathbb{M}, g)$ be a complete and n-dimensional Riemannian manifold such that **Ricci** $\geq \rho$ where $\rho > 0$. We assume $n > 2$. As we already know, we have $\mu(\mathbb{M}) < +\infty$, but as we already stressed we do not want to use Bonnet-Myers theorem, since one of our goals will be to recover it by using heat kernel techniques. Without loss of generality, and to simplify the constants, we assume that $\mu(\mathbb{M}) = 1$. Our goal is to prove the following sharp result:

Theorem 3.48 *For every* $1 \leq p \leq \frac{2n}{n-2}$ *and* $f \in C_0^{\infty}(\mathbb{M})$,

$$\frac{n\rho}{(n-1)(p-2)} \left(\left(\int_{\mathbb{M}} |f|^p d\mu \right)^{2/p} - \int_{\mathbb{M}} f^2 d\mu \right) \leq \int_{\mathbb{M}} \Gamma(f) d\mu.$$

We observe that for $p = 1$, the inequality becomes

$$\frac{n\rho}{(n-1)} \left(\int_{\mathbb{M}} f^2 d\mu - \left(\int_{\mathbb{M}} |f| d\mu \right)^2 \right) \leq \int_{\mathbb{M}} \Gamma(f) d\mu.$$

which is the Poincaré inequality with optimal Lichnerowicz constant. For $p = 2$, we get the log-Sobolev inequality

$$\frac{n\rho}{2(n-1)} \left(\int_{\mathbb{M}} f^2 \ln f^2 d\mu - \int_{\mathbb{M}} f^2 d\mu \ln \int_{\mathbb{M}} f^2 d\mu \right) \leq \int_{\mathbb{M}} \Gamma(f) d\mu.$$

We prove our Sobolev inequality in several steps.

Lemma 3.49 *For every* $1 \le p \le \frac{2n}{n-2}$, *there exists a constant* $C_p \neq 0$ *such that for every* $f \in C_0^\infty(\mathbb{M})$,

$$C_p \left(\left(\int_{\mathbb{M}} |f|^p d\mu \right)^{2/p} - \int_{\mathbb{M}} f^2 d\mu \right) \le \int_{\mathbb{M}} \Gamma(f) d\mu.$$

Proof Using Jensen's inequality, it is enough to prove the result for $p = \frac{2n}{n-2}$. We already proved the following Li-Yau inequality: For $f \in C_0^\infty(\mathbb{M})$, $f \neq 0$, $t > 0$, and $x \in \mathbb{M}$,

$$\|\nabla \ln P_t f(x)\|^2 \le e^{-\frac{2\rho t}{3}} \frac{L P_t f(x)}{P_t f(x)} + \frac{n\rho}{3} \frac{e^{-\frac{4\rho t}{3}}}{1 - e^{-\frac{2\rho t}{3}}}.$$

As a consequence we have

$$\frac{L P_t f(x)}{P_t f(x)} \ge -\frac{n\rho}{3} \frac{e^{-\frac{2\rho t}{3}}}{1 - e^{-\frac{2\rho t}{3}}},$$

which yields,

$$\int_t^{+\infty} \partial_t \ln P_s f(x) ds \ge -\frac{n\rho}{3} \int_t^{+\infty} \frac{e^{-\frac{2\rho s}{3}}}{1 - e^{-\frac{2\rho s}{3}}} ds.$$

We obtain then

$$P_t f(x) \le \left(\frac{1}{1 - e^{-\frac{2\rho t}{3}}} \right)^{n/2} \int_{\mathbb{M}} f d\mu.$$

This of course implies the following upper bound on the heat kernel,

$$p(x, y, t) \le \left(\frac{1}{1 - e^{-\frac{2\rho t}{3}}} \right)^{n/2}.$$

Using Varopoulos theorem, we deduce therefore that there is constant C'_p such that for every $f \in C_0^\infty(\mathbb{M})$,

$$\|f\|_p^2 \le C'_p \left(\|\sqrt{\Gamma(f)}\|_2^2 + \|f\|_2^2 \right).$$

We now use the following inequality which is easy to see:

$$\left(\int_{\mathbb{M}} |f|^p d\mu\right)^{2/p} \le \left(\int_{\mathbb{M}} f d\mu\right)^2 + \left(\int_{\mathbb{M}} \left| f - \int_{\mathbb{M}} f d\mu \right|^p d\mu\right)^{2/p}$$

This yields

$$\left(\int_{\mathbb{M}} |f|^p d\mu\right)^{2/p} \le \left(\int_{\mathbb{M}} f d\mu\right)^2 + (p-1)C'_p \left(\|\sqrt{\Gamma(f)}\|_2^2 + \left\| f - \int_{\mathbb{M}} f d\mu \right\|_2^2 \right)$$

We can now bound

$$\left\| f - \int_{\mathbb{M}} f d\mu \right\|_2^2 \le \frac{1}{\lambda_1} \int_{\mathbb{M}} \Gamma(f) d\mu,$$

using the Poincaré inequality. □

Our proof follows now an argument due to Bakry and largely follows the presentation by Ledoux [19] to which we refer for the details. We now want to prove that the optimal C_p in the previous inequality is given by $C_p = \frac{n\rho}{(n-1)(p-2)}$. We assume $p > 2$ and consider the functional

$$\frac{\left(\int_{\mathbb{M}} |f|^p d\mu\right)^{2/p} - \int_{\mathbb{M}} f^2 d\mu}{\int_{\mathbb{M}} \Gamma(f) d\mu}.$$

Classical non linear variational principles on the functional provide then a positive non trivial solution of the equation

$$C_p(f^{p-1} - f) = -Lf.$$

We point out that existence and smoothness results for this non linear PDE are non a trivial issue, we refer to the comments by Ledoux in [19] and the references therein. Set $f = u^r$ where r is a constant to be later chosen. By the chain rule for diffusion operators, we get

$$C_p(u^{r(p-1)} - u^r) = -ru^{r-1}Lu - r(r-1)u^{r-2}\Gamma(u).$$

Multiplying by $u^{-r}\Gamma(u)$ and integrating yields

$$C_p \left(\int_{\mathbb{M}} u^{r(p-2)} \Gamma(u) d\mu - \int_{\mathbb{M}} \Gamma(u) d\mu \right) = -r \int_{\mathbb{M}} \frac{Lu}{u} \Gamma(u) d\mu - r(r-1) \int_{\mathbb{M}} \frac{\Gamma(u)^2}{u^2} d\mu.$$

Now, integrating by parts,

$$\int_{\mathbb{M}} u^{r(p-2)} \Gamma(u) d\mu = -\frac{1}{r(p-2)+1} \int_{\mathbb{M}} u^{r(p-2)+1} Lu d\mu.$$

On the other hand, multiplying

$$C_p(u^{r(p-1)} - u^r) = -ru^{r-1}Lu - r(r-1)u^{r-2}\Gamma(u),$$

by $u^{1-r}Lu$ and integrating with respect to μ yields

$$C_p\left(\int_{\mathbb{M}} u^{r(p-2)+1}Lud\mu - \int_{\mathbb{M}} uLud\mu\right) = -r\int_{\mathbb{M}}(Lu)^2d\mu - r(r-1)\int_{\mathbb{M}}\frac{Lu}{u}\Gamma(u)d\mu.$$

Combining the previous computations gives

$$\begin{aligned}&C_p\,(r(p-2)+1)\int_{\mathbb{M}} u^{r(p-2)}\Gamma(u)d\mu \\ &\quad = r\int_{\mathbb{M}}(Lu)^2d\mu + r(r-1)\int_{\mathbb{M}}\frac{Lu}{u}\Gamma(u)d\mu + C_p\int_{\mathbb{M}}\Gamma(u)d\mu\end{aligned}$$

Hence, we have

$$\begin{aligned}&C_p(p-2)\int_{\mathbb{M}}\Gamma(u)d\mu \\ &= \int_{\mathbb{M}}(Lu)^2d\mu + \int_{\mathbb{M}}\frac{Lu}{u}\Gamma(u)d\mu + (r-1)\,(r(p-2)+1)\int_{\mathbb{M}}\frac{\Gamma(u)^2}{u^2}d\mu\end{aligned}$$

We have from Bochner's inequality,

$$\Gamma_2(u^s) \geq \frac{1}{n}(Lu^s)^2 + \rho\Gamma(u^s).$$

Once again, s is a parameter that will be later decided. Using the chain rule, to rewrite the previous inequality, leads after tedious computations to

$$\begin{aligned}&\Gamma_2(u) + (s-1)\frac{1}{u}\Gamma(u,\Gamma(u)) + (s-1)^2\frac{\Gamma(u)^2}{u^2} \\ &\quad \geq \rho\Gamma(u) + \frac{1}{n}(Lu)^2 + \frac{2}{n}(s-1)\frac{1}{u}Lu\Gamma(u) + \frac{1}{n}(s-1)^2\frac{1}{u^2}\Gamma(u)^2.\end{aligned}$$

After integration and integration by parts, we see that

$$\begin{aligned}\rho\int_{\mathbb{M}}\Gamma(u)d\mu \leq \left(1-\frac{1}{n}\right)\int_{\mathbb{M}}(Lu)^2d\mu - s'\left(1+\frac{2}{n}\right)\int_{\mathbb{M}}\frac{1}{u}Lu\Gamma(u)d\mu \\ + s'\left(1+s'\left(1-\frac{1}{n}\right)\right)\int_{\mathbb{M}}\frac{\Gamma(u)^2}{u^2}d\mu,\end{aligned}$$

where $s' = s - 1$. Combining the previous inequalities we can eliminate the term $\int_{\mathbb{M}} \frac{1}{u} Lu\Gamma(u)d\mu$. Choosing

$$\frac{s'}{r} = (p-1)\frac{n-1}{n+2},$$

we see that the coefficient in front of $\int_{\mathbb{M}} (Lu)^2 d\mu$ is zero and we are left with

$$\left(C_p \frac{(p-2)(n-1)}{n} - \rho\right) \int_{\mathbb{M}} \Gamma(u)d\mu \geq K(s',r) \int_{\mathbb{M}} \frac{\Gamma(u)^2}{u^2} d\mu,$$

for some constant $K(s',r)$ which is seen to be non-negative as soon as $2 < p \leq \frac{2n}{n-2}$. We conclude

$$C_p \frac{(p-2)(n-1)}{n} - \rho \geq 0.$$

3.16 The Sobolev Inequality Proof of the Myer's Diameter Theorem

It is a well-known result that if $\mathbb{M}$ is a complete n-dimensional Riemannian manifold with **Ricci** $\geq \rho$, for some $\rho > 0$, then $\mathbb{M}$ has to be compact with diameter less than $\pi\sqrt{\frac{n-1}{\rho}}$. The proof of this fact can be found in any graduate book about Riemannian geometry and classically relies on the study of Jacobi fields. We propose here an alternative proof of the diameter theorem that relies on the sharp Sobolev inequality proved in the previous section. The beautiful argument goes back to Bakry and Ledoux. We only sketch the main arguments and refer the readers to the original article [1].

The theorem by Bakry and Ledoux is the following:

Theorem 3.50 *Assume that for some $p > 2$, we have the inequality,*

$$\|f\|_p^2 \leq \|f\|_2^2 + A \int_{\mathbb{M}} \Gamma(f)d\mu, \quad f \in C_0^\infty(\mathbb{M}),$$

then $\mathbb{M}$ is compact with diameter less than $\pi \frac{\sqrt{2pA}}{p-2}$.

Combining this with the inequality

$$\frac{n\rho}{(n-1)(p-2)} \left(\left(\int_{\mathbb{M}} |f|^p d\mu \right)^{2/p} - \int_{\mathbb{M}} f^2 d\mu \right) \leq \int_{\mathbb{M}} \Gamma(f)d\mu,$$

that was proved in the previous section gives $\mathbf{diam}(\mathbb{M}) \leq \pi\sqrt{\frac{2p}{p-2}}\sqrt{\frac{n-1}{n\rho}}$. When $n = 2$ we conclude then by letting $p \to \infty$ and when $n > 2$, we conclude by choosing $p = \frac{2n}{n-2}$.

By using a scaling argument it is easy to see that it is enough to prove that if for some $n > 2$,

$$\|f\|^2_{\frac{2n}{n-2}} \leq \|f\|_2^2 + \frac{4}{n(n-2)}\int_{\mathbb{M}} \Gamma(f)d\mu,$$

then $\mathbf{diam}(\mathbb{M}) \leq \pi$.

The main idea is to apply the Sobolev inequality to the functions which are the extremals functions on the sphere. Such extremals are solutions of the fully non linear PDE

$$f^{(n+2)/(n-2)} - f = -\frac{4}{n(n-2)}Lf$$

and on the spheres the extremals are explicitly given by

$$f = (1 + \lambda \sin d)^{1-n/2}$$

where $-1 < \lambda < 1$ and d is the distance to a fixed point. So, on our manifold $\mathbb{M}$, that satisfies the inequality

$$\|f\|^2_{\frac{2n}{n-2}} \leq \|f\|_2^2 + \frac{4}{n(n-2)}\int_{\mathbb{M}} \Gamma(f)d\mu,$$

we consider the functional

$$F(\lambda) = \int_{\mathbb{M}} (1 + \lambda \sin(f))^{2-n} d\mu, \quad -1 < \lambda < 1,$$

where f is a function on $\mathbb{M}$ that satisfies $\|\Gamma(f)\|_\infty \leq 1$. The first step is to prove a differential inequality on F. For $k > 0$, we denote by D_k the differential operator on $(-1, 1)$ defined by

$$D_k = \frac{1}{k}\lambda\frac{\partial}{\partial\lambda} + I.$$

Lemma 3.51 *Denoting $G = D_{n-1}F$, we have*

$$(D_{n-2}G)^{(n-2)/n} + \frac{n-2}{n}(1-\lambda^2)D_{n-2}G \leq \left(1 + \frac{n-2}{n}\right)G.$$

Proof We denote $\alpha = \frac{n-2}{n}$ and $f_\lambda = (1+\lambda \sin f)^{1-n/2}$, $-1 < \lambda < 1$. By the chain-rule and the hypothesis that $\Gamma(f) \le 1$, we get

$$\int_{\mathbb{M}} \Gamma(f_\lambda) d\mu \le \left(\frac{n}{2}-1\right)^2 \int_{\mathbb{M}} (1+\lambda \sin f)^{-n}(1-\sin^2 f) d\mu.$$

From the Sobolev inequality applied to f_λ, we thus have,

$$\left(\int_{\mathbb{M}} (1+\lambda \sin(f))^{-n} d\mu\right)^{\alpha} \le \int_{\mathbb{M}} (1+\lambda \sin(f))^{2-n} d\mu$$
$$+ \alpha\lambda^2 \int_{\mathbb{M}} (1+\lambda \sin f)^{-n}(1-\sin^2 f) d\mu.$$

It is then an easy calculus exercise to deduce our claim. □

The next idea is then to use a comparison theorem to bound F in terms of solutions of the equation

$$(D_{n-2}H)^{(n-2)/n} + \frac{n-2}{n}(1-\lambda^2) D_{n-2}H \le \left(1+\frac{n-2}{n}\right) H.$$

Actually, such solutions are given by

$$H_c(\lambda) = \frac{1}{1+\alpha} U_c(\lambda)^{\frac{2\alpha}{1-\alpha}} + \frac{\alpha}{1+\alpha}(1-\lambda^2) U_c(\lambda)^{\frac{2}{1-\alpha}},$$

where $c \in \mathbb{R}$, $\alpha = \frac{n-2}{n}$ and

$$U_c(\lambda) = \frac{c\lambda + \sqrt{c^2\lambda^2 + (1-\lambda^2)}}{1-\lambda^2}.$$

We have then the following comparison result:

Lemma 3.52 *Let G be such that*

$$(D_{n-2}G)^{(n-2)/n} + \frac{n-2}{n}(1-\lambda^2) D_{n-2}G \le \left(1+\frac{n-2}{n}\right) G,$$

and assume that $G(\lambda_0) < H_c(\lambda_0)$ for some $\lambda_0 \in [0,1)$. Then for every $\lambda_0 \le \lambda < 1$,

$$G(\lambda) \le H_c(\lambda).$$

Using the previous lemma, we see (again we refer to the original article for the details) that $\int_{\mathbb{M}} \sin f d\mu > 0$ implies that $\int_{\mathbb{M}} (1+\sin f)^{n-1} d\mu < \infty$ and $\int_{\mathbb{M}} \sin f d\mu < 0$ implies that $\int_{\mathbb{M}} (1-\sin f)^{n-1} d\mu < \infty$. Iterating this result on the basis of the Sobolev inequality again, we actually have

$$\|(1 \pm \sin f)^{-1}\|_\infty < \infty,$$

from which the conclusion easily follows.

4 The Heat Semigroup on Sub-Riemannian Manifolds and Its Applications

It turns out that many of methods presented in the last section can be extended to some sub-Riemannian manifolds, provided that a correct generalization of curvature-dimension inequality is used. This generalized curvature dimension condition was first introduced in [6] in the context of sub-Riemannian manifolds with transverse symmetries. In [6], it has been shown that this generalized curvature dimension condition implies a Li-Yau inequality and a Gaussian upper bound for the heat kernel. Then, in [5], it was proved that the curvature dimension inequality implies a lower bound for the heat kernel. Combining those results the volume doubling property and the 2-Poincaré inequality on balls were deduced. In this section we sketch the results obtained in those two papers without entering into details. We refer to the survey [3] and the forthcoming book [7] for further applications of the sub-Riemannian curvature dimension condition (like Bonnet-Myers theorems and eigenvalue estimates), see also [15, 16].

4.1 Framework

Hereafter in this section, $\mathbb{M}$ will be a C^∞ connected manifold endowed with a smooth measure μ and a second-order diffusion operator L on $\mathbb{M}$ with real coefficients, locally subelliptic, satisfying $L1 = 0$ and

$$\int_{\mathbb{M}} fLg d\mu = \int_{\mathbb{M}} gLf d\mu, \qquad \int_{\mathbb{M}} fLf d\mu \le 0,$$

for every $f, g \in C_0^\infty(\mathbb{M})$, where $C_0^\infty(\mathbb{M})$ denotes the space of compactly supported functions. A distance d is constructed as follows (see Sect. 1):

$$d(x, y) = \sup\left\{|f(x) - f(y)| \mid f \in C^\infty(\mathbb{M}), \|\Gamma(f)\|_\infty \le 1\right\}, \quad x, y \in \mathbb{M}, \tag{4.1}$$

where for a function g on $\mathbb{M}$ we have let $||g||_\infty = \operatorname{ess\,sup}_{\mathbb{M}}|g|$. This distance will often be referred to as the subelliptic or sub-Riemannian distance. As before, the quadratic functional $\Gamma(f) = \Gamma(f, f)$, where

$$\Gamma(f, g) = \frac{1}{2}(L(fg) - fLg - gLf), \quad f, g \in C^\infty(\mathbb{M}), \tag{4.2}$$

is known as *le carré du champ*. Given any point $x \in \mathbb{M}$ there exists an open set $x \in U \subset \mathbb{M}$ in which the operator L can be written as

$$L = -\sum_{i=1}^{m} X_i^* X_i, \tag{4.3}$$

where the vector fields X_i have Lipschitz continuous coefficients in U, and X_i^* indicates the formal adjoint of X_i in $L^2(\mathbb{M}, d\mu)$. We remark that such local representation of L is not unique.

In addition to the differential form (4.2), we assume that $\mathbb{M}$ be endowed with another smooth bilinear differential form, indicated with Γ^Z, satisfying for $f, g \in C^\infty(\mathbb{M})$

$$\Gamma^Z(fg, h) = f\Gamma^Z(g, h) + g\Gamma^Z(f, h),$$

and $\Gamma^Z(f) = \Gamma^Z(f, f) \geq 0$. We assume that given any point $x \in \mathbb{M}$ there exists an open set $x \in U \subset \mathbb{M}$ in which the operator Γ^Z can be written as

$$\Gamma^Z(f, g) = \sum_{i=1}^{p} (Z_i f)(Z_i g),$$

where the vector fields Z_i have Lipschitz continuous coefficients in U.

Given the first-order bilinear forms Γ and Γ^Z on $\mathbb{M}$, we now introduce the following second-order differential forms:

$$\Gamma_2(f, g) = \frac{1}{2}\big[L\Gamma(f, g) - \Gamma(f, Lg) - \Gamma(g, Lf)\big], \tag{4.4}$$

$$\Gamma_2^Z(f, g) = \frac{1}{2}\big[L\Gamma^Z(f, g) - \Gamma^Z(f, Lg) - \Gamma^Z(g, Lf)\big]. \tag{4.5}$$

As for Γ and Γ^Z, we will use the notations $\Gamma_2(f) = \Gamma_2(f, f)$, $\Gamma_2^Z(f) = \Gamma_2^Z(f, f)$.

We make the following assumptions:

(H.1) There exists an increasing sequence $h_k \in C_0^\infty(\mathbb{M})$ such that $h_k \nearrow 1$ on $\mathbb{M}$, and

$$||\Gamma(h_k)||_\infty + ||\Gamma^Z(h_k)||_\infty \to 0, \quad \text{as } k \to \infty.$$

(H.2) For any $f \in C^\infty(\mathbb{M})$ one has

$$\Gamma(f, \Gamma^Z(f)) = \Gamma^Z(f, \Gamma(f)).$$

(H.3) The heat semigroup generated by L, which will denoted P_t throughout the paper, is stochastically complete that is, for $t \geq 0$, $P_t 1 = 1$ and for every $f \in C_0^\infty(\mathbb{M})$ and $T \geq 0$, one has

$$\sup_{t\in[0,T]} \|\Gamma(P_t f)\|_\infty + \|\Gamma^Z(P_t f)\|_\infty < +\infty.$$

(H.4) Given any two points $x, y \in \mathbb{M}$, there exist a subunit curve joining them.
(H.5) The metric space $(\mathbb{M}, d)$ is complete.

A large class of examples where all these assumptions are satisfied arises in the context of totally geodesic Riemannian foliations and sub-Riemannian manifolds with transverse symmetries (see [3, 6, 8] for a detailed proof of these assumptions).

Definition 4.1 (See [6]) We shall say that $\mathbb{M}$ satisfies the generalized curvature-dimension inequality CD$(\rho_1, \rho_2, \kappa, n)$ if there exist constants $\rho_1 \in \mathbb{R}$, $\rho_2 > 0$, $\kappa \geq 0$, and $n > 0$ such that the inequality

$$\Gamma_2(f) + \nu\Gamma_2^Z(f) \geq \frac{1}{n}(Lf)^2 + \left(\rho_1 - \frac{\kappa}{\nu}\right)\Gamma(f) + \rho_2\Gamma^Z(f) \tag{4.6}$$

holds for every $f \in C^\infty(\mathbb{M})$ and every $\nu > 0$.

4.2 Li-Yau Inequality and Volume Doubling Properties for the Subelliptic Distance

Throughout the section we assume that $\mathbb{M}$ satisfies the generalized curvature-dimension inequality CD$(\rho_1, \rho_2, \kappa, n)$, and we show how to obtain the Li-Yau estimate for the subelliptic operator L.

Henceforth, we will indicate $C_b^\infty(\mathbb{M}) = C^\infty(\mathbb{M}) \cap L^\infty(\mathbb{M})$ and by P_t the semigroup generated by L. A key lemma is the following.

Lemma 4.2 *Let $f \in C_b^\infty(\mathbb{M})$, $f > 0$ and $T > 0$, and consider the functions*

$$\phi_1(x,t) = (P_{T-t}f)(x)\Gamma(\ln P_{T-t}f)(x),$$

$$\phi_2(x,t) = (P_{T-t}f)(x)\Gamma^Z(\ln P_{T-t}f)(x),$$

which are defined on $\mathbb{M} \times [0, T)$. We have

$$L\phi_1 + \frac{\partial\phi_1}{\partial t} = 2(P_{T-t}f)\Gamma_2(\ln P_{T-t}f)$$

and

$$L\phi_2 + \frac{\partial \phi_2}{\partial t} = 2(P_{T-t}f)\Gamma_2^Z(\ln P_{T-t}f).$$

Proof This is direct computation without trick. Let us just point out that the formula

$$L\phi_2 + \frac{\partial \phi_2}{\partial t} = 2(P_{T-t}f)\Gamma_2^Z(\ln P_{T-t}f).$$

uses the fact that $\Gamma(g, \Gamma^Z(g)) = \Gamma^Z(g, \Gamma(g))$. □

We will need the following lemma already used before.

Lemma 4.3 *Let* $T > 0$. *Let* $u, v : \mathbb{M} \times [0, T] \to \mathbb{R}$ *be smooth functions such that for every* $T > 0$, $\sup_{t\in[0,T]} \|u(\cdot, t)\|_\infty < \infty$, $\sup_{t\in[0,T]} \|v(\cdot, t)\|_\infty < \infty$*; If the inequality*

$$Lu + \frac{\partial u}{\partial t} \geq v$$

holds on $\mathbb{M} \times [0, T]$, *then we have*

$$P_T(u(\cdot, T))(x) \geq u(x, 0) + \int_0^T P_s(v(\cdot, s))(x)ds.$$

We now show how to prove the Li-Yau estimates for the semigroup P_t.

Theorem 4.4 *Let* $\alpha > 2$. *For* $f \in C_0^\infty(\mathbb{M})$, $f \geq 0$, $f \neq 0$, *the following inequality holds for* $t > 0$:

$$\begin{aligned}
&\Gamma(\ln P_t f) + \frac{2\rho_2}{\alpha} t\Gamma^Z(\ln P_t f) \\
&\leq \left(1 + \frac{\alpha\kappa}{(\alpha-1)\rho_2} - \frac{2\rho_1}{\alpha}t\right)\frac{LP_t f}{P_t f} + \frac{n\rho_1^2}{2\alpha}t - \frac{\rho_1 n}{2}\left(1 + \frac{\alpha\kappa}{(\alpha-1)\rho_2}\right) \\
&\quad + \frac{n(\alpha-1)^2\left(1 + \frac{\alpha\kappa}{(\alpha-1)\rho_2}\right)^2}{8(\alpha-2)t}.
\end{aligned}$$

Proof We fix $T > 0$ and consider two functions $a, b : [0, T] \to \mathbb{R}_{\geq 0}$ to be chosen later. Let $f \in C^\infty(\mathbb{M})$, $f \geq 0$. Consider the function

$$\phi(x, t) = a(t)(P_{T-t}f)(x)\Gamma(\ln P_{T-t}f)(x) + b(t)(P_{T-t}f)(x)\Gamma^Z(\ln P_{T-t}f)(x).$$

Applying Lemma 4.2 and the curvature-dimension inequality, we obtain

$$\begin{aligned}
&L\phi+\frac{\partial \phi}{\partial t}\\
=&a'(P_{T-t}f)\Gamma(\ln P_{T-t}f)+b'(P_{T-t}f)\Gamma^Z(\ln P_{T-t}f)+2a(P_{T-t}f)\Gamma_2(\ln P_{T-t}f)\\
&+2b(P_{T-t}f)\Gamma_2^Z(\ln P_{T-t}f)\\
\ge&\left(a'+2\rho_1 a-2\kappa\frac{a^2}{b}\right)(P_{T-t}f)\Gamma(\ln P_{T-t}f)+(b'+2\rho_2 a)(P_{T-t}f)\Gamma^Z(\ln P_{T-t}f)\\
&+\frac{2a}{n}(P_{T-t}f)(L(\ln P_{T-t}f))^2.
\end{aligned}$$

But, for any function $\gamma:[0,T]\to\mathbb{R}$

$$(L(\ln P_{T-t}f))^2\ge 2\gamma L(\ln P_{T-t}f)-\gamma^2,$$

and from chain rule

$$L(\ln P_{T-t}f)=\frac{LP_{T-t}f}{P_{T-t}f}-\Gamma(\ln P_{T-t}f).$$

Therefore, we obtain

$$\begin{aligned}
L\phi+\frac{\partial \phi}{\partial t}\ge&\left(a'+2\rho_1 a-2\kappa\frac{a^2}{b}-\frac{4a\gamma}{n}\right)(P_{T-t}f)\Gamma(\ln P_{T-t}f)\\
&+(b'+2\rho_2 a)(P_{T-t}f)\Gamma^Z(\ln P_{T-t}f)+\frac{4a\gamma}{n}LP_{T-t}f-\frac{2a\gamma^2}{n}P_{T-t}f.
\end{aligned}$$

The idea is now to chose a, b, γ such that

$$\begin{cases} a'+2\rho_1 a-2\kappa\frac{a^2}{b}-\frac{4a\gamma}{n}=0\\ b'+2\rho_2 a=0\end{cases}$$

With this choice we get

$$L\phi+\frac{\partial \phi}{\partial t}\ge\frac{4a\gamma}{n}LP_{T-t}f-\frac{2a\gamma^2}{n}P_{T-t}f \tag{4.7}$$

We wish to apply Lemma 4.3. We take now $f\in C_0^\infty(\mathbb{M})$ and apply the previous inequality with $f_\varepsilon=f+\varepsilon$ instead of f, where $\varepsilon>0$. If moreover $a(T)=b(T)=0$, we end up with the inequality

$$\begin{aligned}
&a(0)(P_T f_\varepsilon)(x)\Gamma(\ln P_T f_\varepsilon)(x)+b(0)(P_T f)(x)\Gamma^Z(\ln P_T f_\varepsilon)(x)\\
\le&-\int_0^T\frac{4a\gamma}{n}dt LP_T f_\varepsilon(x)+\int_0^T\frac{2a\gamma^2}{n}dt P_T f_\varepsilon(x)
\end{aligned} \tag{4.8}$$

If we now chose $b(t) = (T - t)^\alpha$ and b, γ such that

$$\begin{cases} a' + 2\rho_1 a - 2\kappa \frac{a^2}{b} - \frac{4a\gamma}{n} = 0 \\ b' + 2\rho_2 a = 0 \end{cases}$$

the result follows by a simple computation and sending then $\varepsilon \to 0$. □

Observe that if $\rho_1 \geq 0$, then we can take $\rho_1 = 0$ and the estimate simplifies to

$$\Gamma(\ln P_t f) + \frac{2\rho_2}{\alpha} t \Gamma^Z(\ln P_t f) \leq \left(1 + \frac{\alpha\kappa}{(\alpha-1)\rho_2}\right) \frac{LP_t f}{P_t f} + \frac{n(\alpha-1)^2 \left(1 + \frac{\alpha\kappa}{(\alpha-1)\rho_2}\right)^2}{8(\alpha-2)t}.$$

By adapting the classical method of Li and Yau and integrating this last inequality on subunit curves leads to a parabolic Harnack inequality (details are in [6]). For $\alpha > 2$, we denote

$$D_\alpha = \frac{n(\alpha-1)^2 \left(1 + \frac{\alpha\kappa}{(\alpha-1)\rho_2}\right)}{4(\alpha-2)}. \tag{4.9}$$

The minimal value of D_α is difficult to compute, depends on κ, ρ_2 and does not seem relevant because the constants we get are anyhow not optimal. We just point out that the choice $\alpha = 3$ turns out to simplify many computations and is actually optimal when $\kappa = 4\rho_2$.

Corollary 4.5 *Let us assume that $\rho_1 \geq 0$. Let $f \in L^\infty(\mathbb{M})$, $f \geq 0$, and consider $u(x,t) = P_t f(x)$. For every $(x,s), (y,t) \in \mathbb{M} \times (0,\infty)$ with $s < t$ one has with D_α as in* (4.9)

$$u(x,s) \leq u(y,t) \left(\frac{t}{s}\right)^{\frac{D_\alpha}{2}} \exp\left(\frac{D_\alpha}{n} \frac{d(x,y)^2}{4(t-s)}\right).$$

Here $d(x,y)$ is the sub-Riemannian distance between x and y.

It is classical since the work by Li and Yau and not difficult to prove that a parabolic Harnack inequality implies a Gaussian upper bound on the heat kernel. With the curvature dimension inequality in hand, it is actually also possible, but much more difficult, to prove a lower bound. The final result proved in [5] is:

Theorem 4.6 *Let us assume that $\rho_1 \geq 0$, then for any $0 < \varepsilon < 1$ there exists a constant $C(\varepsilon) = C(n, \kappa, \rho_2, \varepsilon) > 0$, which tends to ∞ as $\varepsilon \to 0^+$, such that for every $x, y \in \mathbb{M}$ and $t > 0$ one has*

$$\frac{C(\varepsilon)^{-1}}{\mu(B(x,\sqrt{t}))} \exp\left(-\frac{D_\alpha d(x,y)^2}{n(4-\varepsilon)t}\right) \leq p_t(x,y) \leq \frac{C(\varepsilon)}{\mu(B(x,\sqrt{t}))} \exp\left(-\frac{d(x,y)^2}{(4+\varepsilon)t}\right),$$

where $p_t(x,y)$ is the heat kernel of L.

From the equivalence between Gaussian estimates for the heat kernel and volume doubling properties and Poincaré inequalities (see [21, 22]), this theorem implies the following important result:

Theorem 4.7 *Let us assume that* $\rho_1 \geq 0$. *Then, the metric measure space* $(\mathbb{M}, d, \mu)$ *satisfies the global volume doubling property and supports a scale invariant 2-Poincaré inequality on balls.*

References

1. D. Bakry, M. Ledoux, Sobolev inequalities and Myers's diameter theorem for an abstract Markov generator. Duke Math. J. **85**(1), 253–270 (1996)
2. D. Bakry, I. Gentil, M. Ledoux, *Analysis and Geometry of Markov Diffusion Operators*. Grundlehren der Mathematischen Wissenschaften [Fundamental Principles of Mathematical Sciences], vol. 348 (Springer, Cham, 2014), xx+552 pp.
3. F. Baudoin, *Sub-Laplacians and Hypoelliptic Operators on Totally Geodesic Riemannian Foliations*. Geometry, analysis and dynamics on sub-Riemannian manifolds, vol. 1 (EMS Ser. Lect. Math., Eur. Math. Soc., Zürich, 2016), pp. 259–321
4. F. Baudoin, *Diffusion Processes and Stochastic Calculus*. EMS Textbooks in Mathematics. (European Mathematical Society (EMS), Zürich, 2014). xii+276 pp.
5. F. Baudoin, M. Bonnefont, N. Garofalo, A sub-Riemannian curvature-dimension inequality, volume doubling property and the Poincare inequality. Math. Ann. **358**(3–4), 833–860(2014)
6. F. Baudoin, N. Garofalo, Curvature-dimension inequalities and Ricci lower bounds for sub-Riemannian manifolds with transverse symmetries. J. Eur. Math. Soc. (JEMS) **19**(1), 151–219 (2017)
7. F. Baudoin, N. Garofalo, *Curvature-Dimension Inequalities in Riemannian and Sub-Riemannian Geometry*, In preparation
8. F. Baudoin, B. Kim, J. Wang, Transverse Weitzenböck formulas and curvature dimension inequalities on Riemannian foliations with totally geodesic leaves. Commun. Anal. Geom. **24**(5), 913–937 (2016)
9. F. Baudoin, A. Vatamanelu, A note on lower bounds estimates for the Neumann eigenvalues of manifolds with positive Ricci curvature. Potential Anal. **37**(1), 91–101 (2012)
10. M. Bramanti, *An Invitation to Hypoelliptic Operators and Hörmander's Vector Fields*. SpringerBriefs in Mathematics (Springer, Cham, 2014), xii+150 pp.
11. I. Chavel, *Riemannian Geometry: A Modern Introduction*. Cambridge Tracts in Mathematics, vol. 108 (Cambridge University Press, Cambridge, 1993)
12. C. Fefferman, D.H. Phong, *Subelliptic Eigenvalue Problems*. Conference on Harmonic Analysis in Honor of Antoni Zygmund, vol. I, II (Chicago, IL, 1981), Wadsworth Math. Ser. (Wadsworth, Belmont, CA, 1983), pp. 590–606
13. G. Folland, *Introduction to Partial Differential Equations*. 2nd edn. (Princeton University Press, Princeton, NJ, 1995), xii+324 pp.
14. A. Grigor'yan, *Heat Kernel and Analysis on Manifolds*. AMS/IP Studies in Advanced Mathematics, vol. 47 (American Mathematical Society, Providence, RI; International Press, Boston, MA, 2009). xviii+482 pp.
15. E. Grong, A. Thalmaier, Curvature-dimension inequalities on sub-Riemannian manifolds obtained from Riemannian foliations: Part I. Math. Z. **282**(1–2), 99–130 (2016)
16. E. Grong, A. Thalmaier, Curvature-dimension inequalities on sub-Riemannian manifolds obtained from Riemannian foliations: Part II. Math. Z. **282**(1–2), 131–164 (2016)
17. L. Hörmander, Hypoelliptic second order differential equations. Acta Math. **119**, 147–171 (1967)

18. D. Jerison, A. Sanchez-Calle, Subelliptic differential operators, in Complex Analysis III, ed by. C. Berenstein. Lecture Notes in Mathematics, vol. 1277 (Springer, Berlin, 1987), pp. 46–77
19. M. Ledoux, The geometry of Markov diffusion generators. Probab. Theory Ann. Fac. Sci. Toulouse Math. (6) **9**(2), 305–366 (2000)
20. P. Petersen, *Riemannian Geometry*. Graduate Texts in Mathematics, vol. 171 (Springer, 2006
21. K.-T. Sturm, On the geometry of metric measure spaces I. Acta Math. **196**(1), 65–131 (2006)
22. K.-T. Sturm, On the geometry of metric measure spaces II. Acta Math. **196**(1), 133–177 (2006)

Differentiation of Measures in Metric Spaces

Séverine Rigot

Abstract The theory of differentiation of measures originates from works of Besicovitch in the 1940s. His pioneering works, as well as subsequent developments of the theory, rely as fundamental tools on suitable covering properties. The first aim of these notes is to recall nowadays classical results about differentiation of measures in the metric setting together with the covering properties on which they are based. We will then focus on one of these covering properties, called in the present notes the weak Besicovitch covering property, which plays a central role in the characterization of (complete separable) metric spaces where the differentiation theorem holds for every (locally finite Borel regular) measure. We review in the last part of these notes recent results about the validity or non validity of this covering property.

1 Introduction

The theory of differentiation of measures originates from works of Besicovitch in the 1940s. His pioneering works, as well as subsequent developments of the theory, rely as fundamental tools on suitable covering properties. Our aim in these notes is twofold. In the first sections, we explain classical results about differentiation of measures in metric spaces together with the covering properties on which they are based. Such covering results are more generally useful tools that can be used to deduce global properties of a measure, and, for some of them, of the ambient metric space, from local ones. Our presentation, which does not aim at exhaustivity on the subject, in mainly based on [5, 8, 13, 15]. In the last section, we focus on one of these covering properties, namely, the weak Besicovitch covering property in the terminology of the present notes, which plays a central role in the characterization of (complete separable) metric spaces where the differentiation theorem holds for

S. Rigot (✉)
Université Côte d'Azur, CNRS, LJAD, Nice, France
e-mail: severine.RIGOT@univ-cotedazur.fr

L. Ambrosio et al. (eds.), *New Trends on Analysis and Geometry in Metric Spaces*,
C.I.M.E. Foundation Subseries 2296, https://doi.org/10.1007/978-3-030-84141-6_3

every (locally finite Borel regular) measure. We review in the last part of these notes recent results about the validity or non validity of this covering property that are mainly taken from [10–12].

Throughout these notes, a measure means a nonnegative, monotonic, countably subadditive set function defined on all subsets of a topological space X, vanishing for the empty set. All measures are furthermore assumed to be Borel regular, which means that open sets are measurable and every set is contained in a Borel set with the same measure. We say that a measure is locally finite if every point has a neighborhood with finite measure. We refer for instance to [13, Chapter 1] for detailed definitions.

We denote by (X, d) a space X metrized by a distance d. Balls in (X, d) are assumed to be closed. Namely, a ball denotes a set of the form $B(x, r) := \{y \in X : d(x, y) \leq r\}$ for some $x \in X$ and some $0 < r < \infty$.

We say that the differentiation theorem holds for a measure λ over (X, d) if

$$\lim_{r \downarrow 0} \frac{1}{\lambda(B(x, r))} \int_{B(x,r)} f(y)\, d\lambda y = f(x) \quad \text{for } \lambda\text{-a.e. } x \in X \tag{1.1}$$

for every $f \in L^1_{loc}(\lambda)$, where the latter means that every point has a neighborhood where f is λ-integrable.

In Sect. 2, we consider a classical Vitali type covering property for a measure λ that implies the validity of the differentiation theorem for λ under mild additional assumptions on (X, d) and/or λ. In Sect. 3, we prove that doubling measures are of Vitali type, and hence form a large class of measures for which the results of Sect. 2 apply. Section 4 is devoted to Radon measures over the Euclidean space. We recall in this section that the Euclidean distance satisfies a strong Besicovitch covering property which implies that every Radon measure over the Euclidean space is of Vitali type. Motivated by the example of the Euclidean space, we slightly change our point of view in Sect. 5. We investigate covering properties for metrics over an ambient separable space that imply the validity of the differentiation theorem for every locally finite Borel regular measure. We end this section with a characterization of complete separable metric spaces for which the differentiation theorem holds for every locally finite Borel regular measure. Section 6 is devoted to a closer study of the weak Besicovitch covering property introduced in the previous sections. We give in this last section recent results about the validity or non validity of this covering property.

2 Vitali Type Measures

We present in this section a classical Vitali type covering property for a measure λ over a metric space (X, d) that implies the validity of the differentiation theorem for λ under mild additional assumptions on (X, d) and/or λ. Our presentation follows closely [5, 2.9] although our setting is slightly different. We shall work here with locally finite Borel regular measures over separable metric spaces. Results in this section hold also in slighlty different contexts with minor modifications of the arguments presented here, see Remark 2.7.

Definition 2.1 (Vitali Type Measures) A Borel regular measure λ over a metric space (X, d) is said to be of *Vitali type with respect to d* if for every $A \subset X$ and every family $\mathcal{B}$ of balls in (X, d) such that each point of A is the center of arbitrarily small balls of $\mathcal{B}$, that is,

$$\inf\{r > 0 : B(x, r) \in \mathcal{B}\} = 0 \quad \text{for } x \in A,$$

there exists a countable disjointed subfamily $\mathcal{F} \subset \mathcal{B}$ such that the balls in $\mathcal{F}$ cover λ-almost all of A, namely,

$$\lambda\left(A \setminus \bigcup_{B \in \mathcal{F}} B\right) = 0.$$

We denote by M the class of all locally finite Borel regular measures over (X, d). Recall that if (X, d) is separable, every measure $\lambda \in M$ has the following useful properties. First, there exists a sequence $U_1, U_2, \cdots$ of open sets such that $X = \cup_{j \geq 1} U_j$ and $\lambda(U_j) < \infty$ for every $j \geq 1$. In particular λ is σ-finite. Second, λ is outer regular, which means that for every λ-measurable set A and every $\epsilon > 0$, there is an open set U such that $A \subset U$ and $\lambda(U \setminus A) < \epsilon$.

For a given measure $\lambda \in M$, we associate to $\mu \in M$ the following set function,

$$\mu_\lambda(A) := \inf\{\mu(C) : C \text{ is a Borel set and } \lambda(A \setminus C) = 0\}$$

for $A \subset X$. Evidently $\mu_\lambda \leq \mu$ as set functions. Elementary facts and the relationship between μ and μ_λ are given in Theorem 2.2 below. We recall that, given measures λ and μ over a space X, we say that μ is absolutely continuous with respect to λ, and we write $\mu \ll \lambda$, if $\lambda(A)$ implies $\mu(A) = 0$ for $A \subset X$. We say that μ and λ are mutually singular, and we write $\mu \perp \lambda$, if there exists a set $A \subset X$ such that $\mu(A) = 0 = \lambda(X \setminus A)$.

Theorem 2.2 *Assume that (X, d) is separable. Let $\lambda \in M$. Then, for each measure $\mu \in M$, there exists a borel set A such that*

$$\mu_\lambda = \mu \llcorner A \quad \text{and} \quad \lambda(X \setminus A) = 0.$$

Therefore $\mu_\lambda \in M$, $\mu_\lambda \ll \lambda$, and

$$\mu = \mu_\lambda + \mu_s,$$

where $\mu_s := \mu \llcorner (X \setminus A)$ and λ are mutually singular. Furthermore $\mu \ll \lambda$ iff $\mu = \mu_\lambda$.

Proof For a proof we refer to [5, 2.9.2] and Remark 2.7. □

We also recall that the pair (μ_λ, μ_s) given by Theorem 2.2 is the unique pair of measures in M such that $\mu = \mu_\lambda + \mu_s$ with $\mu_\lambda \ll \lambda$ and $\mu_s \perp \lambda$. It is called the Lebesgue decomposition of μ relative to λ.

Let $\lambda, \mu \in M$. The upper and lower derivatives of μ with respect to λ at a point $x \in X$ are defined by

$$\overline{D}(\mu, \lambda, x) := \limsup_{r \to 0} \frac{\mu(B(x,r))}{\lambda(B(x,r))},$$

$$\underline{D}(\mu, \lambda, x) := \liminf_{r \to 0} \frac{\mu(B(x,r))}{\lambda(B(x,r))}.$$

At points x where the limit exists we define the derivative of μ with respect to λ by

$$D(\mu, \lambda, x) := \overline{D}(\mu, \lambda, x) = \underline{D}(\mu, \lambda, x).$$

Here we interpret $0/0 = 0$.

The main result in this section is given in the following theorem.

Theorem 2.3 *Assume that (X, d) is separable. Let $\lambda \in M$ be of Vitali type with respect to d. Then for every $\mu \in M$,*

$$0 \leq D(\mu, \lambda, x) < \infty \text{ for } \lambda\text{-a.e. } x \in X, \tag{2.1}$$

$$D(\mu, \lambda, \cdot) \text{ is } \lambda\text{-measurable}, \tag{2.2}$$

$$\mu_\lambda(A) = \int_A D(\mu, \lambda, x)\, d\lambda x \text{ for every } \lambda\text{-measurable set } A. \tag{2.3}$$

Before proving Theorem 2.3, we start with a useful lemma that has its own interest. It allows to deduce for two given measures $\alpha, \beta \in M$ a global comparison between $\alpha_\lambda(A)$ and $\beta_\lambda(A)$ if some pointwise estimate of $\underline{D}(\alpha, \beta, x)$ is available on A. Apart from Lemma 2.4, the assumption that λ is of Vitali type with respect to d will not be used elsewhere in the proof of Theorem 2.3.

Lemma 2.4 *Assume that (X, d) is separable. Let $\lambda \in M$ be of Vitali type with respect to d. Let $\alpha, \beta \in M$ and $0 < c < \infty$. Let $A \subset X$ be such that $\underline{D}(\alpha, \beta, x) < c$ for every $x \in A$. Then $\alpha_\lambda(A) \leq c\beta_\lambda(A)$.*

Proof For each $\epsilon > 0$, one can apply the definition of β_λ in conjunction with the regularity of β to find an open set W such that $\lambda(A \setminus W) = 0$ and $\beta(W) \leq \beta_\lambda(A) + \epsilon$. Since λ is of Vitali type, one can find countably many disjointed balls $B_i \subset W$ such that $\lambda((A \cap W) \setminus \cup_i B_i) = 0$ and $\alpha(B_i) < c\beta(B_i)$ for each i. Since $\alpha_\lambda \ll \lambda$, we get

$$\alpha_\lambda(A) \le \alpha_\lambda\left(\bigcup_i B_i\right) \le \alpha\left(\bigcup_i B_i\right)$$
$$\le \sum_i \alpha(B_i) \le c\sum_i \beta(B_i) = c\beta\left(\bigcup_i B_i\right) \le c\beta(W) \le c(\beta_\lambda(A) + \epsilon).$$

The conclusion follows since this holds for every $\epsilon > 0$. □

Corollary 2.5 *Assume that (X, d) is separable. Let $\lambda \in M$ be of Vitali type with respect to d. Let $\mu \in M$ and $0 < c < \infty$. Then*

$$\underline{D}(\mu, \lambda, x) < c \text{ for every } x \in A \Rightarrow \mu_\lambda(A) \le c\lambda(A) \tag{2.4}$$

$$\overline{D}(\mu, \lambda, x) > c \text{ for every } x \in A \Rightarrow \mu_\lambda(A) \ge c\lambda(A) \tag{2.5}$$

Proof Since $\lambda_\lambda = \lambda$, (2.4) follows from Lemma 2.4. For (2.5), we have $\overline{D}(\mu, \lambda, x) > c \Rightarrow \underline{D}(\lambda, \mu, x) < 1/c$ and it follows from Lemma 2.4 that $\lambda(A) = \lambda_\lambda(A) \le \mu_\lambda(A)/c$. □

Proof of Theorem 2.3 To prove (2.1) let us first consider $A \subset X$ such that $\lambda(A) < \infty$ and $\mu(A) < \infty$. Let $0 < s < t < \infty$ and set

$$A_{s,t} := \{x \in A : \underline{D}(\mu, \lambda, x) < s < t < \overline{D}(\mu, \lambda, x)\}.$$

By Corollary 2.5,

$$t\lambda(A_{s,t}) \le \mu_\lambda(A_{s,t}) \le s\lambda(A_{s,t}) < \infty.$$

Since $s < t$, this implies $\lambda(A_{s,t}) = 0$ and hence

$$\lambda\left(\bigcup_{\substack{0<s<t<\infty \\ s,t\in\mathbb{Q}}} A_{s,t}\right) = 0.$$

Let $n \ge 1$ be a positive integer and set

$$A_n := \{x \in A : \overline{D}(\mu, \lambda, x) > n\} \text{ and } A_\infty := \{x \in A : \overline{D}(\mu, \lambda, x) = \infty\}.$$

By (2.5), we have $n\lambda(A_n) \le \mu_\lambda(A_n) \le \mu(A) < \infty$ hence

$$\lambda(A_\infty) = \lim_{n\to+\infty} \lambda(A_n) = 0.$$

Then (2.1) follows noting that

$$A \setminus \{x \in A : 0 \le D(\mu, \lambda, x) < \infty\} \subset A_\infty \bigcup \left(\bigcup_{\substack{0<s<t<\infty \\ s,t \in \mathbb{Q}}} A_{s,t} \right)$$

and recalling that X can be written as a countable union of sets with finite λ and μ measure.

For a proof of (2.2), we refer to [5, 2.9.6] and Remark 2.7.

To prove (2.3), let A be λ-measurable. Then A is μ_λ-measurable, see [5, 2.9.7] and Remark 2.7. Let $1 < t < \infty$. For $p \in \mathbb{Z}$, set

$$A_p := \{x \in A : t^p \le D(\mu, \lambda, x) < t^{p+1}\}.$$

By (2.1), we have

$$\int_A D(\mu, \lambda, x)\, d\lambda x = \sum_p \int_{A_p} D(\mu, \lambda, x)\, d\lambda x \le \sum_p t^{p+1} \lambda(A_p)$$

and $t^p \lambda(A_p) \le \mu_\lambda(A_p)$ by (2.5). Hence

$$\int_A D(\mu, \lambda, x)\, d\lambda x \le t \sum_p \mu_\lambda(A_p) \le t \mu_\lambda(A). \tag{2.6}$$

On the other hand, since $\mu_\lambda \ll \lambda$, (2.1) and (2.4) imply that $\mu_\lambda(\{x \in A : D(\mu, \lambda, x) = 0\}) = 0$. Hence it follows from (2.4) that

$$\begin{aligned} \mu_\lambda(A) &= \sum_p \mu_\lambda(A_p) \le t^{p+1} \sum_p \lambda(A_p) \\ &\le t \sum_p \int_{A_p} D(\mu, \lambda, x)\, d\lambda x = t \int_A D(\mu, \lambda, x)\, d\lambda x. \end{aligned} \tag{2.7}$$

Then (2.3) follows from (2.6) and (2.7) since these hold for every $1 < t < \infty$. □

We stress that, in addition to the classical Radon-Nikodym theorem, Theorem 2.3 gives a concrete representation of the density of the absolutely continuous part of the Lebesgue decomposition of μ relative to λ in terms of the derivative $D(\mu, \lambda, \cdot)$.

Furthermore, the validity of the differentation theorem for Vitali type measures comes as a direct consequence of Theorem 2.3.

Corollary 2.6 *Assume that (X, d) is separable. Let $\lambda \in M$ be of Vitali type with respect to d. Then the differentiation theorem holds for λ.*

Proof It is sufficient to prove that (1.1) holds for nonnegative $f \in L^1_{loc}(\lambda)$. For such an f, we define $\mu \in M$ by

$$\mu(A) := \int_A^* f(x)\, d\lambda x$$

for $A \subset X$. We infer from Theorem 2.3 that

$$\int_A f(x)\, d\lambda x = \mu(A) = \mu_\lambda(A) = \int_A D(\mu, \lambda, x)\, d\lambda x$$

for every λ-measurable set A. Hence

$$f(x) = D(\mu, \lambda, x) \text{ for } \lambda\text{-a.e. } x \in X$$

and it follows that

$$f(x) = \lim_{r \downarrow 0} \frac{1}{\lambda(B(x,r))} \int_{B(x,r)} f(y)\, d\lambda y \text{ for } \lambda\text{-a.e. } x \in X.$$

□

Remark 2.7 In [5, 2.9], (X, d) is not assumed to be separable and measures under consideration are Borel regular measures λ such that every bounded set has finite λ measure. Namely, the main results in [5, 2.9] are gathered in the following theorem.

Theorem 2.8 *Let (X, d) be a metric space. Let λ, μ be Borel regular measures over X such that every bounded set has finite λ and μ measure. Assume that λ is of Vitali type with respect to d. Then the conclusions of Theorems 2.2 and 2.3 hold true. Furthermore, if f is an $\overline{\mathbb{R}}$-valued λ-measurable function such that*

$$\int_A |f(x)|\, d\lambda x < \infty$$

for every bounded λ-measurable set A, then

$$\lim_{r \downarrow 0} \frac{1}{\lambda(B(x,r))} \int_{B(x,r)} f(y)\, d\lambda y = f(x) \quad \textit{for } \lambda\textit{-a.e. } x \in X.$$

It is not difficult to see that minor modifications of the arguments in [5, 2.9] can be used to handle slightly different settings. For locally finite Borel regular measures over separable metric spaces, the minor modifications we used in the proof of Theorem 2.3 are based on the fact that locally finite Borel regular measures over separable metric spaces are σ-finite and outer regular. Similarly one can also prove that Theorems 2.2, 2.3 and Corollary 2.6 hold true when considering Radon measures over locally compact separable metric spaces.

3 Doubling Measures

An important class of Vitali type measures, see Definition 2.1, is given by doubling measures. A classical proof of this result relies on a general covering theorem often called basic $5r$ covering theorem in the literature, see Theorem 3.2. Our presentation in the present section follows closely [8, Chapter 1].

Let (X, d) be a metric space. For a ball B with center x and radius $0 < r < \infty$ in (X, d) and $\tau > 0$, we denote by $\tau B := B(x, \tau r)$ the concentric ball with radius τr.

A Borel regular measure λ over X is said to be *doubling with respect to d* if there is a constant $C \geq 1$ such that

$$\lambda(2B) \leq C\lambda(B) \quad \text{for every ball } B,$$

and λ is nondegenerate in the sense that $\lambda(B_1) > 0$ and $\lambda(B_2) < \infty$ for some balls B_1 and B_2. Note that this implies in particular that $0 < \lambda(B) < \infty$ for every ball B.

Theorem 3.1 *Doubling measures over a metric space (X, d) are of Vitali type with respect to d.*

A classical proof of Theorem 3.1 relies on the following general covering theorem.

Theorem 3.2 *Every family $\mathcal{B}$ of balls with uniformly bounded diameter, that is, such that $\sup\{\operatorname{diam} B : B \in \mathcal{B}\} < \infty$, in a metric space contains a disjointed subfamily $\mathcal{F} \subset \mathcal{B}$ such that*

$$\bigcup_{B \in \mathcal{B}} B \subset \bigcup_{B \in \mathcal{F}} 5B$$

and every ball B in $\mathcal{B}$ meets a ball in $\mathcal{F}$ with radius at least half that of B.

Proof The proof relies on Zorn's lemma, see e.g. [8, Theorem 1.2]. Note that $\mathcal{F}$ is not asserted to be countable but in applications it often will be. Constructive proofs are possible under mild additional assumptions on X, see for instance [13, Theorem 2.1] where closed balls are assumed to be compact. □

Proof of Theorem 3.1 We assume first that A is bounded. Let $\mathcal{B}$ be a family of balls such that $\inf\{r > 0 : B(x, r) \in \mathcal{B}\} = 0$ for $x \in A$. We may assume with no loss of generality that the balls in $\mathcal{B}$ are centered on A and have uniformly bounded diameter. By Theorem 3.2 we can find a disjointed subfamily $\mathcal{F} \subset \mathcal{B}$ such that the balls $5B$, $B \in \mathcal{F}$, cover A. Since the union $\cup_{\mathcal{F}} B$ is contained in some fixed ball, and balls have positive and finite λ measure, the family $\mathcal{F} = \{B_1, B_2, \cdots\}$ is necessarily countable. We infer from the doubling property of λ that

$$\sum_{i \geq 1} \lambda(5B_i) \leq C \sum_{i \geq 1} \lambda(B_i) = C\lambda\left(\bigcup_{i \geq 1} B_i\right) < \infty$$

for some constant $C \geq 1$, hence

$$\sum_{i>N} \lambda(5B_i) \to 0$$

as $N \to \infty$. Therefore it is sufficient to show that

$$A \setminus \bigcup_{i=1}^{N} B_i \subset \bigcup_{j>N} 5B_j.$$

Let $a \in A \setminus \cup_{i \leq N} B_i$. Since balls in $\mathcal{B}$ are closed, we can find a ball $B(a,r) \in \mathcal{B}$ that does not meet $\cup_{i \leq N} B_i$. On the other hand, by Theorem 3.2, one can choose the family $\mathcal{F}$ in such a way that $B(a,r)$ meets some ball $B_j \in \mathcal{F}$ with radius at least $r/2$. Thus $j > N$ and $B(a,r) \subset 5B_j$ as required.

To get rid of the assumption that A is bounded, we fix some $x \in X$ and an increasing sequence of positive numbers $R_0, R_1, \cdots \to \infty$ so that $\lambda(\{y \in X : d(x,y) = R_k\}) = 0$ for every $k \geq 0$. This is possible since $\lambda(\{y \in X : d(x,y) = R\}) = 0$ for a.e. $R > 0$. We set $U_0 := \{y \in X : d(x,y) < R_0\}$ and $U_k := \{y \in X : R_{k-1} < d(y,x) < R_k\}$ for $k \geq 1$. We apply the above construction to the sets $A \cap U_k$ for $k \geq 0$ to find disjointed countable families of balls $\mathcal{F}_k$ such that the balls in $\mathcal{F}_k$ are contained in U_k and cover λ-almost all of $A \cap U_k$. Then $\mathcal{F} = \cup_k \mathcal{F}_k$ gives the conclusion. □

Remark 3.3 Theorem 3.1 holds more generally for asymptotically doubling measures, that is, when the doubling condition is relaxed into

$$\limsup_{r \to 0} \frac{\lambda(B(x,2r))}{\lambda(B(x,r))} < \infty \quad \text{for } \lambda\text{-a.e. } x \in X,$$

see [5, 2.8.17].

4 Radon Measures in Euclidean Spaces

The theory of differentiation of measures originates from works of Besicovitch in the Euclidean space [2, 3]. Besicovitch proved that every locally finite Borel regular measure, or equivalently Radon measure, over the Euclidean space is of Vitali type and hence satisfies the differentiation theorem. His proof relies on a geometric property of the Euclidean distance given in Lemma 4.1 below and called weak Besicovitch covering property (WBCP) in the present notes. Lemma 4.1 turns out to imply two stronger covering properties for the Euclidean distance, namely, Theorems 4.2 (i) and 4.2 (ii), called respectively Besicovitch covering property (BCP) and strong Besicovitch covering property in our terminology. The fact that every Radon measure over $\mathbb{R}^n$ is of Vitali type with respect to the Euclidean distance can then be obtained as a rather simple consequence of Theorem 4.2 (ii), see Theorem 4.3. BCP and WBCP will be studied in more details in the general metric setting in Sects. 5 and 6. Our presentation in the present section follows [13, Chapter 2].

We start with a lemma which is quite elementary in the Euclidean setting. However, as we shall see in Sect. 5, it turns out to play a central role for the theory of differentiation of arbitrary measures when rephrased in the metric setting.

Lemma 4.1 (Weak Besicovitch Covering Property for the Euclidean Distance) *Let $\mathbb{R}^n$ be equipped with the Euclidean distance. There exists an integer $K \geq 1$ with the following property. Assume that there exist k points $x_1, \dots, x_k$ in $\mathbb{R}^n$ and k positive numbers $r_1, \dots, r_k$ such that*

$$x_i \notin B(x_j, r_j) \text{ for } j \neq i, \text{ and } \bigcap_{i=1}^{k} B(x_i, r_i) \neq \emptyset.$$

Then $k \leq K$.

Proof Without loss of generality, we may assume that $x_i \neq 0$ for $i = 1, \dots, k$, and $0 \in \cap_{i=1}^{k} B(x_i, r_i)$. Then $\|x_i\| \leq r_i < \|x_i - x_j\|$ for $i \neq j$ where $\|\cdot\|$ denotes the Euclidean norm. It follows from elementary geometric arguments that the angle between x_i and x_j for $i \neq j$ is at least 60°, that is,

$$\left\| \frac{x_i}{\|x_i\|} - \frac{x_j}{\|x_j\|} \right\| \geq 1 \quad \text{for } i \neq j,$$

see [13, Lemma 2.5] for more details. Then the conclusion follows by compactness of the unit Euclidean sphere. □

The next theorem, usually known as Besicovitch's covering theorem for the Euclidean space in the literature, comes as a consequence of Lemma 4.1.

Theorem 4.2 *Let $\mathbb{R}^n$ be equipped with the Euclidean distance. There are integers $N \geq 1$ and $Q \geq 1$ with the following properties. Let A be a bounded subset of $\mathbb{R}^n$ and $\mathcal{B}$ be a family of balls such that each point of A is the center of some ball of $\mathcal{B}$.*

(i) (Besicovitch covering property) There is a countable subfamily $\mathcal{F} \subset \mathcal{B}$ such that the balls in $\mathcal{F}$ cover A and every point in $\mathbb{R}^n$ belongs to at most N balls in $\mathcal{F}$.

(ii) (Strong Besicovitch covering property) There are countable subfamilies $\mathcal{B}_1, \dots, \mathcal{B}_Q \subset \mathcal{B}$ covering A such that each $\mathcal{B}_i$ is disjointed, namely,

$$A \subset \bigcup_{i=1}^{Q} \bigcup_{B \in \mathcal{B}_i} B$$

and $B \cap B' = \emptyset$ for $B, B' \in \mathcal{B}_i$ with $B \neq B'$.

For a proof of Theorem 4.2 we refer for instance to [13, Theorem 2.7]. See also Sect. 5 for the connections between Lemma 4.1 and Theorem 4.2 (i) in the general metric setting.

We explain now how the fact that every Radon measure over $\mathbb{R}^n$ is of Vitali type with respect to the Euclidean distance can be obtained as a rather simple consequence of the strong Besicovitch covering property given by Theorem 4.2 (ii).

Theorem 4.3 *Radon measures over $\mathbb{R}^n$ are of Vitali type with respect to the Euclidean distance.*

As a consequence of Theorem 4.3 together with the results of Sect. 2, we get the following corollary.

Corollary 4.4 *The conclusions of Theorem 2.3 and Corollary 2.6 hold true for every Radon measure over the Euclidean space.*

Proof of Theorem 4.3 Let λ be a Radon measure over $\mathbb{R}^n$. Let $A \subset \mathbb{R}^n$ and let $\mathcal{B}$ be a family of Euclidean balls such that each point of A is the center of arbitrarily small balls of $\mathcal{B}$. We may assume that $\lambda(A) > 0$. Let us assume first that A is bounded and hence has finite λ measure. By outer regularity of λ, there is an open set U such that $A \subset U$ and

$$\lambda(U) \leq (1 + (4Q)^{-1})\, \lambda(A)$$

where Q is given by Theorem 4.2. By Theorem 4.2 (ii) we can find countable subfamilies $\mathcal{B}_1, \ldots, \mathcal{B}_Q \subset \mathcal{B}$ such that each $\mathcal{B}_i$ is disjointed and

$$A \subset \bigcup_{i=1}^{Q} \bigcup_{B \in \mathcal{B}_i} B \subset U.$$

Then

$$\lambda(A) \leq \sum_{i=1}^{Q} \lambda\left(\bigcup_{B \in \mathcal{B}_i} B\right)$$

hence there is an i such that

$$\lambda(A) \leq Q\, \lambda\left(\bigcup_{B \in \mathcal{B}_i} B\right).$$

Further we can find finitely many disjointed balls $B_1, \ldots, B_{k_1} \in \mathcal{B}_i$ such that

$$\lambda(A) \leq 2Q\, \lambda\left(\bigcup_{j=1}^{k_1} B_j\right).$$

Letting

$$A_1 := A \setminus \bigcup_{j=1}^{k_1} B_j,$$

we get

$$\lambda(A_1) \le \lambda\left(U \setminus \bigcup_{j=1}^{k_1} B_j\right) = \lambda(U) - \lambda\left(\bigcup_{j=1}^{k_1} B_j\right) \le u\lambda(A)$$

with $u := 1 - (4Q)^{-1} < 1$. Now A_1 is contained in the open set $\mathbb{R}^n \setminus \cup_{j=1}^{k_1} B_j$ and therefore we can find an open set U_1 such that $A_1 \subset U_1 \subset \mathbb{R}^n \setminus \cup_{j=1}^{k_1} B_j$ and

$$\lambda(U_1) \le (1 + (4Q)^{-1})\,\lambda(A_1).$$

As before there are finitely many disjointed balls $B_{k_1+1}, \dots, B_{k_2}$ for which $B_j \subset U_1$ and

$$\lambda(A_2) \le u\lambda(A_1) \le u^2\lambda(A)$$

where

$$A_2 := A_1 \setminus \bigcup_{j=k_1+1}^{k_2} B_j = A \setminus \bigcup_{j=1}^{k_2} B_j.$$

Clearly, the balls $B_1, \dots, B_{k_2}$ are disjointed. After m steps, we get

$$\lambda\left(A \setminus \bigcup_{j=1}^{k_m} B_j\right) \le u^m \lambda(A)$$

with the balls $B_1, \dots, B_{k_m} \in \mathcal{B}$ that are disjointed and the fact that λ is of Vitali type with respect to the Euclidean distance follows since $u < 1$.

To get rid of the assumption that A is bounded, we write $\mathbb{R}^n$ as the union of a countable collection of closed cubes $\overline{Q_i}$ such that the corresponding open cubes Q_i are disjointed and such that $\lambda(\mathbb{R}^n \setminus \cup_{i\ge 1} Q_i) = 0$. This is possible since $\lambda(V)$ can be positive for at most countably many parallel hyperplanes V. Applying the above arguments to the sets $A \cap Q_i$ and noting that $\lambda(A \setminus \cup_{i\ge 1} Q_i) = 0$, we get that λ is of Vitali type with respect to the Euclidean distance. □

Remark 4.5 The conclusions of Lemma 4.1 and Theorem 4.2 hold for families of open balls. However, Theorem 4.3 does not hold in its full generality when considering family of open balls in the definition of Vitali type measures, see [1, Example 2.20].

Remark 4.6 More general metric spaces satisfy the strong Besicovitch covering property, see Theorem 4.2 (ii) for the statement in the Euclidean case. We refer for instance to [5, 2.8.9] for the notion of directionally limited metrics and to [5, 2.8.14] for an extension of Theorem 4.2 (ii) for such metrics. Finite dimensional normed vector spaces fit in particular into this more general setting. We shall however not go further in that direction in these notes. In the next sections, we shall instead investigate in more details the geometric content of the Besicovitch and weak Besicovitch covering properties and their connections with the validity of the differentiation theorem for every locally finite Borel regular measure in the metric setting.

5 σ-Finite Dimensional Metrics

We present in this section the notion of σ-finite dimentional metrics introduced by Preiss in [15]. This notion characterizes complete separable metric spaces on which the differentiation theorem holds for every locally finite Borel regular measure, see Theorem 5.10.

We start with a model case to explain without to much technicalities how one can get from such kind of notions the validity of the differentiation theorem for every locally finite Borel regular measure.

Definition 5.1 (Besicovitch Covering Property) We say that a distance d on a space X satisfies the *Besicovitch covering property* (BCP) if there exists an integer $N \geq 1$ such that the following holds. Let A be a bounded subset of X and $\mathcal{B}$ be a family of balls such that each point of A is the center of some ball of $\mathcal{B}$. Then there is a countable subfamily $\mathcal{F} \subset \mathcal{B}$ such that the balls in $\mathcal{F}$ cover A and every point in X belongs to at most N balls in $\mathcal{F}$, namely,

$$\chi_A \leq \sum_{B \in \mathcal{F}} \chi_B \leq N$$

where χ_A denotes the characteristic function of the set A.

With this terminology, Theorem 4.2 (i) exactly says that the Euclidean distance satisfies BCP. We show in the next theorem that the validity of BCP is sufficient to imply the validity of the differentiation theorem for every locally finite Borel regular measure over a separable metric space. Recall that, given a metric space (X, d), we denote by M the class of all locally finite Borel regular measures over (X, d).

Theorem 5.2 *Let (X, d) be a separable metric space. Assume that d satisfies BCP. Then the differentiation theorem holds for every $\lambda \in M$.*

Proof Every $\lambda \in M$ satisfies the assumption of Theorem 5.3 below with $C = 1$ and N given by the validity of BCP for d. □

Theorem 5.3 *Let (X, d) be a separable metric space and let $\lambda \in M$. Assume there exist constants $C > 0$ and $N > 0$ such that the following holds. For every bounded set $A \subset X$ and every family $\mathcal{B}$ of balls in (X, d) such that each point of A is the center of some ball of $\mathcal{B}$, there is a countable subfamily $\mathcal{F} \subset \mathcal{B}$ such that*

$$\lambda(A) \leq C\lambda\left(\bigcup_{B \in \mathcal{F}} B\right) \quad \text{and} \quad \sum_{B \in \mathcal{F}} \chi_B \leq N.$$

Then the differentiation theorem holds for λ.

Proof The proof given here is based on a weak type $(1, 1)$ inequality for the maximal operator. We refer for instance to [4] or [17] for more specialized results in that direction. First, it is sufficient to prove (1.1) for $f \in L^1(\lambda)$ multiplying our original function by the characteristic function of an open set where f is λ-integrable and then exhausting X by a countable union of such open sets.

Next, for $f \in L^1(\lambda)$ and $x \in X$, set

$$Mf(x) := \sup_{r>0} \frac{1}{\lambda(B(x, r))} \int_{B(x,r)} |f(y)|\, d\lambda y.$$

Let A be a bounded subset of X. For $\alpha > 0$, set

$$A_\alpha := \{x \in A : Mf(x) > \alpha\}.$$

For each $x \in A_\alpha$, we can find a ball $B(x, r_x)$ such that

$$\lambda(B(x, r_x)) < \frac{1}{\alpha} \int_{B(x,r_x)} |f(y)|\, d\lambda y.$$

By assumption, one can extract from this family of balls a sequence $B_1, B_2, \ldots$ such that

$$\begin{aligned}\lambda(A_\alpha) &\leq C\lambda\left(\bigcup_{j\geq 1} B_j\right) \leq C\sum_{j\geq 1} \lambda(B_j) \\ &\leq \frac{C}{\alpha}\sum_{j\geq 1}\int_{B_j} |f(y)|\, d\lambda y = \frac{C}{\alpha}\int_X \sum_{j\geq 1} \chi_{B_j}(y)|f(y)|\, d\lambda y \leq \frac{CN}{\alpha}\|f\|_1.\end{aligned}$$

Since this holds for every bounded set A, we get the following weak type (1,1) inequality for the maximal operator,

$$\lambda(\{x \in X : Mf(x) > \alpha\}) \leq \frac{CN}{\alpha}\|f\|_1. \tag{5.1}$$

For $g \in L^1(\lambda)$ and $r > 0$, denote by g_r the function defined by

$$g_r(x) := \frac{1}{\lambda(B(x,r))} \int_{B(x,r)} g(y)\, d\lambda y$$

and set

$$\Omega g(x) := \limsup_{r\to 0} |g_r(x) - g(x)|.$$

We have $\Omega g(x) \leq Mg(x) + g(x)$. Therefore, using (5.1) and Chebychev's inequality, we get

$$\begin{aligned}
&\lambda(\{x \in X : \Omega g(x) > \alpha\}) \\
&\quad \leq \lambda\left(\left\{x \in X : Mg(x) > \frac{\alpha}{2}\right\}\right) + \lambda\left(\left\{x \in X : |g(x)| > \frac{\alpha}{2}\right\}\right) \\
&\quad \leq \frac{C'}{\alpha} \|g\|_1
\end{aligned}$$

for every $\alpha > 0$ and for some constant $C' > 0$.

To conclude the proof of (1.1), let $f \in L^1(\lambda)$. Let $h \in L^1(\lambda)$ and assume that h is continuous. Then Ωh is identically zero on the support of λ and hence λ-a.e. Next, we have $\Omega f \leq \Omega(f - h) + \Omega h = \Omega(f - h)$, therefore

$$\lambda(\{x \in X : \Omega f(x) > \alpha\}) \leq \lambda(\{x \in X : \Omega(f - h)(x) > \alpha\}) \leq \frac{C'}{\alpha} \|f - h\|_1.$$

By density of continuous functions in $L^1(\lambda)$, we get

$$\lambda(\{x \in X : \Omega f(x) > \alpha\}) = 0$$

for every $\alpha > 0$. Hence $\Omega f(x) = 0$ for λ-a.e. $x \in X$ which proves (1.1). □

Remark 5.4 Note that every doubling measure satisfies the assumption of Theorem 5.3. This can be deduced from Theorem 3.2 and is left to the reader as an easy exercise. Hence one can also recover from Theorem 5.3 the validity of the differentiation theorem for doubling measures over separable metric spaces.

More generally, results about derivatives of measures similar to those presented in Sect. 2 can be obtained with the assumption of Theorem 5.3. Namely,

Theorem 5.5 *Let (X, d) be a separable metric space. Let $\lambda \in M$ and assume that the assumption of Theorem 5.3 holds. Then for every $\mu \in M$, we have*

$$\mu_\lambda(A) = \int_A D(\mu, \lambda, x)\, d\lambda x \quad \textit{for every } \lambda\textit{-measurable set } A, \tag{5.2}$$

where μ_λ denotes the absolute continuous part of the Lebesgue decomposition of μ relative to λ.

In particular, if d satisfies BCP, then (5.2) holds for every $\lambda, \mu \in M$.

Proof The fact that there exists $f \in L^1_{loc}(\lambda)$ such that

$$\mu_\lambda(A) = \int_A f(x)\, d\lambda x$$

for every λ-measurable set A follows from Radon-Nikodym theorem for σ-finite measures. Theorem 5.3 implies the validity of the differentiation theorem for λ and in particular $f(x) = D(\mu_\lambda, \lambda, x)$ for a.e. $x \in X$. Then (5.2) follows from Lemma 5.6 below. We apply it to $\nu = \mu_s$, where μ_s denotes the singular part of the Lebesgue decomposition of μ relative to λ, to get that $D(\mu_s, \lambda, x) = 0$, and hence $f(x) = D(\mu_\lambda, \lambda, x) = D(\mu, \lambda, x)$, for λ-a.e. $x \in X$. □

Lemma 5.6 *Let (X, d) be a separable metric space. Let $\lambda \in M$ and assume that the assumption of Theorem 5.3 holds. Let $\nu \in M$ be such that $\nu \perp \lambda$. Then $D(\nu, \lambda, x) = 0$ for λ-a.e. $x \in X$.*

Proof The proof is very similar to the proof of Theorem 5.3. Let A be a Borel set such that $\nu(A) = 0 = \lambda(X \setminus A)$. Let $\varepsilon > 0$ and let U be an open set such that $A \subset U$ and $\nu(U) \leq \varepsilon$. Let V be a bounded subset of A. For $\alpha > 0$, set

$$V_\alpha := \{x \in V : \overline{D}(\nu, \lambda, x) > \alpha\}.$$

For each $x \in V_\alpha$, we can find a ball $B(x, r_x) \subset U$ such that

$$\lambda(B(x, r_x)) < \frac{1}{\alpha}\, \nu(B(x, r_x)).$$

By assumption we can extract from this family of balls of countable subfamily $\mathcal{F}$ such that

$$\lambda(V_\alpha) \leq C\lambda\left(\bigcup_{B \in \mathcal{F}} B\right) \quad \text{and} \quad \sum_{B \in \mathcal{F}} \chi_B \leq N.$$

Arguing in a similar way as in the proof of Theorem 5.3, it follows that

$$\lambda(V_\alpha) \leq \frac{CN}{\alpha}\, \nu(U) \leq \frac{CN}{\alpha}\, \varepsilon.$$

Since this holds for every $\varepsilon > 0$, we get $\lambda(V_\alpha) = 0$ for every $\alpha > 0$. Therefore

$$\lambda(\{x \in V : \overline{D}(\nu, \lambda, x) > 0\}) = 0$$

and the conclusion follows since the latter holds for every bounded subset of A and since $\lambda(X \setminus A) = 0$. □

We introduce now a slightly weaker geometric condition than BCP. We say that a family $\mathcal{B}$ of balls in a metric space (X, d) is a *Besicovitch family of balls* if, first, for every ball $B \in \mathcal{B}$ with center x_B, we have $x_B \notin B'$ for all $B' \in \mathcal{B}$, $B \neq B'$, and, second, $\bigcap_{B \in \mathcal{B}} B \neq \emptyset$.

Definition 5.7 (Weak Besicovitch Covering Property) We say that a distance d on a space X satisfies the *weak Besicovitch covering property* (WBCP) if there is an integer $K \geq 1$ such that $\operatorname{Card} \mathcal{B} \leq K$ for every Besicovitch family $\mathcal{B}$ of balls in (X, d).

Going back to the Euclidean setting, Lemma 4.1 exactly shows that the Euclidean distance satisfies WBCP.

It can easily be seen that BCP implies WBCP. But, as the terminology suggests, WBCP is in general strictly weaker than BCP, see [12, Example 3.4], and the comment after Proposition 5.8 for the more precise relationship between BCP and WBCP.

There is however an important class of metric spaces, namely, doubling metric spaces, where BCP and WBCP are equivalent. Recall that a metric space is said to be doubling if there is an integer $D \geq 1$ such that for each $0 < r < \infty$, each ball in (X, d) of radius $2r$ can be covered by a family of at most D balls of radius r.

Proposition 5.8 *Let (X, d) be a doubling metric space. Then d satisfies BCP if and only if d satisfies WBCP.*

For a proof of Proposition 5.8, we refer for instance to [12, Proposition 3.7] which follows actually closely the arguments of the proof of [13, Theorem 2.7] where the validity of BCP for the Euclidean distance, namely, Theorem 4.2 (i), is deduced from the validity of WBCP.

Although a distance d that satisfies WBCP may not satisfy BCP in general, it satisfies a weak form of BCP that can be stated as follows. There is an integer $N \geq 1$ such that the following holds. Let A be a bounded subset of X. Let $\mathcal{B}$ be a family of balls such that each point of A is the center of some ball of $\mathcal{B}$ and such that either $\sup\{r : \ B(x, r) \in \mathcal{B}\} = +\infty$ or $B(x, r) \in \mathcal{B} \mapsto r$ attains only an isolated set of values in $(0, +\infty)$. Then there is a countable subfamily $\mathcal{F} \subset \mathcal{B}$ such that the balls in $\mathcal{F}$ cover A and every point in X belongs to at most N balls in $\mathcal{F}$. Mild modifications of the arguments presented in this section can be used to prove the validity of the differentiation theorem for every locally finite Borel regular measure from this weak form of Besicovitch covering property.

We conclude this section with the notion of σ-finite dimensional metric that generalizes WBCP and is due to Preiss, see [15]. This notion involves the decomposition of the ambient space into countably many pieces on which an ad hoc version of WBCP holds.

Definition 5.9 (σ-Finite Dimensional Metric) Let (X, d) be a metric space. We say that d is finite dimensional on a subset $Y \subset X$ if there exist constants $K \geq 1$

and $0 < r \leq \infty$ such that Card $\mathcal{B} \leq K$ for every Besicovitch family $\mathcal{B}$ of balls in (X, d) centered on Y with radius $< r$. We say that d is *σ-finite dimensional* if X can be written as a countable union of subsets on which d is finite dimensional.

With this terminology, the fact that WBCP holds on a metric space (X, d) exactly means that the distance d is finite dimensional on X for some constant $K \geq 1$ and with $r = \infty$ in the previous definition. Examples of σ-finite dimensional metrics are Riemannian metrics over Riemannian manifolds of class ≥ 2, see [5, Chapter 2]. On the contrary, (non-Riemannian) sub-Riemannian distances are not σ-finite dimensional, see [12, Theorem 7.5]. We end this section with the following characterization whose proof can be found in [15].

Theorem 5.10 ([15, Theorem 1]) *Let (X, d) be a complete separable metric space. The differentiation theorem holds for every locally finite Borel regular measure over (X, d) if and only if d is σ-finite dimensional.*

6 Weak Besicovitch Covering Property

We devote this last section to a closer study of the weak Besicovitch covering property, see Definition 5.7. Besides the fact that it implies the validity of the differentiation theorem for every locally finite Borel regular measures over a separable metric space, WBCP is a useful tool that can be used to deduce global properties of a metric space from local ones, and can for instance be used to study arbitrary measures.

We begin with a couple of general facts. We first observe that WBCP is a property for family of balls in a metric space (X, d) that may not be preserved under (even arbitrarily small) perturbations of the metric.

Theorem 6.1 ([11, Theorem 1.6]) *Let d be a distance on a space X. Assume that there exists an accumulation point in (X, d). Then, for every $0 < c < 1$, there exists a distance d_c on X that does not satisfy WBCP and such that $cd \leq d_c \leq d$.*

See also [15, Theorem 3] for a version of this result about σ-finite dimensional metrics. Conversely, there are examples of metric spaces (X, d) that do not satisfy WBCP but for which there exists for every $\epsilon > 0$, a distance d_ϵ on X that satisfies WBCP and is such that $d \leq d_\epsilon \leq (1 + \epsilon)d$. See the example of the stratified first Heisenberg group, Example 6.6 below. WBCP is in particular far from being preserved by a bi-Lipschitz change of metric.

On the other hand, let d and ρ be two distances on a space X and assume that every ball with respect to ρ is a ball with respect to d, with the same center but possibly a different radius. Then the validity of WBCP for d implies the validity of WBCP for ρ. For instance if WBCP holds for a distance d, then, for every $0 < s < 1$, the snowflake distance d^s satisfies WBCP as well. Note that on the contrary it is well known that a metric space (X, d) and its snowflakes (X, d^s), $0 < s < 1$, have for many other purposes significantly different behaviour.

Our next observation concerns product of metric spaces. Given two metric spaces (X, d_X) and (Y, d_Y), there are many ways to define distances on their product. If (X, d_X) and (Y, d_Y) both satisfy WBCP, then WBCP may fail for classical choices of distances on $X \times Y$ such as their l^p-mean for $1 \leq p < \infty$. Such examples are the following ones. Given $1 \leq p < \infty$ and $s > p$, let d be the distance on $\mathbb{R} \times \mathbb{R}$ defined by $d((x, y), (x', y')) := (|x' - x|^p + |y' - y|^{p/s})^{1/p}$. It is the l^p-mean of two distances on $\mathbb{R}$ that satisfy WBCP, but WBCP does not hold on $(\mathbb{R} \times \mathbb{R}, d)$, see [10, Lemma 3.2]. However, considering the max distance on a product of metric spaces preserves the validity of WBCP. Namely,

Theorem 6.2 ([12, Theorem 3.16]) *Let (X, d_X) and (Y, d_Y) be two metric spaces. Assume that d_X and d_Y satisfy WBCP on X and Y respectively. Then the max distance*

$$d_{X \times Y}((x, y), (x', y')) := \max(d_X(x, x'), d_Y(y, y'))$$

satisfies WBCP on $X \times Y$.

The proof of Theorem 6.2 relies on tools from graph theory, namely, Ramsey's Theorem. We refer to [12] for a proof. We also refer to the latter paper, and in particular to its Sect. 3, for further general results about WBCP.

Until recently, there were only few known examples of metric spaces satisfying WBCP. As far as we know, finite dimensional normed vector spaces, see Remark 4.6, were the main known such examples. We end this section with recent results that enlarge this picture.

The setting we are considering in the rest of these notes is the one of homogeneous groups equipped with homogeneous distances. Roughly speaking, a homogeneous group is a Lie group equipped with an appropriate family of dilations. A homogeneous distance on a homogeneous group is a left-invariant distance that is one-homogeneous with respect to the family of dilations, see below for more detailed definitions. To some extend, homogeneous groups equipped with homogeneous distances generalize naturally finite dimensional normed vector spaces. Due to the presence of translations and dilations, they provide for instance a setting where many aspects of classical analysis and geometry can be carried out. However, it is also well known that, for many purposes, their behaviour can be significantly different from the behaviour of finite dimensional normed vector spaces. The discussion below about the validity or non validity of WBCP will give some more evidence about these differences. We also recall that, beyond such a priori considerations, homogeneous groups equipped with homogeneous distances form an important framework because of their occurrences in many settings, see for instance [6], or the introduction in [12] and the references therein.

Before stating the results, we introduce some definitions and terminology. We refer to [12] for a complete presentation, see also Examples 6.5, 6.6 and 6.8 below for some explicit examples.

A positive grading of a finite dimensional real Lie algebra $\mathfrak{g}$ is a family $(V_t)_{t \in (0, +\infty)}$ of vector subspaces of $\mathfrak{g}$ where all but finitely many of the V_t's are

$\{0\}$ and such that

$$\mathfrak{g} = \oplus_{t>0} V_t \quad \text{with } [V_s, V_t] \subset V_{s+t} \quad \text{for all } s, t > 0.$$

Here $[V, W] := \text{span}\{[X, Y] : X \in V, Y \in W\}$. Given a positive grading $(V_t)_{t>0}$ of a Lie algebra and $t > 0$, the subspace V_t is called the degree t layer of the grading. We say that a Lie algebra is positively graduable if it admits a positive grading. In general, a positively graduable Lie algebra admits several positive gradings that are not isomorphic as graded Lie algebras. We say that a Lie algebra is graded if it is positively graduable and endowed with a fixed positive grading.

For a graded Lie algebra $\mathfrak{g} = \oplus_{t>0} V_t$ and $r > 0$, the associated dilation of factor r is defined as the unique linear map $\delta_r : \mathfrak{g} \to \mathfrak{g}$ such that $\delta_r(X) = r^t X$ for $X \in V_t$. The family of associated dilations $(\delta_r)_{r>0}$ is a one-parameter group of Lie algebra automorphisms.

We say that a connected and simply connected Lie group G is graded if its Lie algebra is graded. For a graded group G and $r > 0$, we define, with a slight abuse of notation and terminology, the associated dilation $\delta_r : G \to G$ of factor r as the unique Lie group automorphism such that $\delta_r \circ \exp = \exp \circ \delta_r$ where $\exp : \mathfrak{g} \to G$ denotes the exponential map. Recall that graded groups are connected and simply connected nilpotent Lie groups hence the exponential map is a diffeomorphism from $\mathfrak{g}$ to G.

We say that a distance d on a graded group G with associated family of dilations $(\delta_r)_{r>0}$ is homogeneous if it is left-invariant, that is, $d(p \cdot q, p \cdot q') = d(q, q')$ for all $p, q, q' \in G$, and one-homogeneous with respect to the associated family of dilations $(\delta_r)_{r>0}$, that is, $d(\delta_r(p), \delta_r(q)) = rd(p, q)$ for all $p, q \in G$ and all $r > 0$.

Homogeneous distances do exist on a graded group if and only if, for all $t < 1$, degree t layers of the associated positive grading are $\{0\}$. We call such groups *homogeneous*. For general graded groups, one can consider homogeneous quasi-distances. For simplicity, we restrict ourselves in these notes to homogeneous groups and we refer to [12] for the more general case of graded groups.

Note that we do not require topological assumptions in the definition of homogeneous distances. It can indeed be proved that homogeneous distances on a homogeneous group induce the manifold topology, see [12, Proposition 2.26]. Note also that a homogeneous group equipped with a homogeneous distance is doubling, hence the validity of WBPC is equivalent to the validity of BCP in this setting. Furthermore, it can be proved that a homogeneous distance on a homogeneous group is σ-finite dimensional if and only if it satisfies WBCP, see [12, Proposition 6.1].

The following theorem gives a characterization of homogeneous groups that admit homogeneous distances for which WBCP holds. We say that a graded group G with associated positive grading of its Lie algebra given by $\mathfrak{g} = \oplus_{t>0} V_t$ has *commuting different layers* if $[V_t, V_s] = \{0\}$ for all $t, s > 0$ such that $t \neq s$.

Theorem 6.3 ([12, Corollary 1.3]) *Let G be a homogeneous group. There exist homogeneous distances on G for which WBCP holds if and only if G has commuting different layers.*

Note that, given a homogeneous group with commuting different layers, a layer of its associated positive grading may not commute with itself, and there are indeed examples of homogeneous groups with commuting different layers beyond the Abelian case, see below for few of them.

A large class of homogeneous groups is given by stratified groups, also known as Carnot groups in the literature (although the latter terminology might sometimes more specifically be used when such groups are equipped with sub-Riemannian or sub-Finsler distances). We recall that a stratification of step s of a finite dimensional real Lie algebra $\mathfrak{g}$ is a positive grading of the form

$$\mathfrak{g} = V_1 \oplus V_2 \oplus \cdots \oplus V_s \quad \text{where } [V_1, V_j] = V_{j+1} \text{ for all } 1 \leq j \leq s$$

for some integer $s \geq 1$ and where $V_s \neq \{0\}$ and $V_{s+1} = \{0\}$. Equivalently, a stratification is a positive grading whose degree one layer generates $\mathfrak{g}$ as a Lie algebra. We say that a Lie algebra is stratified of step s if it is graded with associated positive grading that is a stratification of step s. We say that a connected and simply connected Lie group is stratified of step s if its Lie algebra is stratified of step s. It can easily be seen that a stratified group has commuting different layers if and only if it is of step 1 or 2. Hence Theorem 6.3 has the following corollary.

Corollary 6.4 ([12, Corollary 1.4]) *Let G be a stratified group. There exist homogeneous distances on G for which WBCP holds if and only if G is of step 1 or 2.*

Let us now give some examples together with an outline of the main ideas involved in the proof of Theorem 6.3.

Example 6.5 Stratified groups of step 1 equipped with homogeneous distances can be identified with finite dimensional normed vector spaces. As already explained, in such a case the validity of WBCP was known for a long time, see Remark 4.6.

The simplest example of non Abelian positively graduable Lie algebra is given by the first Heisenberg Lie algebra.

The first Heisenberg Lie algebra $\mathfrak{h}$ is the 3-dimensional Lie algebra that admits a basis (X, Y, Z) where the only non trivial bracket relation is $[X, Y] = Z$. With no loss of generality, let us fix such a basis of $\mathfrak{h}$.

The first Heisenberg group $\mathbb{H}$ is the connected and simply connected Lie group whose Lie algebra is $\mathfrak{h}$. Using exponential coordinates of the first kind, we write $p \in \mathbb{H}$ as $p = \exp(xX + yY + zZ)$ and we identify p with $(x, y, z) \in \mathbb{R}^3$. Using the Baker-Campbell-Hausdorff formula, the group law is given by

$$(x, y, z) \cdot (x, y, z) = (x + x', y + y', z + z' + \frac{1}{2}(xy' - yx'))\ .$$

Example 6.6 (Stratified First Heisenberg Group) The first Heisenberg Lie algebra $\mathfrak{h}$ is stratifiable of step 2. Namely,

$$\mathfrak{h} = V_1 \oplus V_2 \quad \text{where } V_1 := \text{span}\{X, Y\}, \; V_2 := \text{span}\, Z,$$

is a stratification that we call the *standard stratification*. Note that stratifications are unique up to Lie algebra automorphisms of graded Lie algebras. Hence there is no loss of generality here to work with the standard stratification of $\mathfrak{h}$ rather than with other possible stratifications.

We call *stratified first Heisenberg group* the first Heisenberg group viewed as a stratified group whose Lie algebra is equipped with the standard stratification. Associated dilations on the stratified first Heisenberg group are given by

$$\delta_r(x, y, z) := (rx, ry, r^2 z).$$

There are several classical examples of homogeneous distances on the stratified first Heisenberg group, such as the Korányi (see (6.2)) or sub-Riemannian distances. The following class of examples is due to Hebisch et Sikora. For $\gamma > 0$, set

$$A_\gamma := \{(x, y, z) \in \mathbb{H} : x^2 + y^2 + z^2 \leq \gamma^2\}$$

and, for $p, q \in \mathbb{H}$,

$$d_\gamma(p, q) := \inf\{r > 0 : \delta_{1/r}(p^{-1} \cdot q) \in A_\gamma\}. \tag{6.1}$$

It is proved in [7] that there is $\gamma^* \geq 2$, such that, for every $0 < \gamma \leq \gamma^*$, d_γ defines a homogeneous distance on the stratified first Heisenberg group. Note that the unit ball centered at the origin for such a distance is given by the set A_γ, that is, a Euclidean ball centered at the origin with a small enough radius. We refer to [7] for a complete statement about existence of such homogeneous distances on arbitrary homogeneous groups.

Going back to the validity of WBCP, the main result in [11] is the validity of WBCP for the homogeneous distances d_γ on the stratified first Heisenberg group.

Theorem 6.7 ([11, Theorem 1.14]) *Let $\gamma > 0$ be such that d_γ (see* (6.1)*) is a homogeneous distance on the stratified first Heisenberg group. Then WBCP holds on* $(\mathbb{H}, d_\gamma)$.

When $\gamma = 2$, d_2 defines a homogeneous distance on the stratified first Heisenberg group that is related to the Korányi distance d_K via the following formula

$$d_2(0, (x, y, z)) = (2\sqrt{2})^{-1} \left((x^2 + y^2) + d_K(0, (x, y, z))^2\right)^{1/2},$$

where the Korányi distance from the origin is given by

$$d_K(0, (x, y, z)) := ((x^2 + y^2)^2 + 16z^2)^{1/4}. \tag{6.2}$$

A proof that, for every $\epsilon > 0$, the homogeneous distance d_ϵ given by

$$d_\epsilon(0,(x,y,z)) := (\epsilon(x^2+y^2) + d_K(0,(x,y,z))^2)^{1/2}$$

satisfies WBCP on the stratified first Heisenberg group is given in [14, Theorem 1.7]. This gives in particular a sequence of homegeneous distances that are as close as one wants to the Korányi distance and that satisfy WBCP. However, it was noticed independently and at the same time in [9] and [16] that the Korányi distance itself does not satisfy WBCP.

Theorem 6.7 has been extended to stratified free nilptotent groups of step 2 in [12, Theorem 4.5]. The existence of homogeneous distances for which WBCP holds on finite dimensional normed vector spaces and on stratified free nilptotent groups of step 2 is the first crucial geometric step in the proof of the "if" part in Theorem 6.3. This implication can indeed be deduced from these two model cases using the algebraic structure of graded Lie algebras with commuting different layers and ad hoc submetries, together with some of the general results stated at the beginning of this section. We refer to [10, Section 2] for the definition of submetries and their relationship with the validity of WBCP, and to [12, Section 4] for a complete proof of the "if" part in Theorem 6.3.

Example 6.8 (Non-Standard First Heisenberg Groups) The first Heisenberg Lie algebra $\mathfrak{h}$ admits positive gradings that are not stratifications. Namely, for $\alpha > 1$, we call *non standard grading of exponent* α the positive grading of $\mathfrak{h}$ given by

$$\mathfrak{h} = W_1 \oplus W_\alpha \oplus W_{\alpha+1}$$

where $W_1 := \operatorname{span}\{X\}$, $W_\alpha := \operatorname{span}\{Y\}$, $W_{\alpha+1} = \operatorname{span}\{Z\}$. Dilations associated to the non standard grading of exponent α are given by

$$\delta_r(x,y,z) := (rx, r^\alpha y, r^{\alpha+1} z)\ .$$

We call *non standard first Heisenberg group of exponent* α the first Heisenberg group viewed as a graded group whose Lie algebra is endowed with the non standard grading of exponent α. We have $[W_1, W_\alpha] = W_{\alpha+1}$ hence non standard Heisenberg groups do not have commuting different layers.

The non existence of continuous homogeneous quasi-distances for which WBCP holds on non standard first Heisenberg groups, see [12, Theorem 5.6], is the first crucial geometric step in the proof of the "only if" part in Theorem 6.3. The general case of homogeneous groups that do not have commuting different layers can indeed be obtained from this model case using an algebraic relationship between positive grading of Lie algebras that do not have commuting different layers and non standard gradings of the first Heisenberg Lie algebra, together with ad hoc submetries. We refer to [12, Section 5] for a complete proof of the "only if" part in Theorem 6.3.

References

1. L. Ambrosio, N. Fusco, D. Pallara, *Functions of Bounded Variation and Free Discontinuity Problems*. Oxford Mathematical Monographs (The Clarendon Press, Oxford University Press, New York, 2000), xviii+434 pp.
2. A.S. Besicovitch, A general form of the covering principle and relative differentiation of additive functions. Proc. Camb. Phil. Soc. **41**, 103–110 (1945)
3. A.S. Besicovitch, A general form of the covering principle and relative differentiation of additive functions. II. Proc. Camb. Phil. Soc. **42**, 1–10 (1946)
4. M. de Guzmán, *Real Variable Methods in Fourier Analysis*, North-Holland Mathematics Studies, vol. 46. Notas de Matemtica [Mathematical Notes], vol. 75 (North-Holland Publishing, Amsterdam-New York, 1981), xiii+392 pp.
5. H. Federer, *Geometric Measure Theory*. Die Grundlehren der mathematischen Wissenschaften, Band 153 (Springer-Verlag New York Inc., New York, 1969), xiv+676 pp.
6. G.B. Folland, E.M. Stein, *Hardy Spaces on Homogeneous Groups*. Mathematical Notes, vol. 28 (Princeton University Press, Princeton, NJ; University of Tokyo Press, Tokyo, 1982), xii+285 pp.
7. W. Hebisch, A. Sikora, A smooth subadditive homogeneous norm on a homogeneous group. Studia Math. **96**(3), 231–236 (1990)
8. J. Heinonen, *Lectures on Analysis on Metric Spaces*. Universitext (Springer, New York, 2001), x+140 pp.
9. A. Korányi, H.M. Reimann, Foundations for the theory of quasiconformal mappings on the Heisenberg group. Adv. Math. **111**(1), 1–87 (1995)
10. E. Le Donne, S. Rigot, Remarks about the Besicovitch covering property in Carnot groups of step 3 and higher. Proc. Am. Math. Soc. **144**(5), 2003–2013 (2016)
11. E. Le Donne, S. Rigot, Besicovitch covering property for homogeneous distances on the Heisenberg groups. J. Eur. Math. Soc. (JEMS) **19**(5), 1589–1617 (2017)
12. E. Le Donne, S. Rigot, Besicovitch covering property on graded groups and applications to measure differentiation. J. Reine Angew. Math. **750**, 241–297 (2019)
13. P. Mattila, *Geometry of Sets and Measures in Euclidean Spaces. Fractals and Rectifiability*. Cambridge Studies in Advanced Mathematics, vol. 44 (Cambridge University Press, Cambridge, 1995), xii+343 pp.
14. S. Golo, S. Rigot, The Besicovitch covering property in the Heisenberg group revisited. J. Geom. Anal. **29**(4), 3345–3383 (2019)
15. D. Preiss, *Dimension of Metrics and Differentiation of Measures*. General Topology and Its Relations to Modern Analysis and Algebra, V (Prague, 1981)
16. R.L. Sawyer, R.L. Wheeden, Weighted inequalities for fractional integrals on Euclidean and homogeneous spaces. Am. J. Math. **114**(4), 813–874 (1992)
17. E.M. Stein, *Singular Integrals and Differentiability Properties of Functions*. Princeton Mathematical Series, vol. 30 (Princeton University Press, Princeton, NJ, 1970), xiv+290 pp.

Sobolev Spaces in Extended Metric-Measure Spaces

Giuseppe Savaré

Abstract These lecture notes contain an extended version of the material presented in the C.I.M.E. summer course in 2017. The aim is to give a detailed introduction to the metric Sobolev theory.

The notes are divided in four main parts. The first one is devoted to a preliminary and detailed study of the underlying topological, metric, and measure-theoretic aspects needed for the development of the theory in a general *extended metric-topological measure space* $\mathbb{X} = (X, \tau, \mathsf{d}, \mathfrak{m})$.

The *second part* is devoted to the construction of the Cheeger energy, initially defined on a distinguished unital algebra $\mathscr{A}$ of bounded, τ-continuous and d-Lipschitz functions.

The *third part* deals with the basic tools needed for the dual characterization of the Sobolev spaces: the notion of p-Modulus of a collection of (nonparametric) rectifiable arcs and its duality with the class of nonparametric dynamic plans, i.e. Radon measures on the space of rectifiable arcs with finite q-barycentric entropy with respect to $\mathfrak{m}$.

The *final part* of the notes is devoted to the dual/weak formulation of the Sobolev spaces $W^{1,p}(\mathbb{X})$ in terms of nonparametric dynamic plans and to their relations with the Newtonian spaces $N^{1,p}(\mathbb{X})$ and with the spaces $H^{1,p}(\mathbb{X})$ obtained by the Cheeger construction. In particular, when (X, d) is complete, a new proof of the equivalence between these different approaches is given by a direct duality argument.

A substantial part of these Lecture notes relies on well established theories. New contributions concern the extended metric setting, the role of general compatible algebras of Lipschitz functions and their density w.r.t. the Sobolev energy, a

Partially supported by the Institute of Advanced Study of the Technical University of Munich, by the MIUR-PRIN 2017 project *Gradient flows, Optimal Transport and Metric Measure Structures* and by the IMATI-CNR, Pavia.

G. Savaré (✉)
Department of Decision Sciences and BIDSA, Bocconi University, Milan, Italy
e-mail: giuseppe.savare@unibocconi.it

L. Ambrosio et al. (eds.), *New Trends on Analysis and Geometry in Metric Spaces*,
C.I.M.E. Foundation Subseries 2296, https://doi.org/10.1007/978-3-030-84141-6_4

general embedding/compactification trick, the study of reflexivity and infinitesimal Hilbertianity inherited from the underlying space, and the use of nonparametric dynamic plans for the definition of weak upper gradients.

1 Introduction

These lecture notes contain an extended version of the material presented in the C.I.M.E. summer course in 2017. The aim is to give a detailed introduction to the metric Sobolev theory, trying to unify at least two of the main approaches leading to the construction of the Sobolev spaces in general metric-measure spaces.

The notes are divided in four main parts. The first one is devoted to a preliminary and detailed study of the underlying topological, metric, and measure-theoretic aspects needed for the development of the general theory. In order to cover a wide class of examples, including genuinely infinite dimensional cases, we consider a general *extended metric-topological measure space* [3] $\mathbb{X} = (X, \tau, \mathsf{d}, \mathfrak{m})$, where d is an extended distance on X, τ is an auxiliary weaker topology compatible with d and $\mathfrak{m}$ is a Radon measure in (X, τ). The simplest example is a complete and separable metric space (X, d) where τ is the topology induced by the distance, but more general situations as duals of separable Banach spaces or Wiener spaces can be included as well.

The use of an auxiliary weaker (usually Polish or Souslin) topology τ has many technical advantages: first of all, it is easier to check the Radon property of the finite Borel measure $\mathfrak{m}$, one of our crucial structural assumptions. A second advantage is to add more flexibility in the choice of well behaved sub-algebras of Lipschitz functions and to allow for a powerful compactification method. As a reward, roughly speaking, many results which can be proved for a compact topology, can be extended to the case of a complete metric space (X, d) without too much effort. Therefore, for a first reading, it would not be too restrictive to assume compactness of the underlying topology, in order to avoid cumbersome technicalities.

The first part also includes a careful analysis of the topological-metric properties of the path space (Sect. 2.2) in particular concerning invariant properties with respect to parametrizations. We first recall the compact-open topology of $\mathrm{C}([0, 1]; (X, \tau))$ and the induced quotient space of arcs, obtained by identifying two curves $\gamma_1, \gamma_2 \in \mathrm{C}([0, 1]; (X, \tau))$ if there exist continuous, nondecreasing and surjective maps $\sigma_1, \sigma_2 : [0, 1] \to [0, 1]$ such that $\gamma_1 \circ \sigma_1 = \gamma_2 \circ \sigma_2$. This provides the natural quotient topology for the space of *continuous and* d-*rectifiable arcs* $\mathrm{RA}(X)$, for which the length

$$\ell(\gamma) := \sup \left\{ \sum_{j=1}^{N} \mathsf{d}(\gamma(t_j), \gamma(t_{j-1})) : \{t_j\}_{j=0}^{N} \subset [0, 1], \quad t_0 < t_1 < \cdots < t_N \right\} \tag{1.1}$$

is finite. It results a natural metric-topological structure for the space $\mathrm{RA}(X)$, where the distance d characterizes the length and the integrals, whereas the topology τ induces the appropriate notion of convergence. This analysis plays a crucial role, since one of the main tools for studying Sobolev spaces involves *dynamic plans*, i.e. Radon measures on $\mathrm{RA}(X)$. It is also the natural setting to study the properties of length-conformal distances (Sect. 2.3).

The *second part* is devoted to the construction of the Cheeger energy [8, 22] (Sect. 3.1), the $L^p(X, \mathfrak{m})$-relaxation of the energy functional

$$\int_X \big(\operatorname{lip} f(x)\big)^p \,\mathrm{d}\mathfrak{m}(x), \quad \operatorname{lip} f(x) := \limsup_{y,z\to x} \frac{|f(y)-f(z)|}{\mathsf{d}(y,z)}, \tag{1.2}$$

initially defined on a distinguished unital algebra $\mathscr{A}$ of bounded, τ-continuous and d-Lipschitz functions satisfying the approximation property

$$\mathsf{d}(x,y) = \sup\Big\{f(x) - f(y) : f \in \mathscr{A},\ |f(x') - f(y')| \le \mathsf{d}(x', y') \text{ for every } x', y' \in X\Big\} \tag{1.3}$$

for every pair of points $x, y \in X$. This gives raise to the Cheeger energy

$$\mathsf{CE}_{p,\mathscr{A}}(f) := \inf\Big\{\int_X \big(\operatorname{lip} f_n\big)^p \,\mathrm{d}\mathfrak{m} : f_n \in \mathscr{A},\ f_n \to f \text{ in } L^p(X, \mathfrak{m})\Big\}, \tag{1.4}$$

whose proper domain characterizes the strongest Sobolev space (deeply inspired by the Cheeger approach [22])

$$H^{1,p}(\mathbb{X}, \mathscr{A}) := \Big\{f \in L^p(X, \mathfrak{m}) : \mathsf{CE}_{p,\mathscr{A}}(f) < \infty\Big\}. \tag{1.5}$$

We will discuss various useful properties of the Cheeger energy, in particular its local representation in terms of the minimal relaxed gradient $|\mathrm{D}f|_{\star,\mathscr{A}}$ as

$$\mathsf{CE}_{p,\mathscr{A}}(f) = \int_X |\mathrm{D}f|^p_{\star,\mathscr{A}}(x)\,\mathrm{d}\mathfrak{m}(x), \tag{1.6}$$

the non-smooth first-order calculus properties of $|\mathrm{D}f|_{\star,\mathscr{A}}$, and the invariance properties of $\mathsf{CE}_{p,\mathscr{A}}$ with respect to measure-preserving isometric imbeddings of X.

A first, non obvious, important result is the independence of $\mathsf{CE}_{p,\mathscr{A}}$ with respect to $\mathscr{A}$, at least when (X, τ) is compact. It is a consequence of a delicate and powerful approximation method based on the metric Hopf-Lax flow

$$\mathsf{Q}_t f(x) := \inf_{y\in X} f(y) + \frac{1}{q t^{q-1}} \mathsf{d}^q(x, y), \quad t > 0, \tag{1.7}$$

which we will discuss in great detail in Sect. 3.2.

The *third part* of these notes deals with the basic tools needed for the dual characterization of the Sobolev spaces. First of all the notion of p-Modulus [30, 44, 46, 57] (Sect. 4.1) of a collection of (nonparametric) rectifiable arcs $\Gamma \subset \mathrm{RA}(X)$,

$$\mathrm{Mod}_p(\Gamma) := \inf\Big\{\int_X f^p\,\mathrm{d}\mathfrak{m} : \ f : X \to [0,\infty] \text{ Borel}, \quad \int_\gamma f \ge 1 \text{ for every } \gamma \in \Gamma\Big\}, \tag{1.8}$$

which is mainly used to give a precise meaning to negligible sets. It will be put in duality with the class $\mathcal{B}_q$ of Radon measures $\boldsymbol{\pi}$ on $\mathrm{RA}(X)$ with finite q-barycentric entropy $\mathrm{Bar}_q(\boldsymbol{\pi})$ with respect to $\mathfrak{m}$ [2] (Sect. 4.2). Every $\boldsymbol{\pi} \in \mathcal{B}_q$ induces a measure $\mu_{\boldsymbol{\pi}} = h_{\boldsymbol{\pi}}\mathfrak{m}$ with density $h_{\boldsymbol{\pi}} \in L^q(X,\mathfrak{m})$ such that for every bounded Borel function $\zeta : X \to \mathbb{R}$

$$\int_{\mathrm{RA}(X)}\int_\gamma \zeta\,\mathrm{d}\boldsymbol{\pi}(\gamma) = \int_X \zeta\,\mathrm{d}\mu_{\boldsymbol{\pi}} = \int_X \zeta\, h_{\boldsymbol{\pi}}\,\mathrm{d}\mathfrak{m}, \quad \mathrm{Bar}_q^q(\boldsymbol{\pi}) := \int_X h_{\boldsymbol{\pi}}^q\,\mathrm{d}\mathfrak{m}. \tag{1.9}$$

At least when (X,d) is complete, we will show (Sect. 4.3) that the Modulus of a Borel subset $\Gamma \subset \mathrm{RA}(X)$ can be essentially identified with the conjugate of the q-barycentric entropy:

$$\frac{1}{p}\mathrm{Mod}_p(\Gamma) = \sup_{\boldsymbol{\pi}\in\mathcal{B}_q} \boldsymbol{\pi}(\Gamma) - \frac{1}{q}\mathrm{Bar}_q^q(\boldsymbol{\pi}). \tag{1.10}$$

The duality formula shows that a Borel set $\Gamma \subset \mathrm{RA}(X)$ is Mod_p-negligible if and only if it is $\boldsymbol{\pi}$-negligible for every dynamic plan $\boldsymbol{\pi}$ with finite q-barycentric entropy.

The *final part* of the notes is devoted to the dual/weak formulation of the Sobolev spaces $W^{1,p}(\mathbb{X})$ and to their relations with the spaces $H^{1,p}(\mathbb{X})$ obtained by the Cheeger construction. The crucial concept here is the notion of *upper gradient* [22, 44, 46] of a function $f : X \to \mathbb{R}$: it is a nonnegative Borel function $g : X \to [0,+\infty]$ such that

$$|f(\gamma_1) - f(\gamma_0)| \le \int_\gamma g \tag{1.11}$$

for every rectifiable arc $\gamma \in \mathrm{RA}(X)$; γ_0 and γ_1 in (1.11) denote the initial and final points of γ. As suggested by the theory of Newtonian spaces [57], it is possible to adapt the notion of upper gradient to Sobolev functions by asking that $g \in L^p(X,\mathfrak{m})$, by selecting a corresponding notion of "exceptional" or "negligible" sets of rectifiable arcs in $\mathrm{RA}(X)$, and by imposing that the set of curves where (1.11) does not hold is exceptional.

According to the classic approach leading to Newtonian spaces, a subset $\Gamma \subset \mathrm{RA}(X)$ is negligible if $\mathrm{Mod}_p(\Gamma) = 0$. This important notion, however, is not invariant with respect to modification of f and g in $\mathfrak{m}$-negligible sets. Here we present a different construction, based on the new class $\mathcal{T}_q$ of dynamic plans $\boldsymbol{\pi}$ with finite q-barycentric entropy and with finite q-entropy of the initial and final distribution of points,

$$\boldsymbol{\pi} \in \mathcal{T}_q \quad \Leftrightarrow \quad \mathrm{Bar}_q(\boldsymbol{\pi}) < \infty, \quad (\mathsf{e}_i)_\sharp \boldsymbol{\pi} = h_i \mathfrak{m} \quad \text{for some } h_i \in L^q(X, \mathfrak{m}), \quad i = 0, 1, \tag{1.12}$$

where $\mathsf{e}_i(\gamma) = \gamma_i$. The last condition requires that there exist functions $h_i \in L^q(X, \mathfrak{m})$ such that

$$\int_{\mathrm{RA}(X)} \zeta(\gamma_i)\,\mathrm{d}\boldsymbol{\pi}(\gamma) = \int_X \zeta(x)\,h_i\,\mathrm{d}\mathfrak{m} \quad \text{for every bounded Borel function } \zeta : X \to \mathbb{R}. \tag{1.13}$$

A collection $\Gamma \subset \mathrm{RA}(X)$ is $\mathcal{T}_q$-negligible if it is $\boldsymbol{\pi}$-negligible for every $\boldsymbol{\pi} \in \mathcal{T}_q$. The Sobolev space $W^{1,p}(\mathbb{X}, \mathcal{T}_q)$ precisely contains all the functions $f \in L^p(X, \mathfrak{m})$ with a $\mathcal{T}_q$-weak upper gradient $g \in L^p(X, \mathfrak{m})$, so that

$$|f(\gamma_1) - f(\gamma_0)| \le \int_\gamma g \quad \text{for } \mathcal{T}_q\text{-a.e. arc } \gamma \in \mathrm{RA}(X). \tag{1.14}$$

Among all the $\mathcal{T}_q$-weak upper gradients g of f it is possible to select the minimal one, denoted by $|\mathrm{D}f|_{w,\mathcal{T}_q}$, such that $|\mathrm{D}f|_{w,\mathcal{T}_q} \le g$ for every $\mathcal{T}_q$-weak upper gradient g. The norm of $W^{1,p}(\mathbb{X}, \mathcal{T}_q)$ is then given by

$$\|f\|^p_{W^{1,p}(\mathbb{X},\mathcal{T}_q)} := \int_X \Big(|f|^p + |\mathrm{D}f|^p_{w,\mathcal{T}_q}\Big)\,\mathrm{d}\mathfrak{m}. \tag{1.15}$$

Differently from the Newtonian weak upper gradient, the notion of $\mathcal{T}_q$-weak upper gradient is invariant w.r.t. modifications of f and g in $\mathfrak{m}$-negligible sets; moreover it is possible to prove that functions in $W^{1,p}(\mathbb{X}, \mathcal{T}_q)$ are Sobolev along $\mathcal{T}_q$-a.e. arc γ with distributional derivative bounded by $g \circ \gamma$. The link with the Newtonian theory appears more clearly by a further property of functions in $W^{1,p}(\mathbb{X}, \mathcal{T}_q)$, at least when (X, d) is complete: for every function $f \in W^{1,p}(\mathbb{X}, \mathcal{T}_q)$ it is possible to find a "good representative" $\tilde{f}$ (i.e., $\{\tilde{f} \neq f\}$ is $\mathfrak{m}$-negligible), so that the modified function $\tilde{f}$ is absolutely continuous along Mod_p-a.e. rectifiable arc. In this way, the a-priori weaker approach by $\mathcal{T}_q$-dynamic plans is equivalent to the Newtonian one and it is possible to identify $W^{1,p}(\mathbb{X}, \mathcal{T}_q)$ with $N^{1,p}(\mathbb{X})$. We will also show that the approach by nonparametric dynamic plans is equivalent to the definition by parametric q-test plans of [7, 8].

A further main identification result is stated in Sect. 5.2: when (X, d) is complete, we can show that $W^{1,p}(\mathbb{X}, \mathcal{T}_q)$ coincides with $H^{1,p}(\mathbb{X}, \mathscr{A})$. This fact (originally proved by [22, 57] in the case of doubling-Poincaré spaces) can be interpreted as a density result of a compatible algebra of functions $\mathscr{A}$ in $W^{1,p}(\mathbb{X}, \mathcal{T}_q)$ and has important consequences, some of them recalled in the last section of the notes. Differently from other recent approaches [7, 8] the proof arises from a direct application of the Von Neumann min-max principle and relies on two equivalent characterizations of the dual Cheeger energy $\mathsf{CE}_p^*(h)$ for functions $h \in L^q(X, \mathfrak{m})$

$$\frac{1}{q}\mathsf{CE}_p^*(h) = \sup_{h \in H^{1,p}(\mathbb{X})} \int_X fh \,\mathrm{d}\mathfrak{m} - \frac{1}{p}\mathsf{CE}_p(f) \quad h \in L^q(X, \mathfrak{m}), \ \int_X h \,\mathrm{d}\mathfrak{m} = 0. \tag{1.16}$$

When (X, τ) is compact we can prove that

$$\mathsf{CE}_p^*(h) = \sup\Big\{ \mathrm{Bar}_q^q(\boldsymbol{\pi}) : (\mathsf{e}_0)_\sharp \boldsymbol{\pi} = h_-\mathfrak{m}, \quad (\mathsf{e}_1)_\sharp \boldsymbol{\pi} = h_+\mathfrak{m}\Big\}, \tag{1.17}$$

h_-, h_+ being the negative and positive parts of h, and

$$\frac{1}{p}\mathsf{CE}_p^*(h) = \sup\Big\{\mathsf{K}_{\mathsf{d}_g}(h_-\mathfrak{m}, h_+\mathfrak{m}) - \frac{1}{q}\int_X g^q \,\mathrm{d}\mathfrak{m} : g \in \mathrm{C}_b(X), \ \inf_X g > 0\Big\}, \tag{1.18}$$

where $\mathsf{K}_{\mathsf{d}_g}$ is the Kantorovich-Rubinstein distance induced by the cost

$$\mathsf{d}_g(x_0, x_1) := \inf\Big\{\int_\gamma g : \gamma \in \mathrm{RA}(X), \ \gamma_0 = x_0, \ \gamma_1 = x_1\Big\}. \tag{1.19}$$

Thanks to the identification Theorem and the compactification method, we obtain that for a general complete space (X, d)

$$H^{1,p}(\mathbb{X}, \mathscr{A}) = H^{1,p}(\mathbb{X}) = W^{1,p}(\mathbb{X}, \mathfrak{T}_q) = N^{1,p}(\mathbb{X}) \tag{1.20}$$

(the last identity holds up to the selection of a good representative) with equality of the corresponding minimal gradients. As a consequence, all the approaches lead to one canonical object and this property does not rely on the validity of doubling properties or Poincaré inequalities for $\mathfrak{m}$.

In the last Sect. 5.3 we will show various invariance properties of the Cheeger energy and of the metric Sobolev spaces. In particular, when the underlying space X has a linear structure, we show that the metric approach coincides with more classic definitions of Sobolev spaces (e.g. the weighted Sobolev spaces in $\mathbb{R}^d$ [43] or the Sobolev spaces associated to a log-concave measure in a Banach-Hilbert space), obtaining the reflexivity (resp. the Hilbertianity) of $W^{1,p}(\mathbb{X})$ whenever X is a reflexive Banach (resp. Hilbert) space.

A substantial part of these Lecture notes relies on well established theories: our main sources have been [7, 8, 11] (for the parts concerning the Cheeger energy, the weak upper gradients, and the properties of the Hopf-Lax flow), [2, 16, 45] (for the notion of the p-Modulus and the Newtonian spaces), [2] (for the notion of nonparametric dynamic plans in $\mathcal{B}_q$, the dual characterization of the p-Modulus and the selection of a good representative of a Sobolev function), [3] (for the extended metric-topological structures), [6, 64] (for the results involving the Kantorovich-Rubinstein distances of Optimal Transport), [56] for the theoretic aspects of Radon measures. Further bibliographical notes are added to each Section with more detailed comments. We also refer to the overviews and lecture notes [4, 14, 35, 42].

New contributions concern the role of general compatible algebras of Lipschitz functions and their invariance in the construction of the Cheeger energy, the embedding/compactification tricks, the use of nonparametric dynamic plans for the definition of weak upper gradients, the characterization of the dual Cheeger energy and the proof of the identification theorem $H = W$ by a direct duality argument.

Of course, there are many important aspects that we did not include in these notes: just to name a few of them at the level of the Sobolev construction we quote

- the Hajłasz's Sobolev spaces [41],
- the theoretical aspects related to the doubling and to the Poincaré inequality assumptions [16, 45],
- the point of view of *parametric* dynamic plans (i.e. Radon measures on the space of parametric curves with finite q-energy) [7, 8] (but see the discussion in Sect. 5.1.5),
- the properties of the L^2-gradient flow of the Cheeger energy,
- the original proof of the "$H = W$" Theorem [7, 8] by a dynamic approach based on the identification of the L^2-gradient flow of the Cheeger energy with the Kantorovich-Wasserstein gradient flow of the Shannon-Rény entropies,
- the approach [4, 25] by derivations and integration by parts,
- the Gigli's nonsmooth differential structures [34–36] (see also [37, 38]),
- the applications to metric measure spaces satisfying a lower Ricci curvature bounds [5, 10–12, 29, 50, 61, 62].

1.1 Main Notation

(X, τ)	Hausdorff topological space
(X, τ, d)	Extended metric-topological (e.m.t.) space, see Sect. 2.1.2 and Definition 2.1.3
$\mathbb{X} = (X, \tau, \mathsf{d}, \mathfrak{m})$	Extended metric-topological measure (e.m.t.m.) space, see Sect. 2.1.2
$\mathcal{M}(X),\ \mathcal{M}_+(X)$	Signed and positive Radon measures on a Hausdorff topological space X
$\mathcal{P}(X)$	Radon probability measures on X
$\mathscr{F}(X),\ \mathscr{K}(X), \mathscr{B}(X), \mathscr{S}(X)$	Closed, compact, Borel and Souslin subsets of X
$\mathrm{supp}(\mu)$	Support of a Radon measure, see p. 126
$f_\sharp\mu$	Push forward of $\mu \in \mathcal{M}(X)$ by a (Lusin μ-measurable) map $f : X \to Y$, (2.1.12)

$\mathrm{C}_b(X,\tau),\ \mathrm{C}_b(X)$	τ-continuous and bounded real functions on X
$\mathrm{B}_b(X,\tau),\ \mathrm{B}_b(X)$	Bounded τ-Borel real functions
$\mathrm{Lip}(f, A, \delta)$	Lipschitz constant of f on A w.r.t. the extended semidistance δ, (2.1.14)
$\mathrm{Lip}_b(X,\tau,\delta)$	Bounded, τ-continuous and δ-Lipschitz real functions on X, (2.1.15)
$\mathrm{Lip}_{b,\kappa}(X,\tau,\delta)$	Functions in $\mathrm{Lip}_b(X,\tau,\delta)$ with Lipschitz constant bounded by κ, (2.1.16)
$\mathrm{lip}_\delta f(x)$	Asymptotic Lipschitz constant of f at a point x, Sect. 2.1.5
$\mathscr{A}, \mathscr{A}_1$	Compatible unital sub-algebra of $\mathrm{Lip}_b(X,\tau,\mathsf{d})$, Sect. 2.1.6
$\mathcal{L}^p(X,\mathfrak{m})$	Space of p-summable Borel functions
$\mathscr{L}^q(\gamma\|\mu)$	Entropy functionals on Radon measures
$\mathsf{K}_\delta(\mu_1,\mu_2)$	Kantorovich-Rubinstein extended distance in $\mathcal{M}_+(X)$, Sect. 2.1.4
$\mathrm{C}([a,b];(X,\tau)),\ \mathrm{C}([a,b];X)$	τ-continuous curves defined in $[a,b]$ with values in X, Sect. 2.2.1
$\tau_{\mathrm{C}},\ \mathsf{d}_{\mathrm{C}}$	Compact open topology and extended distance on $\mathrm{C}([a,b];X)$, Sect. 2.2.1
$\mathrm{A}(X,\tau),\ \mathrm{A}(X)$	Space of arcs, classes of curves equivalent up to a reparametrization, Sect. 2.2.2
$\mathrm{A}(X,\mathsf{d})$	Space of arcs with a d-continuous reparametrization, Sect. 2.2.2
$\tau_{\mathrm{A}},\ \mathsf{d}_{\mathrm{A}}$	Quotient topology and extended distance on $\mathrm{A}(X,\mathsf{d})$, Sect. 2.2.2
$\mathrm{BV}([a,b];(X,\delta))$	Curves $\gamma:[a,b]\to X$ with finite total variation w.r.t. δ, Sect. 2.2.3
$\mathrm{BVC}([a,b];(X,\mathsf{d}))$	Continuous curves in $\mathrm{BV}([a,b];(X,\mathsf{d}))$, Sect. 2.2.3
$\mathrm{RA}(X,\mathsf{d}),\ \mathrm{RA}(X)$	Continuous and rectifiable arcs, Sect. 2.2.3
R_γ	Arc-length reparametrization of a rectifiable arc γ, Sect. 2.2.3
$\int_\gamma f$	Integral of a function f along a rectifiable curve (or arc) γ, Sect. 2.2.3
$\ell(\gamma)$	length of γ, Sect. 2.2.3
ν_γ	Radon measure in $\mathcal{M}_+(X)$ induced by integration along a rectifiable arc γ, Sect. 2.2.3
$\mathsf{d}_\ell,\ \mathsf{d}_g$	Length and conformal distances generated by d, Sect. 2.3
$\mathsf{pCE}_p,\ \mathsf{CE}_{p,\mathscr{A}}$	(pre)Cheeger energy, Definition 3.1.1
$H^{1,p}(\mathbb{X},\mathscr{A})$	Metric Sobolev space induced by the Cheeger energy, Definition 3.1.3
$\|\mathrm{D}f\|_{\star,\mathscr{A}}$	Minimal $(p,\mathscr{A})$-relaxed gradient, Sect. 3.1.1

$\mathsf{Q}_t^{K,\delta}(f)$, $\mathsf{Q}_t(f)$	(Generalized) Hopf-Lax flow, Sect. 3.2.1
$\mathsf{Mod}_p(\Sigma)$, $\mathsf{Mod}_{p,c}(\Sigma)$	p-Modulus of a collection of measures $\Sigma \subset \mathcal{M}_+(X)$, (4.1.1) and (4.1.2)
$\mathrm{Mod}_p(\Gamma)$, $\widetilde{\mathrm{Mod}}_p(\Gamma)$	p-Moduli of a collection $\Gamma \subset \mathrm{RA}(X)$, (4.1.11) and (4.1.13)
$\mathrm{Bar}_q(\boldsymbol{\pi})$	q-barycentric entropy of a dynamic plan, Definition 4.2.2
$\mathcal{B}_q$	Plans with barycenter in $L^q(X, \mathfrak{m})$, Definition 4.2.2
$\mathrm{Cont}_p(\Gamma)$	p-Content of a family of arcs, Definition 4.2.6
$\mathcal{T}_q$, $\mathcal{T}_q^*$	nonparametric q-test plans, Definition 5.1.1
$\|\mathrm{D}f\|_w$, $\|\mathrm{D}f\|_{w,\mathcal{T}_q}$	Minimal $\mathcal{T}_q$-weak upper gradient, Definition 5.1.23
$\mathsf{wCE}_{p,\mathcal{T}_q}$, $W^{1,p}(\mathbb{X}, \mathcal{T}_q)$	Weak $(p, \mathcal{T}_q)$-energy and weak Sobolev space, Definition 5.1.24
$\mathscr{D}_q(\mu_0, \mu_1)$	Dual dynamic cost, (5.2.2)

2 Topological and Metric-Measure Structures

2.1 Metric-Measure Structures

In this section we will recall the main notion and facts we will use in the sequel. Our main ingredients are

- a Hausdorff topological space (X, τ),
- an extended distance $\mathsf{d} : X \times X \to [0, \infty]$,
- a finite Radon measure $\mathfrak{m}$ on (X, τ),
- an algebra $\mathscr{A}$ of τ-continuous, d-Lipschitz, bounded real functions defined in X.

All these objects will satisfy suitable compatibility conditions, which we are going to explain. We will call the system

$$\mathbb{X} = (X, \tau, \mathsf{d}, \mathfrak{m}), \text{ an } \textit{extended metric-topological measure (e.m.t.m.) space}. \tag{2.1.1}$$

The choice of $\mathscr{A}$ will play a role in the construction of the Cheeger energy.

Let us first consider the topological and measurable side of this structure.

2.1.1 Topological and Measure Theoretic Notions

Let (X, τ) be a Hausdorff topological space. We will denote by $\mathrm{C}_b(X, \tau)$ (resp. $\mathrm{B}_b(X, \tau)$) the space of τ-continuous (resp. Borel) and bounded real functions defined on X. $\mathscr{B}(X,\tau)$ is the collection of the Borel subsets of X. For every $x \in X$, $\mathscr{U}_x$ will denote the system of neighborhoods of x. We will often omit the explicit indication of the topology τ, when it will be clear from the context.

We will always deal with a *completely regular* topology, i.e.

$$\begin{gathered}\text{for any closed set } F \subset X \text{ and any } x_0 \in X \setminus F\\ \text{there exists } f \in \mathrm{C}_b(X,\tau) \text{ with } f(x_0) > 0 \text{ and } f \equiv 0 \text{ on } F.\end{gathered} \tag{2.1.2}$$

We can always assume that f takes values in $[0,1]$ and $f(x_0) = 1$. An immediate consequence of (2.1.2) is that for every open subset $G \subset X$ its characteristic function χ_G can be represented as

$$\chi_G(x) = \sup\Big\{\varphi(x) : \varphi \in \mathrm{C}_b(X,\tau),\ 0 \le \varphi \le \chi_G\Big\}, \tag{2.1.3}$$

and the same representation holds for every nonnegative lower semicontinuous (l.s.c.) $f : X \to [0,+\infty]$:

$$f(x) = \sup\Big\{\varphi(x) : \varphi \in \mathrm{C}_b(X,\tau),\ 0 \le \varphi \le f\Big\}, \qquad f : X \to [0,+\infty] \text{ l.s.c.}. \tag{2.1.4}$$

Definition 2.1.1 (Radon Measures [56, Chap. I, Sect. 2]) A finite Radon measure $\mu : \mathscr{B}(X,\tau) \to [0,+\infty)$ is a Borel nonnegative σ-additive finite measure satisfying the following inner regularity property:

$$\forall\, B \in \mathscr{B}(X,\tau) : \quad \mu(B) = \sup\Big\{\mu(K) : K \subset B,\ K \text{ compact}\Big\}. \tag{2.1.5}$$

A finite Radon measure μ is also outer regular:

$$\forall\, B \in \mathscr{B}(X,\tau) : \quad \mu(B) = \inf\Big\{\mu(O) : O \subset X,\ O \text{ open}\Big\}. \tag{2.1.6}$$

We will denote by $\mathcal{M}_+(X)$ (resp. $\mathcal{P}(X)$) the collection of all finite (resp. Probability) Radon measures on X. By the very definition of Radon topological space [56, Ch. II, Sect. 3], every Borel measure in a Radon space is Radon: such class of spaces includes locally compact spaces with a countable base of open sets, Polish, Lusin and Souslin spaces. In particular the notation of $\mathcal{P}(X)$ is consistent with the standard one adopted e.g. in [6, 9, 63], where Polish or second countable locally compact spaces are considered.

Equation (2.1.5) implies in particular that a Radon measure is tight:

$$\forall\, \varepsilon > 0 \quad \exists\, K_\varepsilon \subset X \quad \text{compact such that} \quad \mu(X \setminus K_\varepsilon) \le \varepsilon. \tag{2.1.7}$$

We can also define in the usual way the support $\operatorname{supp}\mu$ of a Radon measure as the set of points $x \in X$ such that every neighborhood $U \in \mathscr{U}_x$ has strictly positive measure $\mu(U) > 0$. Thanks to (2.1.5), one can verify that $\mu(X \setminus \mathrm{supp}(\mu)) = 0$ [56, p. 60].

Radon measures have stronger additivity and continuity properties in connection with open sets and lower semicontinuous functions; in particular we shall use this version of the monotone convergence theorem (see [18, Lemma 7.2.6])

$$\lim_{i \in I} \int f_i \, \mathrm{d}\mu = \int \lim_{i \in I} f_i \, \mathrm{d}\mu \tag{2.1.8}$$

valid for Radon measures μ and for nondecreasing nets $i \mapsto f_i$, $i \in I$, of τ-lower semicontinuous and equibounded functions $f_i : X \to [0, \infty]$. Here I is a directed set with a partial order $\preceq$ satisfying $i \preceq j \Rightarrow f_i \le f_j$, see the Appendix A.1.

The weak (or narrow) topology $\tau_{\mathcal{M}_+}$ on $\mathcal{M}_+(X)$ can be defined as the coarsest topology for which all maps

$$\mu \mapsto \int h \, \mathrm{d}\mu \qquad \text{from } \mathcal{M}_+(X) \text{ into } \mathbb{R} \tag{2.1.9}$$

are continuous as $h : X \to \mathbb{R}$ varies in $\mathrm{C}_b(X, \tau)$ [56, p. 370, 371].

Prokhorov Theorem provides a sufficient condition for compactness w.r.t. the weak topology: [56, Theorem 3, p. 379].

Theorem 2.1.2 (Prokhorov) *Let (X, τ) be a completely regular Hausdorff topological space. Assume that a collection $\mathcal{K} \subset \mathcal{M}_+(X)$ is uniformly bounded and equi-tight, i.e.*

$$\sup_{\mu \in \mathcal{K}} \mu(X) < \infty, \tag{2.1.10}$$

$$\text{for every } \varepsilon > 0 \text{ there exists a compact set } K_\varepsilon \subset X \text{ such that} \quad \sup_{\mu \in \mathcal{K}} \mu(X \setminus K_\varepsilon) \le \varepsilon\,. \tag{2.1.11}$$

Then $\mathcal{K}$ has limit points in the class $\mathcal{M}_+(X)$ w.r.t. the weak topology.

Recall that a set $A \subset X$ is $\mathfrak{m}$-measurable if there exist Borel sets $B_1, B_2 \in \mathscr{B}(X, \tau)$ such that $B_1 \subset A \subset B_2$ and $\mathfrak{m}(B_2 \setminus B_1) = 0$. $\mathfrak{m}$-measurable sets form a σ-algebra $\mathscr{B}_{\mathfrak{m}}(X)$. A set is called *universally (Radon) measurable* if it is μ-measurable for every Radon measure $\mu \in \mathcal{M}_+(X)$.

Let (Y, τ_Y) be a Hausdorff topological space. A map $f : X \to Y$ is Borel (resp. Borel $\mathfrak{m}$-measurable) if for every $B \in \mathscr{B}(Y)$ $f^{-1}(B) \in \mathscr{B}(X)$ (resp. $f^{-1}(B)$ is $\mathfrak{m}$-measurable). f is Lusin $\mathfrak{m}$-measurable if for every $\varepsilon > 0$ there exists a compact set $K_\varepsilon \subset X$ such that $\mathfrak{m}(X \setminus K_\varepsilon) \le \varepsilon$ and the restriction of f to K_ε is continuous. A map $f : X \to Y$ is called *universally measurable* if it is Lusin μ-measurable for every Radon measure $\mu \in \mathcal{M}_+(X)$.

Every Lusin $\mathfrak{m}$-measurable map is also Borel $\mathfrak{m}$-measurable; the converse is true if, e.g., the topology τ_Y is metrizable and separable [56, Chap. I Section 1.5,

Theorem 5]. Whenever f is Lusin $\mathfrak{m}$-measurable, its push-forward

$$f_\sharp\mathfrak{m} \in \mathcal{M}_+(Y), \quad f_\sharp\mathfrak{m}(B) := \mathfrak{m}(f^{-1}(B)) \quad \text{for every Borel subset } B \subset \mathscr{B}(Y) \tag{2.1.12}$$

induces a Radon measure in Y.

Given a power $p \in (1, \infty)$ and a Radon measure $\mathfrak{m}$ in (X, τ) we will denote by $L^p(X, \mathfrak{m})$ the usual Lebesgue space of class of p-summable $\mathfrak{m}$-measurable functions defined up to $\mathfrak{m}$-negligible sets. We will also set

$$\mathcal{L}^p_+(X, \mathfrak{m}) := \Big\{ f : X \to [0, \infty] \ : \ f \text{ is Borel}, \ \int_X f^p \,\mathrm{d}\mathfrak{m} < \infty \Big\}; \tag{2.1.13}$$

this space is not quotiented under any equivalence relation. We will keep using the notation

$$\|f\|_p = \|f\|_{L^p(X,\mathfrak{m})} := \Big(\int_X |f|^p \,\mathrm{d}\mathfrak{m} \Big)^{1/p}$$

as a seminorm on $\mathcal{L}^p_+(X, \mathfrak{m})$ and a norm in $L^p(X, \mathfrak{m})$.

2.1.2 Extended Metric-Topological (Measure) Spaces

Let (X, τ) be a Hausdorff topological space.

An extended semidistance is a symmetric map $\delta : X \times X \to [0, \infty]$ satisfying the triangle inequality; δ is an extended distance if it also satisfies the property $\delta(x, y) = 0$ iff $x = y$ in X: in this case, we call (X, δ) an extended metric space. We will omit the adjective "extended" if δ takes real values.

Whenever $f : X \to \mathbb{R}$ is a given function, $A \subset X$, and δ is an extended semidistance on X, we set

$$\mathrm{Lip}(f, A, \delta) := \inf \Big\{ L \in [0, \infty] : |f(y) - f(z)| \le L\delta(y, z) \quad \text{for every } y, z \in A \Big\}. \tag{2.1.14}$$

We adopt the convention to omit the set A when $A = X$. We consider the class of τ-continuous and δ-Lipschitz functions

$$\mathrm{Lip}_b(X, \tau, \delta) := \Big\{ f \in \mathrm{C}_b(X, \tau) : \mathrm{Lip}(f, \delta) < \infty \Big\}, \tag{2.1.15}$$

and for every $\kappa > 0$ we will also consider the subsets

$$\mathrm{Lip}_{b,\kappa}(X, \tau, \delta) := \Big\{ f \in \mathrm{C}_b(X, \tau) : \mathrm{Lip}(f, \delta) \le \kappa \Big\}. \tag{2.1.16}$$

A particular role will be played by $\mathrm{Lip}_{b,1}(X, \tau, \delta)$. We will sometimes omit to indicate the explicit dependence on τ and δ whenever it will be clear from the context. It is easy to check that $\mathrm{Lip}_b(X, \tau, \delta)$ is a real and commutative sub-algebras of $\mathrm{C}_b(X, \tau)$ with unit.

According to [3, Definition 4.1], an extended metric-topological space (e.m.t. space) (X, τ, d) is characterized by a Hausdorff topology τ and an extended distance d satisfying a suitable compatibility condition.

Definition 2.1.3 (Extended Metric-Topological Spaces) Let (X, d) be an extended metric space, let τ be a Hausdorff topology in X. We say that (X, τ, d) is an extended metric-topological (e.m.t.) space if:

(X1) the topology τ is generated by the family of functions $\mathrm{Lip}_b(X, \tau, \mathsf{d})$ (see the Appendix A.2);
(X2) the distance d can be recovered by the functions in $\mathrm{Lip}_{b,1}(X, \tau, \mathsf{d})$ through the formula

$$\mathsf{d}(x, y) = \sup_{f \in \mathrm{Lip}_{b,1}(X,\tau,\mathsf{d})} |f(x) - f(y)| \quad \text{for every } x, y \in X. \tag{2.1.17}$$

We will say that (X, τ, d) is *complete* if d-Cauchy sequences are d-convergent. All the other topological properties (as compactness, separability, metrizability, Borel, Polish-Lusin-Souslin, etc) usually refers to (X, τ).

The previous assumptions guarantee that (X, τ) is completely regular, according to (2.1.2) (see the Appendix A.2). As in (2.1.1), when an e.m.t. space (X, τ, d) is provided by a positive Radon measure $\mathfrak{m} \in \mathcal{M}_+(X, \tau)$ we will call the system $\mathbb{X} = (X, \tau, \mathsf{d}, \mathfrak{m})$ an extended metric-topological measure (e.m.t.m.) space.

Definition 2.1.3 yields two important properties linking d and τ: first of all

$$\mathsf{d} \text{ is } \tau \times \tau\text{-lower semicontinuous in } X \times X, \tag{2.1.18}$$

since it is the supremum of a family of continuo us maps by (2.1.17). On the other hand, every d-converging net $(x_j)_{j \in J}$ indexed by a directed set J is also τ-convergent:

$$\lim_{j \in J} \mathsf{d}(x_j, x) = 0 \quad \Rightarrow \quad \lim_{j \in J} x_j = x \quad \text{w.r.t. } \tau. \tag{2.1.19}$$

It is sufficient to observe that τ is the initial topology generated by $\mathrm{Lip}_b(X,\tau,\mathsf{d})$ so that a net (x_j) is convergent to a point x if and only if

$$\lim_{j\in J} f(x_j) = f(x) \quad \text{for every } f \in \mathrm{Lip}_b(X,\tau,\mathsf{d}). \tag{2.1.20}$$

A basis of neighborhoods for the τ-topology at a point $x \in X$ is given by the sets of the form

$$U_{F,\varepsilon}(x) := \Big\{ y \in X : \sup_{f\in F} |f(y) - f(x)| < \varepsilon \Big\} \quad F \subset \mathrm{Lip}_{b,1}(X,\tau,\mathsf{d}) \text{ finite}, \ \varepsilon > 0. \tag{2.1.21}$$

Definition 2.1.3 is in fact equivalent to other seemingly stronger assumptions, as we discuss in the following Lemma.

Lemma 2.1.4 (Monotone Approximations of the Distance) *Let (X,τ,d) be an e.m.t. space, let us denote by Λ the collection of all the finite subsets in $\mathrm{Lip}_{b,1}(X,\tau,\mathsf{d})$, a directed set ordered by inclusion, and let us define*

$$\mathsf{d}_\lambda(x,y) := \sup_{f\in\lambda} |f(x) - f(y)|, \quad \lambda \in \Lambda, \ x, y \in X. \tag{2.1.22}$$

The family $(\mathsf{d}_\lambda)_{\lambda\in\Lambda}$ is a monotone collection of τ continuous and bounded semidistances on X generating the τ-topology and the extended distance d, in the sense that for every net $(x_j)_{j\in J}$ in X

$$x_j \xrightarrow{\tau} x \quad \Leftrightarrow \quad \lim_{j\in J} \mathsf{d}_\lambda(x_j, x) = 0 \quad \textit{for every } \lambda \in \Lambda, \tag{2.1.23a}$$

and

$$\mathsf{d}(x,y) = \sup_{\lambda\in\Lambda} \mathsf{d}_\lambda(x,y) = \lim_{\lambda\in\Lambda} \mathsf{d}_\lambda(x,y) \quad \textit{for every } x, y \in X. \tag{2.1.23b}$$

Conversely, suppose that $(\mathsf{d}_i)_{i\in I}$ is a directed family of real functions on $X \times X$ satisfying

$$\mathsf{d}_i : X \times X \to [0,+\infty) \quad \textit{is a bounded and continuous semidistance for every } i \in I, \tag{2.1.24a}$$

$$i \preceq j \quad \Rightarrow \quad \mathsf{d}_i \le \mathsf{d}_j, \tag{2.1.24b}$$

$$x_j \xrightarrow{\tau} x \quad \Leftrightarrow \quad \lim_{j\in J} \mathsf{d}_i(x_j, x) = 0 \quad \textit{for every } i \in I, \tag{2.1.24c}$$

$$\mathsf{d}(x,y) = \sup_{i\in I} \mathsf{d}_i(x,y) = \lim_{i\in I} \mathsf{d}_i(x,y) \quad \textit{for every } x, y \in X, \tag{2.1.24d}$$

then (X,τ,d) is an extended metric-topological space.

Proof Equations (2.1.23a) and (2.1.23b) are immediate consequence of the Definition 2.1.3. In order to prove the second statement, we simply observe that the collection of functions $\mathcal{F} := \{\mathsf{d}_i(y, \cdot) : i \in I,\ y \in X\}$ is included in $\mathrm{Lip}_{b,1}(X, \tau, \mathsf{d})$ and generates the topology τ thanks to (2.1.24c). A fortiori, $\mathrm{Lip}_{b,1}(X, \tau, \mathsf{d})$ satisfies conditions (X1) and (X2) of Definition 2.1.3. □

We will often use the following simple and useful property involving a directed family of semidistances $(\mathsf{d}_i)_{i\in I}$ satisfying (2.1.24a,b,c,d): whenever $\mathrm{i} : J \to I$ is a subnet and x_j, y_j, $j \in J$, are τ-converging to x, y respectively, we have

$$\liminf_{j\in J} \mathsf{d}_{\mathrm{i}(j)}(x_j, y_j) \geq \mathsf{d}(x, y). \tag{2.1.25}$$

It follows easily by the continuity of d_i and (2.1.24b), since for every $i \in I$

$$\liminf_{j\in J} \mathsf{d}_{\mathrm{i}(j)}(x_j, y_j) \geq \liminf_{j\in J} \mathsf{d}_i(x_j, y_j) \geq \mathsf{d}_i(x, y);$$

(2.1.25) then follows by taking the supremum w.r.t. $i \in I$.

Remark 2.1.5 Notice that if K is a Souslin subset of X (in particular $K = X$ if (X, τ) is Souslin) then $K \times K$ is Souslin as well, so that by Lemma A.4(b) there exists a countable collection $\mathcal{F} = (f_n)_{n\in\mathbb{N}} \subset \mathrm{Lip}_{b,1}(X, \tau, \mathsf{d})$ such that

$$\mathsf{d}(x, y) = \sup_{n\in\mathbb{N}} |f_n(x) - f_n(y)| \quad \text{for every } x, y \in K. \tag{2.1.26}$$

If τ' is the initial topology generated by $\mathcal{F}$, (K, τ', d) is an e.m.t. space whose topology τ' is coarser than τ. τ' is also metrizable and separable: it is sufficient to choose an increasing 1-Lipschitz homeomorphism $\vartheta : \mathbb{R} \to]0, 1/12[$ and setting $f_n' := \vartheta \circ f_n$; the family $(f_n')_{n\in\mathbb{N}}$ induces the same topology τ', it separates the points of K, and the distance

$$\mathsf{d}'(x, y) := \sum_{n=1}^{\infty} 2^{-n} |f_n'(x) - f_n'(y)| \tag{2.1.27}$$

is a bounded τ-continuous semidistance dominated by d whose restriction to $K \times K$ is a distance inducing the topology τ'. If K is also compact, than τ coincides with the topology induced by d'.

Let us recap a useful property discussed in the previous Remark.

Definition 2.1.6 (Auxiliary Topologies) Let (X, τ, d) be an e.m.t. space. We say that τ' is an *auxiliary* topology if there exist a countable collection $\mathcal{F} = (f_n)_{n\in\mathbb{N}} \subset \mathrm{Lip}_b(X, \tau, \mathsf{d})$ such that τ' is generated by $\mathcal{F}$ and

$$\mathsf{d}(x, y) = \sup_{n\in\mathbb{N}} |f_n(x) - f_n(y)|. \tag{2.1.28}$$

Equivalently

(A1) τ' is coarser than τ,
(A2) τ' is separable and metrizable by a bounded τ-continuous distance $\mathsf{d}' \le \mathsf{d}$,
(A3) there exists a sequence $f_n \in \mathrm{Lip}_b(X, \tau', \mathsf{d})$ such that (2.1.28) holds.

In particular, (X, τ', d) is an e.m.t. space.

If τ' is generated by a countable collection $\mathcal{F} \subset \mathrm{Lip}_b(X, \tau, \mathsf{d})$ satisfying (2.1.28) then properties (A1,2,3) obviously hold by the discussion of Remark 2.1.5. Conversely, if τ' satisfies (A1,2,3) then one can consider the countable collection $\mathcal{F}$ resulting by the union of $(f_n)_{n\in\mathbb{N}}$ given in (A3) and the set $\{\mathsf{d}'(x_n, \cdot)\}_{n\in\mathbb{N}}$ where $(x_n)_{n\in\mathbb{N}}$ is a τ' dense subset of X and d' is given by (A2). It is clear that τ' is the initial topology of $\mathcal{F}$ and (2.1.28) holds.

By setting

$$\mathsf{d}_n(x, y) := \sup_{1\le k\le n} |f_k(x) - f_k(y)|$$

one can easily see that (A3) is in fact equivalent to

(A3') There exists an increasing sequence of τ' continuous and bounded (semi) distances $(\mathsf{d}_n)_{n\in\mathbb{N}}$ such that

$$\mathsf{d}(x, y) = \sup_{n\in\mathbb{N}} \mathsf{d}_n(x, y) = \lim_{n\to\infty} \mathsf{d}_n(x, y) \quad \text{for every } x, y \in X. \tag{2.28'}$$

It is also possible to assume $\mathsf{d}_n \ge \mathsf{d}'$ for every $n \in \mathbb{N}$.

As a consequence of Remark 2.1.5 we have:

Corollary 2.1.7 (Auxiliary Topologies for Souslin e.m.t. Spaces) *If (X, τ, d) is a Souslin e.m.t. space (i.e. (X, τ) is Souslin) then it admits an auxiliary topology τ' according to Definition 2.1.6.*

Notice that if τ' is an auxiliary topology of a Souslin e.m.t. space (X, τ, d), (X, τ') is Souslin as well. If $\mathfrak{m}$ is a Radon measure in (X, τ) then it is Radon also w.r.t. τ'. An important consequence of the existence of an auxiliary topology is the following fact:

Lemma 2.1.8 *If (X, τ, d) admits an auxiliary topology τ' then every τ-compact set $K \subset X$ is a Polish space (with the relative topology).*

Proof It is sufficient to note that τ and τ' induces the same topology on K and that τ' is metrizable and separable. □

2.1.3 Examples

Example 2.1.9 (Complete and Separable Metric Spaces) The most important and common example is provided by a complete and separable metric space (X, d); in this case, the canonical choice of τ is the (Polish) topology induced by d. Any positive and finite Borel measure $\mathfrak{m}$ on (X, d) is a Radon measure so that $(X, \mathsf{d}, \mathfrak{m})$ is a Polish metric measure space. This case covers the Euclidean spaces $\mathbb{R}^d$, the complete Riemannian or Finsler manifolds, the separable Banach spaces and their closed subsets.

In some situation, however, when (X, d) is not separable or d takes the value $+\infty$, it could be useful to distinguish between the topological and the metric aspects. This will particularly important when a measure will be involved, since the Radon property with respect to a coarser topology is less restrictive.

Example 2.1.10 (Dual of a Banach Space) A typical example is provided by the dual $X = B'$ of a separable Banach space: in this case the distance d is induced by the dual norm $\|\cdot\|_{B'}$ of B' (which may not be separable) and the topology τ is the weak* topology of B', which is Lusin [56, Corollary 1, p. 115]. All the functions of the form $f(x; y, v, \mathrm{t}) := \mathrm{t}(\langle x - y, v\rangle)$ where $y \in B'$, $\mathrm{t} : \mathbb{R} \to \mathbb{R}$ is a bounded 1-Lipschitz map and $v \in B$ with $\|v\|_B \le 1$ clearly belong to $\mathrm{Lip}_b(X, \tau, \mathsf{d})$ and are sufficient to recover the distance d since

$$\begin{aligned}\mathsf{d}(x, y) = \|x - y\|_{B'} &= \sup_{\|v\|_B \le 1} \langle x - y, v\rangle \\ &= \sup_{\|v\| \le 1, M>0} f(x; y, v, \mathrm{t}_M), \quad \mathrm{t}_M(r) := 0 \vee r \wedge M,\end{aligned}$$

e.g. by chosing $M > 2\|x - y\|$.

A slight modification of the previous setting leads to a somehow universal model: we will see in Sect. 2.1.7 that every e.m.t. space can be isometrically and continuously embedded in such a framework and every metric Sobolev space has an isomorphic representation in this setting (see Corollary 5.3.16).

Example 2.1.11 Let X be weakly* compact subset of a dual Banach space B' endowed with the weak* topology τ and a Radon measure $\mathfrak{m}$. We select a strongly closed and symmetric convex set $L \subset B$ containing 0 and separating the points of B' and we set

$$\theta(z) := \sup_{f \in L} \langle z, f\rangle, \quad \mathsf{d}(x, y) := \theta(x - y). \tag{2.1.29}$$

It is immediate to check that (X, τ, d) is an e.m.t. space. Notice that θ is 1-homogeneous and convex, therefore it is an "extended" norm (possibly assuming the value $+\infty$), so that d is translation invariant. The previous Example 2.1.10 correspond to the case when L is the unit ball of B.

Example 2.1.12 (Abstract Wiener Spaces) Let $(X, \|\cdot\|_X)$ be a separable Banach space endowed with a Radon measure $\mathfrak{m}$ and let $(W, |\cdot|_W)$ be a reflexive Banach space (in particular an Hilbert space) densely and continuously included in X, so that there exists a constant $C > 0$ such that

$$\|h\|_X \le C|h|_W \quad \text{for every } h \in W. \tag{2.1.30}$$

We call τ the Polish topology of X induced by the Banach norm and for every $x, y, z \in X$ we set

$$\phi(z) := \begin{cases} |z|_W & \text{if } z \in W, \\ +\infty & \text{otherwise,} \end{cases} \qquad \mathsf{d}(x, y) := \phi(x - y) = \begin{cases} |x - y|_W & \text{if } x - y \in W, \\ +\infty & \text{otherwise,} \end{cases} \tag{2.1.31}$$

The functional $z \mapsto \phi(z)$ is 1-homogenous, convex and lower semicontinuous in X (thanks to the reflexivity of W) so that setting

$$L := \Big\{ f \in X' : \langle f, z \rangle \le |z|_B \text{ for every } z \in B \Big\},$$

Fenchel duality yields

$$\phi(z) = \sup_{f \in L} \langle f, z \rangle, \quad \mathsf{d}(x, y) = \sup_{f \in L} \langle f, x - y \rangle; \tag{2.1.32}$$

the same truncation trick of Example 2.1.10 shows that (2.1.17) is satisfied. On the other hand, the distance functions $x \mapsto \|x - z\|_X$, $z \in X$, induced by the norm in X belong to $\mathrm{Lip}_b(X, \tau, \mathsf{d})$ so that the first condition of Definition 2.1.3 is satisfied as well. This setting covers the important case of an abstract Wiener space, when $\mathfrak{m}$ is a Gaussian measure in X and W is the Cameron-Martin space, see e.g. [17].

Example 2.1.13 Let $X := \mathbb{R}^d$ and let $h : X \times \mathbb{R}^d \to [0, +\infty]$ be a lower semicontinuous function such that for every $x \in X$

$$h(x, \cdot) \quad \text{is 1-homogeneous and convex,} \quad h(x, v) > h(x, 0) = 0$$
$$\text{for every } x \in X, \ v \in \mathbb{R}^d \setminus \{0\}.$$

We can define the extended "Finsler" distance

$$\mathsf{d}(x_0, x_1) := \inf \Big\{ \int_0^1 h(\mathrm{x}(t), \mathrm{x}'(t))\, \mathrm{d}t : \mathrm{x} \in \mathrm{Lip}([0, 1], \mathbb{R}^d), \quad \mathrm{x}(i) = x_i, \ i = 0, 1 \Big\} \tag{2.1.33}$$

with the convention that $\mathsf{d}(x_0, x_1) = +\infty$ if there is no Lipschitz curve connecting x_0 to x_1 with a finite cost. When there exist constants $C_0, C_1 > 0$ such that

$$C_0|v| \le h(x, v) \le C_1|v| \quad \text{for every } x, v \in \mathbb{R}^d, \tag{2.1.34}$$

d is the "Finsler" distance induced by the family of norms $\big(h(x, \cdot)\big)_{x\in\mathbb{R}^d}$, inducing the usual topology of $\mathbb{R}^d$.

When

$$h(x, v) = \begin{cases} |v| & \text{if } x \in X_0, \\ +\infty & \text{if } x \in \mathbb{R}^d \setminus X_0,\ v \neq 0, \end{cases} \qquad X_0 \quad \text{is a closed subset of } \mathbb{R}^d, \tag{2.1.35}$$

then d is the "geodesic extended distance" induced by the Euclidean tensor on X_0. When h is expressed in terms of a smooth family of bounded vector fields $(X_j)_{j=1}^J$, $X_j : \mathbb{R}^d \to \mathbb{R}^d$, by the formula

$$h^2(x, v) := \inf\Big\{\sum_{j=1}^J u_j^2 : \sum_{j=1}^J u_j X_j(x) = v\Big\} \tag{2.1.36}$$

we obtain the Carnot-Caratheodory distance induced by the vector fields X_j. In all these cases, we can approximate h by its Yosida regularization:

$$h_\varepsilon^2(x, v) := \inf_{w\in\mathbb{R}^d} h^2(x, w) + \frac{1}{2\varepsilon}|w - v|^2 \quad x, v \in \mathbb{R}^d,\ \varepsilon > 0, \tag{2.1.37}$$

which satisfy

$$0 < h_\varepsilon^2(x, v) \le \frac{1}{2\varepsilon}|v|^2, \quad \lim_{\varepsilon\downarrow 0} h_\varepsilon(x, v) = h(x, v) \quad \text{for every } v \in \mathbb{R}^d \setminus \{0\}. \tag{2.1.38}$$

If we define the Finsler distance d_ε as in (2.1.33) in terms of h_ε we can easily see that

$$0 < \mathsf{d}_\varepsilon^2(x, y) \le \frac{1}{2\varepsilon}|x - y|^2 \quad \lim_{\varepsilon\downarrow 0} \mathsf{d}_\varepsilon(x, y) = \mathsf{d}(x, y) \quad \text{for every } x, y \in \mathbb{R}^d,\ x \neq y. \tag{2.1.39}$$

If τ is the usual Euclidean topology, we obtain that $(\mathbb{R}^d, \tau, \mathsf{d})$ is an extended metric-topological space.

2.1.4 The Kantorovich-Rubinstein Distance

Let (X, τ, d) be an extended metric-topological space. We want to lift the same structure to the space of Radon probability measures $\mathcal{P}(X)$. We introduce the main definitions for pair of measures $\mu_0, \mu_1 \in \mathcal{M}_+(X)$ with the same mass $\mu_0(X) = \mu_1(X)$.

We denote by $\Gamma(\mu_0, \mu_1)$ the collection of plans $\boldsymbol{\mu} \in \mathcal{M}_+(X \times X)$ whose marginals are μ_0 and μ_1 respectively:

$$\Gamma(\mu_0, \mu_1) := \Big\{\boldsymbol{\mu} \in \mathcal{M}_+(X \times X) : \pi^i_\sharp \boldsymbol{\mu} = \mu_i\Big\}, \quad \pi^i(x_0, x_1) := x_i. \tag{2.1.40}$$

It is not difficult to check that $\Gamma(\mu_0, \mu_1)$ is a nonempty (it always contains $\mu_0^{-1}(X)\, \mu_0 \otimes \mu_1$) and compact subset of $\mathcal{M}_+(X \times X)$.

Let $\delta : X \times X \to [0, +\infty]$ be a lower semicontinuous extended semi distance. The Kantorovich formulation of the optimal transport problem with cost δ induces the celebrated Kantorovich-Rubinstein (extended, semi-)distance K_δ in $\mathcal{P}(X)$ [63, Chap. 7]

$$\mathsf{K}_\delta(\mu_0, \mu_1) := \inf\Big\{ \int_{X \times X} \delta(x_0, x_1)\,\mathrm{d}\boldsymbol{\mu}(x_0, x_1) : \boldsymbol{\mu} \in \Gamma(\mu_0, \mu_1)\Big\}. \tag{2.1.41}$$

Proposition 2.1.14 *Let $\mu_0, \mu_1 \in \mathcal{M}_+(X)$ with the same mass.*

(a) *If $\mathsf{K}_\delta(\mu_0, \mu_1)$ is finite then the infimum in* (2.1.41) *is attained. In particular, this holds if δ is bounded.*

(b) [Kantorovich-Rubinstein duality] *If δ is a bounded continuous (semi)-distance in (X, τ) then K_δ is a bounded continuous (semi)-distance in $\mathcal{P}(X)$ and*

$$\mathsf{K}_\delta(\mu_0, \mu_1) = \sup\Big\{ \int \phi_0\,\mathrm{d}\mu_0 - \int \phi_1\,\mathrm{d}\mu_1 : \phi_i \in \mathrm{C}_b(X, \tau), \\ \phi_0(x_0) - \phi_1(x_1) \le \delta(x_0, x_1) \quad \textit{for every } x_o, x_1 \in X\Big\} \tag{2.1.42}$$

$$= \sup\Big\{ \int \phi\,\mathrm{d}(\mu_0 - \mu_1) : \phi \in \mathrm{Lip}_{b,1}(X, \tau, \delta)\Big\} \tag{2.1.43}$$

(c) *If $(\mathsf{d}_i)_{i \in I}$ is a directed collection of bounded continuous semidistances satisfying $\lim_{i \in I} \mathsf{d}_i = \mathsf{d}$ then*

$$\mathsf{K}_\mathsf{d}(\mu_0, \mu_1) = \lim_{i \in I} \mathsf{K}_{\mathsf{d}_i}(\mu_0, \mu_1). \tag{2.1.44}$$

(d) *If* (X, τ, d) *is an extended metric-topological space*

$$\mathsf{K}_{\mathsf{d}}(\mu_0, \mu_1) = \sup\Big\{ \int \phi_0 \,\mathrm{d}\mu_0 - \int \phi_1 \,\mathrm{d}\mu_1 : \phi_i \in C_b(X, \tau),$$

$$\phi_0(x_0) - \phi_1(x_1) \le \mathsf{d}(x_0, x_1) \quad \textit{for every } x_0, x_1 \in X \Big\} \tag{2.1.45}$$

$$= \sup\Big\{ \int \phi \,\mathrm{d}(\mu_0 - \mu_1) : \phi \in \mathrm{Lip}_{b,1}(X, \tau, \mathsf{d}) \Big\} \tag{2.1.46}$$

Proof

(a) follows by the lower semicontinuity of δ and the compactness of Γ.
(b) we refer to [63, Chap. 7].
(c) follows by the property

$$\liminf_{j\in J} \int \mathsf{d}_{i(j)} \,\mathrm{d}\boldsymbol{\pi}_j \ge \int \mathsf{d} \,\mathrm{d}\boldsymbol{\pi} \tag{2.1.47}$$

whenever $(\boldsymbol{\pi}_j)_{j\in J}$ is a net in $\Gamma(\mu_0, \mu_1)$ converging weakly to $\boldsymbol{\pi}$ and $j \mapsto i(j)$ is a subnet in I. See [3, Theorem 5.1].
(d) is an immediate consequence of (2.1.44) and Claim (b), which yields that K_{d} is less or equal than the two expression in the right-hand side of (2.1.45) and (2.1.46). The converse inequality is obvious.

□

Remark 2.1.15 Thanks to the previous proposition, it would not be difficult to check that $(\mathcal{P}(X), \tau_{\mathcal{P}}, \mathsf{K}_{\mathsf{d}})$ is an extended metric-topological space as well.

2.1.5 The Asymptotic Lipschitz Constant

Whenever δ is an extended, τ-lower semicontinuous semidistance, and $f : X \to \mathbb{R}$, we set

$$\mathrm{lip}_\delta f(x) := \lim_{U\in\mathscr{U}_x} \mathrm{Lip}(f, U, \delta) = \inf_{U\in\mathscr{U}_x} \mathrm{Lip}(f, U, \delta) \quad x \in X; \tag{2.1.48}$$

recall that $\mathscr{U}_x$ is the directed set of all the τ-neighborhoods of x. Notice that $\mathrm{Lip}(f, \{x\}) = 0$ and therefore $\mathrm{lip}\, f(x) = 0$ if x is an isolated point of X. We will often omit the index δ when $\delta = \mathsf{d}$. When δ is a distance, we can also define lip_δ as

$$\mathrm{lip}_\delta f(x) = \limsup_{\substack{y,z\to x\\ y\neq z}} \frac{|f(y) - f(z)|}{\delta(y, z)}; \tag{2.1.49}$$

in particular,

$$\mathrm{lip}_\delta f(x) \geq |\mathrm{D}_\delta f|(x) := \limsup_{y\to x} \frac{|f(y)-f(x)|}{\delta(x,y)}. \tag{2.1.50}$$

It is not difficult to check that $x \mapsto \mathrm{lip}_\delta f(x)$ is a τ-upper semicontinuous map and f is locally δ-Lipschitz in X iff $\mathrm{lip}_\delta f(x) < \infty$ for every $x \in X$. When (X, δ) is a length space, $\mathrm{lip}_\delta f$ coincides with the upper semicontinuous envelope of the local Lipschitz constant (2.1.50).

We collect in the next useful lemma the basic calculus properties of $\mathrm{lip}_\delta f$.

Lemma 2.1.16 *For every $f, g, \chi \in \mathrm{C}_b(X)$ with $\chi(X) \subset [0,1]$ we have*

$$\mathrm{lip}_\delta(\alpha f + \beta g) \leq |\alpha|\, \mathrm{lip}_\delta f + |\beta|\, \mathrm{lip}_\delta g \quad \textit{for every } \alpha, \beta \in \mathbb{R}, \tag{2.1.51a}$$

$$\mathrm{lip}_\delta(fg) \leq |f|\, \mathrm{lip}_\delta g + |g|\, \mathrm{lip}_\delta f, \tag{2.1.51b}$$

$$\mathrm{lip}_\delta((1-\chi)f + \chi g) \leq (1-\chi)\, \mathrm{lip}_\delta f + \chi\, \mathrm{lip}_\delta g + \mathrm{lip}_\delta \chi |f-g|. \tag{2.1.51c}$$

Moreover, whenever $\phi \in \mathrm{C}^1(\mathbb{R})$

$$\mathrm{lip}_\delta(\phi \circ f) = |\phi' \circ f|\, \mathrm{lip}_\delta f \tag{2.1.51d}$$

and for every convex and nondecreasing function $\psi : [0,\infty) \to \mathbb{R}$ and every map $\zeta \in \mathrm{C}^1(\mathbb{R})$ with $0 \leq \zeta' \leq 1$, the transformation

$$\tilde f := f + \zeta(g-f), \quad \tilde g := g + \zeta(f-g)$$

satisfies

$$\psi(\mathrm{lip}_\delta \tilde f) + \psi(\mathrm{lip}_\delta \tilde g) \leq \psi(\mathrm{lip}_\delta f) + \psi(\mathrm{lip}_\delta g). \tag{2.1.51e}$$

Proof Equation (2.1.51a) follows by the obvious inequalities

$$\mathrm{Lip}(\alpha f + \beta g, U, \delta) \leq |\alpha|\, \mathrm{Lip}(f, U, \delta) + |\beta|\, \mathrm{Lip}(g, U, \delta)$$

for every subset $U \subset X$. Similarly, for every $y, z \in U$

$$\begin{aligned} |f(y)g(y) - f(z)g(z)| &\leq |(f(y)-f(z))g(y)| + |(g(y)-g(z))f(z)| \\ &\leq \Big(\mathrm{Lip}(f, U, \delta) \sup_U |g| + \mathrm{Lip}(g, U, \delta) \sup_U |f|\Big)\delta(z,y) \end{aligned}$$

and we obtain (2.1.51b) passing to the limit w.r.t. $U \in \mathscr{U}_x$. Setting $\tilde{\chi} := 1 - \chi$, (2.1.51c) follows by

$$
\begin{aligned}
&\big|\tilde{\chi}(y)f(y) + \chi(y)g(y) - \big(\tilde{\chi}(z)f(z) + \chi(z)g(z)\big)\big| \\
&\qquad \le \big|\tilde{\chi}(y)(f(y) - f(z))\big| + \big|\chi(y)(g(y) - g(z))\big| \\
&\qquad \quad + \big|(\chi(y) - \chi(z))(g(z) - f(z))\big| \\
&\qquad \le \Big(\sup_U \tilde{\chi}\ \mathrm{Lip}(f, U, \delta) + \sup_U \chi\ \mathrm{Lip}(g, U, \delta) \\
&\qquad \quad + \sup_U |f - g|\,\mathrm{Lip}(\chi, U, \delta)\Big)\,\delta(y, z)
\end{aligned}
$$

and passing to the limit w.r.t. $U \in \mathscr{U}_x$.

Concerning (2.1.51d), for every $y, z \in U$ we get

$$
|\phi(f(y)) - \phi(f(z))| \le \mathrm{Lip}(\phi, f(U))\ \mathrm{Lip}(f, U, \delta)\delta(y, z)
$$

which easily yields $\mathrm{lip}_\delta\,\big(\phi \circ f\big)(x) \le |\phi'(f(x))|\,\mathrm{lip}_\delta\, f(x)$. If $\phi'(f(x)) \neq 0$, we can find a C^1 function $\psi : \mathbb{R} \to \mathbb{R}$ such that $\psi(\phi(r)) = r$ in a neighborhood of $f(x)$, so that the same property yields $\mathrm{lip}_\delta\, f(x) \le \frac{1}{|\phi'(f(x))|}\,\mathrm{lip}_\delta\,\big(f \circ \phi\big)(x)$ and the identity in (2.1.51d).

Let us eventually consider (2.1.51e). As usual, we consider arbitrary points $y, z \in U$, $U \in \mathscr{U}_x$ obtaining

$$
\begin{aligned}
|\tilde{f}(y) - \tilde{f}(z)| &= |f(y) - f(z) + \zeta(g(y) - f(y)) - \zeta(g(z) - f(z))| \\
&= |f(y) - f(z) + \alpha\big((g(y) - g(z) - (f(y) - f(z))| \\
&= |(1 - \alpha)(f(y) - f(z)) + \alpha(g(y) - g(z))| \\
&\le \Big((1 - \alpha)\,\mathrm{Lip}(f, U, \delta) + \alpha\,\mathrm{Lip}(g, U, \delta)\Big)\delta(y, z)
\end{aligned}
$$

for some $\alpha = \alpha_{y,z} = \zeta'(\theta_{y,z}) \in [0, 1]$, where $\theta_{y,z}$ is a convex combination of $g(y) - f(y)$ and $g(z) - f(z)$. Passing to the limit w.r.t. U and observing that $\alpha \to \zeta'(g(x) - f(x))$ we get

$$
\mathrm{lip}_\delta\, \tilde{f}(x) \le (1 - \zeta'(g(x) - f(x)))\,\mathrm{lip}_\delta\, f(x) + \zeta'(g(x) - f(x))\,\mathrm{lip}_\delta\, g(x).
$$

A similar argument yields

$$
\mathrm{lip}_\delta\, \tilde{g}(x) \le (1 - \zeta'(f(x) - g(x)))\,\mathrm{lip}_\delta\, g(x) + \zeta'(f(x) - g(x))\,\mathrm{lip}_\delta\, f(x).
$$

Since ψ is convex and nondecreasing, we obtain (2.1.51e). □

2.1.6 Compatible Algebra of Functions

We have seen in Sect. 2.1.2 the important role played by the algebra of function $\mathrm{Lip}_b(X, \tau, \mathsf{d})$. In many situations it could be useful to consider smaller subalgebras which are however sufficiently rich to recover the metric properties of an extended metric topological space (X, τ, d).

Definition 2.1.17 (Compatible Algebras of Lipschitz Functions) Let $\mathscr{A}$ be a unital subalgebra of $\mathrm{Lip}_b(X, \tau, \mathsf{d})$ and let us set $\mathscr{A}_\kappa := \mathscr{A} \cap \mathrm{Lip}_{b,\kappa}(X, \tau, \mathsf{d})$.

We say that $\mathscr{A}$ is *compatible* with the metric-topological structure (X, τ, d) if

$$\mathsf{d}(x, y) = \sup_{f \in \mathscr{A}_1} |f(x) - f(y)| \quad \text{for every } x, y \in X. \tag{2.1.52}$$

In particular, $\mathscr{A}$ separates the points of X.
We say that $\mathscr{A}$ is *adapted* to (X, τ, d) if $\mathscr{A}$ is compatible with (X, τ, d) and it generates the topology τ.

If we do not make a different explicit choice, we will always assume that an e.t.m.m. space $\mathbb{X}$ is endowed with the canonical algebra $\mathscr{A}(\mathbb{X}) := \mathrm{Lip}_b(X, \tau, \mathsf{d})$.

Remark 2.1.18 (Coarser Topologies and Countably Generated Algebras) Suppose that $\mathscr{A} \subset \mathrm{Lip}_b(X, \tau, \mathsf{d})$ is an algebra compatible with (X, τ, d) and let $\tau_{\mathscr{A}}$ be the initial topology generated by $\mathscr{A}$ (see Sect. A.2 in the Appendix). Then $(X, \tau_{\mathscr{A}}, \mathsf{d})$ is an e.m.t. space as well and $\mathscr{A}$ is adapted to $(X, \tau_{\mathscr{A}}, \mathsf{d})$; a Radon measure $\mathfrak{m} \in \mathcal{M}_+(X, \tau)$ is also Radon in $(X, \tau_{\mathscr{A}})$.
This property shows that there is some flexibility in the choice of the topology τ, as long as τ-continuous functions are sufficiently rich to generate the distance d. An interesting example occurs when (X, τ) is a Souslin space. By Remark 2.1.5 we can always find a countable collection $(f_n)_{n \in \mathbb{N}}$ of $\mathrm{Lip}_b(X, \tau, \mathsf{d})$ (or of a compatible algebra $\mathscr{A}$) satisfying (2.1.26). If we denote by $\mathscr{A}'$ the algebra generated by the functions f_n, $n \in \mathbb{N}$, we obtain a countably generated algebra and an auxiliary topology $\tau' = \tau_{\mathscr{A}'}$ according to Definition 2.1.6.

Examples

Example 2.1.19 (Cylindrical Functions in Banach Spaces and Their Dual) Let $(X, \|\cdot\|)$ be a Banach space (in particular the space $\mathbb{R}^d$ with any norm) endowed with its weak topology (or the dual of a Banach space B with the weak* topology) and let $\mathscr{A}$ be the set of smooth cylindrical functions: a function $f : X \to \mathbb{R}$ belongs to $\mathscr{A}$ if there exists $\psi \in \mathrm{C}^\infty(\mathbb{R}^d)$ with bounded derivatives of every order and d linear functionals $\mathsf{h}_1, \cdots, \mathsf{h}_d \in X'$ (resp. in B if the weak* topology is considered) such that

$$f(x) = \psi(\langle \mathsf{h}_1, x\rangle, \langle \mathsf{h}_2, x\rangle, \cdots, \langle \mathsf{h}_d, x\rangle). \tag{2.1.53}$$

It is not difficult to check that $\mathscr{A} \subset \mathrm{Lip}_b(X, \tau, \mathsf{d})$. In order to approximate the distance $\mathsf{d}(x, y) = \|x - y\|$ between two points in X we can argue as in Example 2.1.10 by choosing functions of the form $f(x) := \mathsf{t}_\varepsilon(\langle \mathsf{h}, x\rangle - \langle \mathsf{h}, y\rangle)$ where h belongs to the dual (resp. predual) unit ball of X' (resp. B) and $\mathsf{t}_\varepsilon(r)$ is a smooth regularization of $\mathsf{t}(r) := 0 \vee r \wedge M$ (with $M > 0$) coinciding with r in the interval $[\varepsilon, M]$ (it will be enough to choose $M > 2\|x - y\|$). In the case of Example 2.1.11 it is sufficient to choose h in the convex set L.

The same approach can be adapted to the "Wiener" construction of Example 2.1.12: in this case one can use linear functionals in X'.

In the case X is separable (resp. $X = B'$ and B is separable) any Borel (resp. weakly* Borel) measure is Radon.

Example 2.1.20 A compatible algebra is provided by

$$\mathrm{Lip}_b(X, \tau, (\mathsf{d}_i)) := \{f \in \mathrm{C}_b(X, \tau) : \exists i \in I : \mathrm{Lip}(f, \mathsf{d}_i) < \infty\}, \tag{2.1.54}$$

whenever $(\mathsf{d}_i)_{i\in I}$ is a directed family satisfying (2.1.24a,b,c,d). One can also consider the smaller unital algebra of functions generated by the collection of distance functions

$$\Big\{\mathsf{d}_i(\cdot, y) : y \in X,\ i \in I\Big\}. \tag{2.1.55}$$

Example 2.1.21 (Cartesian Products) Let us consider two e.t.m. spaces (X', τ', d') and $(X'', \tau'', \mathsf{d}'')$ with two compatible algebras $\mathscr{A}'$, $\mathscr{A}''$. For every $p \in [1, +\infty]$ we can consider the product space (X, τ, d_p) where $X = X' \times X''$, τ is the product topology of τ' and τ'', and

$$\begin{aligned}
\mathsf{d}_p((x', x''), (y', y'')) &:= \Big(\mathsf{d}'(x', y')^p + \mathsf{d}''(x'', y'')^p\Big)^{1/p} \quad \text{if } p < \infty,\\
\mathsf{d}_\infty((x', x''), (y', y'')) &:= \max\Big(\mathsf{d}'(x', y'), \mathsf{d}''(x'', y'')\Big).
\end{aligned} \tag{2.1.56}$$

The algebra $\mathscr{A} = \mathscr{A}' \otimes \mathscr{A}''$ generated by functions $f' \in \mathscr{A}'$ and $f'' \in \mathscr{A}''$ (an element of $\mathscr{A}$ is a linear combination of functions of the form $f(x', x'') := f'(x')f''(x'')$) is compatible with (X, τ, d_p). In order to prove that (2.1.52) holds, let q be the conjugate exponent of p and let us introduce the convex subset of $\mathbb{R}^2$ $C_q := \{(\alpha, \beta) \in \mathbb{R}^2 : \alpha^q + \beta^q \le 1\}$ (with obvious modification when $q = \infty$). For every pair of points (x', x''), (y', y'') in X we can find $(\alpha, \beta) \in C_q$ such that

$$\mathsf{d}_p((x', x''), (y', y'')) = \alpha\mathsf{d}'(x', y') + \beta\mathsf{d}''(x'', y'').$$

It is easy to check that for every $f' \in \mathscr{A}_1'$ and $f'' \in \mathscr{A}_1''$ the function $f(z', z'') := \alpha f'(z') + \beta f''(z'')$ belongs to $\mathscr{A}_1$. Since $\mathscr{A}'$ and $\mathscr{A}''$ are compatible in the respective spaces, we then get

$$\begin{aligned}\mathsf{d}_p((x',x''),(y',y'')) &= \sup_{f'\in\mathscr{A}'_1,\, f''\in\mathscr{A}''_1} \alpha(f'(x')-f'(y'))+\beta(f''(x'')-f''(y''))\\ &= \sup_{f'\in\mathscr{A}'_1,\, f''\in\mathscr{A}''_1} \alpha f'(x')+\beta f''(x'')-\big(\alpha f'(y')+\beta f''(y'')\big).\end{aligned}$$

Remark 2.1.22 The previous Example 2.1.21 shows in particular that the cartesian product of two e.t.m. spaces is also an e.t.m. space, a property that one can also directly check by using the approximating semidistance functions $(\mathsf{d}'_i)_{i\in I}$, $(\mathsf{d}''_j)_{j\in J}$.

In order to deal with functions in $\mathscr{A}$ it will be useful to have suitable polynomial approximations of the usual truncation maps.

Lemma 2.1.23 (Polynomial Approximation) *Let $c > 0$, $a_i, b_i \in \mathbb{R}$, and $\phi : \mathbb{R} \to \mathbb{R}$ be a Lipschitz function satisfying*

$$a_0 \le \phi \le b_0 \quad \text{in } [-c,c], \qquad a_1 \le \phi' \le b_1 \quad \mathscr{L}^1\text{-a.e. in } [-c,c]. \tag{2.1.57}$$

There exists a sequence $(P_n)_{n\in\mathbb{N}}$ of polynomials such that

$$\lim_{n\to\infty} \sup_{[-c,c]} |P_n-\phi| = 0, \qquad a_0 \le P \le b_0,\ a_1 \le P' \le b_1 \quad \text{in } [-c,c], \tag{2.1.58}$$

and

$$\lim_{n\to\infty} |P'_n(r)-\phi'(r)| = 0 \quad \text{for every } r\in[-c,c] \text{ where } \phi \text{ is differentiable.} \tag{2.1.59}$$

If moreover $\phi \in \mathrm{C}^1([-c,c])$ we also have

$$\lim_{n\to\infty} \sup_{r\in[-c,c]} |P'_n(r)-\phi'(r)| = 0. \tag{2.1.60}$$

Proof In order to prove the first statement of the lemma, it is sufficient to use the Bernstein polynomials of degree $2n$ on the interval $[-c,c]$ given by the formula

$$P_n(r) := \frac{1}{(2c)^n}\sum_{k=-n}^{n} \phi(k/n)\binom{2n}{k+n}(r+c)^{n+k}(c-r)^{n-k} \tag{2.1.61}$$

recalling that P_n uniformly converge to ϕ in $[-c,c]$ as $n\to\infty$ and that formula (2.1.61) preserves the bounds on ϕ and ϕ' [49, Sect. 1.7]. □

Applying the previous Lemma to the the function $\phi(r) := \alpha \vee r \wedge \beta$ (with $a_0 = \alpha$, $b_0 = \beta$, $a_1 = 0$, $a_2 = 1$), we immediately get the following property.

Corollary 2.1.24 *For every interval* $[-c, c]$, $c > 0$, $\alpha, \beta \in \mathbb{R}$ *with* $\alpha < \beta$, *and every* $\varepsilon > 0$ *there exists a polynomial* $P_\varepsilon = P_\varepsilon^{c,\alpha,\beta}$ *such that*

$$|P_\varepsilon(r) - \alpha \vee r \wedge \beta| \le \varepsilon, \quad \alpha \le P_\varepsilon(r) \le \beta, \quad 0 \le P_\varepsilon'(r) \le 1 \quad \textit{for every } r \in [-c, c],$$

$$\lim_{\varepsilon \downarrow 0} P_\varepsilon'(r) = \begin{cases} 1 & \textit{if } \alpha < r < \beta \\ 0 & \textit{if } r < \alpha \textit{ or } r > \beta. \end{cases} \tag{2.1.62}$$

If $\alpha = -\beta$ *we can also find an odd* P_ε, *thus satisfying* $P_\varepsilon(0) = 0$.

A more refined argument yields:

Corollary 2.1.25 *For every interval* $[-c, c] \subset \mathbb{R}$ *and every* $\varepsilon > 0$, *there exists a polynomial* $Q_\varepsilon = Q_\varepsilon^c : \mathbb{R} \times \mathbb{R} \to \mathbb{R}$ *such that*

$$r \wedge s \le Q_\varepsilon(r, s) \le r \vee s, \quad |Q_\varepsilon(r, s) - r \vee s| \le \varepsilon \quad \textit{for every } r, s \in [-c, c], \tag{2.1.63}$$

$$0 \le \partial_r Q_\varepsilon \le 1, \quad 0 \le \partial_s Q_\varepsilon \le 1 \quad \textit{in } [-c, c] \times [-c, c], \tag{2.1.64}$$

$$|Q_\varepsilon(r_2, s_2) - Q_\varepsilon(r_1, s_1)| \le \max\big(|r_2 - r_1|, |s_2 - s_1|\big) \quad \textit{for every } r_i, s_i \in [-c, c]. \tag{2.1.65}$$

Proof We apply Lemma 2.1.23 to the function $\phi(r) := r_+$ in the interval $[-4c, 4c]$ (with $a_0 = a_1 = 0$, $b_0 = 4c$ and $b_1 = 1$) obtaining a polynomial P_ε such that

$$|P_\varepsilon(r) - r_+| \le \varepsilon, \quad 0 \le P_\varepsilon(r) \le 4c, \quad 0 \le P_\varepsilon'(r) \le 1 \quad \text{for every } r \in [-4c, 4c]. \tag{2.1.66}$$

We set $Q_\varepsilon(r, s) := r + P_\varepsilon(s - r) - P_\varepsilon(0)$. Notice that Q_ε is increasing w.r.t. r, s in $[-2c, 2c] \times [-2c, 2c]$ since

$$\partial_r Q_\varepsilon(r, s) = 1 - P_\varepsilon'(s - r) \ge 0, \quad \partial_s Q_\varepsilon(r, s) = P_\varepsilon'(s - r) \ge 0;$$

in particular

$$Q_\varepsilon(r, s) \ge Q_\varepsilon(r \wedge s, r \wedge s) = r \wedge s, \quad Q_\varepsilon(r, s) \le Q_\varepsilon(r \vee s, r \vee s) = r \vee s.$$

By construction, if $r, s \in [-c, c]$ then

$$|Q_\varepsilon(r, s) - r \vee s| = |r + P_\varepsilon(s - r) - (r + (s - r)_+)| = |P_\varepsilon(s - r) - (s - r)_+| \le \varepsilon.$$

Concerning the Lipschitz estimate, let us consider points $(r_1, s_1), (r_2, s_2) \in [-c, c]$. Up to inverting the order of the pairs, it is not restrictive to assume that $Q_\varepsilon(r_1, s_1) \ge Q_\varepsilon(r_2, s_2)$. Setting $r_- := r_1 \wedge r_2$, $r_+ := r_1 \vee r_2$, $s_- := s_1 \wedge s_2$, $s_+ := s_1 \vee s_2$,

$\bar{r} := (r_+ - r_-) \le 2c, \bar{s} = s_+ - s_- \le 2c, \bar{z} := \bar{r} \vee \bar{s} = \max(|r_2 - r_1|, |s_2 - s_1|) \le 2c$, the partial monotonicity of Q_ε yields

$$\begin{aligned}|Q_\varepsilon(r_1, s_1) - Q_\varepsilon(r_2, s_2)| &\le Q_\varepsilon(r_+, s_+) - Q_\varepsilon(r_-, s_-)\\ &\le Q_\varepsilon(r_- + \bar{z}, s_- + \bar{z}) - Q_\varepsilon(r_-, s_-)\\ &= \int_0^{\bar{z}} \Big(\partial_r Q_\varepsilon(r_- + z, s_- + z) + \partial_s Q_\varepsilon(r_- + z, s_- + z)\Big)\,\mathrm{d}z\\ &= \int_0^{\bar{z}} \Big(1 - P'_\varepsilon(s_- - r_-) + P'_\varepsilon(s_- - r_-)\Big)\,\mathrm{d}z = \bar{z}.\end{aligned}$$

□

The next result shows how to obtain good approximations of the maximum of a finite number of functions in $\mathscr{A}$.

Lemma 2.1.26 *Let $f^1, f^2, \cdots, f^M \in \mathscr{A}$ and let $f := \max(f^1, f^2, \cdots, f^M)$. Then for every $\varepsilon > 0$ there exists a sequence $f_\varepsilon \in \mathscr{A}$ such that*

$$\min_m f^m(x) \le f_\varepsilon(x) \le \max_m f^m(x) \quad \text{for every } x \in X, \quad \sup_X |f_\varepsilon - f| \le \varepsilon. \tag{2.1.67}$$

If moreover $\mathrm{Lip}(f^m, A, \delta) \le L$ *for* $1 \le m \le M$ *where* $A \subset X$ *and* δ *is an extended semidistance on* X*, then* $\mathrm{Lip}(f_\varepsilon, A, \delta) \le L$ *for every* $n \in \mathbb{N}$.

Proof We split the proof in two steps.

1. *The thesis of the Lemma holds for $M = 2$.* We set $c > 0$ so that $f^m(X) \subset [-c, c]$ and then we define $f_\varepsilon := Q_\varepsilon(f^1, f^2)$, where Q_ε has been provided by Corollary 2.1.25. Equation (2.1.67) follows immediately by (2.1.63). Equation (2.1.65) yields for every $x, y \in X$

$$\begin{aligned}|f_\varepsilon(x) - f_\varepsilon(y)| &= |Q_\varepsilon(f^1(x), f^2(x)) - Q_\varepsilon(f^1(y), f^2(y))|\\ &\le \max\big(|f^1(x) - f^1(y)|, |f^2(x) - f^2(y)|\big)\end{aligned}$$

so that the composition with Q_ε preserve the Lipschitz constant w.r.t. arbitrary sets and semidistances.

2. *The thesis of the Lemma holds for arbitrary $M \in \mathbb{N}$.* We argue by induction, assuming that the result is true for $M - 1$. We fix a constant c so that $f^m(X) \subset [-c, c]$ for $1 \le m \le M$. We thus find $h_{\varepsilon/2} \subset \mathscr{A}$ satisfying

$$\begin{aligned}&\min_{1 \le m \le M-1} f^m(x) \le h_{\varepsilon/2}(x)\\ &\qquad \le \max_{1 \le m \le M-1} f^m(x) \quad \text{for every } x \in X, \quad \sup_X |h_{\varepsilon/2} - \tilde{f}| \le \varepsilon/2,\end{aligned}$$

where $\tilde{f} := f^1 \vee \cdots \vee f^{M-1}$; in particular $h_{\varepsilon/2}(X) \subset [-c, c]$. We then set $f_\varepsilon := Q_{\varepsilon/2}(h_{\varepsilon/2}, f^M)$; clearly for every $x \in X$

$$\min_{1\le m\le M} f^m(x) \le h_{\varepsilon/2}(x) \wedge f^M(x) \le f_\varepsilon(x) \le h_{\varepsilon/2}(x) \vee f^M(x) \le \max_{1\le m\le M} f^m(x);$$

moreover

$$\begin{aligned}|f_\varepsilon - f| &\le |Q_{\varepsilon/2}(h_{\varepsilon/2}, f^M) - f|\\ &\le |Q_{\varepsilon/2}(h_{\varepsilon/2}, f^M) - Q_{\varepsilon/2}(\tilde{f}, f^M)| + |Q_{\varepsilon/2}(\tilde{f}, f^M) - \tilde{f} \vee f^M|\\ &\le |h_{\varepsilon/2} - \tilde{f}| + \varepsilon/2 \le \varepsilon/2 + \varepsilon/2 \le \varepsilon.\end{aligned}$$

□

We conclude this section by a simple density results that will be useful in the following.

Lemma 2.1.27 (Density of $\mathscr{A}$ in $L^p(X, \mathfrak{m})$) *Let $p \in [1, +\infty)$ and let I a closed (possibly unbounded) interval of $\mathbb{R}$. If $\mathscr{A}$ is a compatible sub-algebra of* $\mathrm{Lip}_b(X, \tau, \mathsf{d})$, *then for every $f \in \mathcal{L}^p(X, \mathfrak{m})$ with values in I there exists a sequence $f_n \in \mathscr{A}$ with values in I such that $\int_X |f - f_n|^p \,\mathrm{d}\mathfrak{m} \to 0$.*

Proof By standard approximation, it is not restrictive to assume that $I = [\alpha, \beta]$ for some $\alpha, \beta \in \mathbb{R}$; we set $\gamma := |\alpha| \vee |\beta|$. Since $\mathfrak{m}$ is Radon, every $\mathfrak{m}$-measurable function f is Lusin $\mathfrak{m}$-measurable: thus for every $\varepsilon > 0$ there exists a compact $K \subset X$ such that $f_{|K}$ is continuous and $\mathfrak{m}(X \setminus K) \le \varepsilon$.

Since $\mathscr{A}$ contains the constants and separates the points of K, the restriction of $\mathscr{A}$ to K is uniformly dense in $C_b(K, \tau)$ by Stone-Weierstrass Theorem: we thus find $\tilde{f}_\varepsilon \in \mathscr{A}$ such that $\sup_{x\in K} |f(x) - \tilde{f}_\varepsilon(x)| \le \varepsilon$. If $c := \sup_X |\tilde{f}_\varepsilon| \vee \gamma$, applying Corollary 2.1.24 we can find a polynomial P_ε satisfying (2.1.62), so that $f_\varepsilon := P_\varepsilon \circ \tilde{f}_\varepsilon$ belongs to $\mathscr{A}$, takes values in $[\alpha, \beta]$, and satisfies

$$|f_\varepsilon - f| \le |P_\varepsilon(\tilde{f}_\varepsilon) - P_\varepsilon(f)| + |P_\varepsilon(f) - f| \le 2\varepsilon \quad \text{in } K, \tag{2.1.68}$$

so that

$$\int_X |f_\varepsilon - f|^p \,\mathrm{d}\mathfrak{m} \le (2\varepsilon)^p \mathfrak{m}(X) + (\beta - \alpha)^p \varepsilon. \tag{2.1.69}$$

Choosing a sequence $f_n := f_{\varepsilon_n}$ corresponding to a vanishing sequence $\varepsilon_n \downarrow 0$ we conclude. □

2.1.7 Embedding and Compactification of Extended Metric-Measure Spaces

Let $\mathbb{X} = (X, \tau, \mathsf{d}, \mathfrak{m})$ and $\mathbb{X}' = (X', \tau', \mathsf{d}', \mathfrak{m}')$ be two extended metric measure spaces, endowed with compatible algebras $\mathscr{A}$, $\mathscr{A}'$ according to Definition 2.1.17.

Definition 2.1.28 (Embedding, Compactification, and Isomorphism) We say that a map $\iota : X \to X'$ is a measure-preserving embedding of $(\mathbb{X}, \mathscr{A})$ into $(\mathbb{X}', \mathscr{A}')$ if

(E1) ι is a continuous and injective map of (X, τ) into (X', τ');
(E2) ι is an isometry, in the sense that

$$\mathsf{d}'(\iota(x), \iota(y)) = \mathsf{d}(x, y) \quad \text{for every } x, y \in X. \tag{2.1.70}$$

(E3) $\iota_\sharp \mathfrak{m} = \mathfrak{m}'$.
(E4) For every $f' \in \mathscr{A}'$ the function $\iota^* f' := f' \circ \iota$ belongs to $\mathscr{A}$.

We say that $(\mathbb{X}', \mathscr{A}')$ is a compactification of $(\mathbb{X}, \mathscr{A})$ if (X', τ') is compact and there exists a measure-preserving embedding of $(\mathbb{X}, \mathscr{A})$ into $(\mathbb{X}', \mathscr{A}')$.

We say that a measure-preserving embedding ι is an isomorphism of $(\mathbb{X}, \mathscr{A})$ onto $(\mathbb{X}', \mathscr{A}')$ if ι is an homeomorphism of (X, τ) onto (X', τ') and $\iota^*(\mathscr{A}') = \mathscr{A}$.

Remark 2.1.29 (Canonical Lipschitz Algebra) When $\mathscr{A} = \mathrm{Lip}_b(X, \tau, \mathsf{d})$ and $\mathscr{A}' = \mathrm{Lip}_b(X', \tau', \mathsf{d}')$ we simply say that ι is a measure preserving embedding of $\mathbb{X}$ into $\mathbb{X}'$. In this case it is sufficient to check conditions (E1-2-3), since condition (E4) is a consequence of (E1-2). In this case ι is an isomorphism of $(\mathbb{X}, \mathscr{A})$ onto $(\iota(X), \tau', \mathsf{d}', \mathfrak{m}')$.

Example 2.1.30 Let us show three simple examples of embeddings involving an e.m.t.m. space $\mathbb{X} = (X, \tau, \mathsf{d}, \mathfrak{m})$.

(a) Let τ' be a weaker topology than τ, such that (X, τ', d) is an e.t.m. space. The identity map provides a measure-preserving embedding of $\mathbb{X}$ into $(X, \tau', \mathsf{d}, \mathfrak{m})$.
(b) Let $\mathscr{A}' \subset \mathscr{A}$ be two compatible sub-algebras of $\mathbb{X}$. The identity map provides a measure-preserving embedding of $(\mathbb{X}, \mathscr{A})$ into $(\mathbb{X}, \mathscr{A}')$ (in particular when $\mathscr{A} = \mathrm{Lip}_b(X, \tau, \mathsf{d})$).
(c) By Corollary 2.1.7, if (X, τ) is a Souslin space, one can always find a measure preserving embedding in the space $(X, \tau', \mathscr{A}', \mathsf{d})$ where $\mathscr{A}' \subset \mathscr{A}$ is countably generated and τ' is an auxiliary topology (thus metrizable and separable, coarser than τ).
(d) Let Y be any $\mathfrak{m}$-measurable subset of X such that $\mathfrak{m}(X \setminus Y) = 0$; we denote by τ_Y the relative topology of Y, d_Y the restriction of d to $Y \times Y$ and $\mathfrak{m}_Y := \mathfrak{m}_{|Y}$. If $\mathscr{A}$ is a compatible algebra for $\mathbb{X}$ we define $\mathscr{A}_Y := \{f_{|Y} : f \in \mathscr{A}\}$ as the algebra obtained by the restriction to Y of the elements of $\mathscr{A}$. It is easy to check that $\mathbb{Y} = (Y, \tau_Y, \mathsf{d}, \mathfrak{m}_{|Y})$ is an e.m.t.m. space with a compatible algebra $\mathscr{A}_Y$ and the inclusion map $\iota : Y \to X$ is a measure-preserving embedding of $(\mathbb{Y}, \mathscr{A}_Y)$ into $(\mathbb{X}, \mathscr{A})$.

Let us collect a few simple results concerning the corresponding between measurable functions induced by a measure-preserving embedding. Whenever $f' : X' \to \mathbb{R}$ we write

$$f = \iota^* f' := f' \circ \iota. \tag{2.1.71}$$

Note that if $\iota(X)$ is τ'-dense in X' the pull back map ι^* is injective. Independently from this property, we will show that ι^* induces an isomorphism between classes of measurable functions identified by $\mathfrak{m}'$ and $\mathfrak{m}$-a.e. equivalence respectively. We write $f_1 \sim_{\mathfrak{m}} f_2$ if $\mathfrak{m}(\{f_1 \neq f_2\}) = 0$ and we will denote by $[f]_{\mathfrak{m}}$ the equivalence class of a $\mathfrak{m}$-measurable map f.

Lemma 2.1.31 *Let $\iota : X \to X'$ be a measure-preserving embedding of $(\mathbb{X}, \mathscr{A})$ into $(\mathbb{X}', \mathscr{A}')$ according to the previous definition.*

(a) *For every $\mathfrak{m}'$-measurable function $f' : X' \to \mathbb{R}$ the function $\iota^* f'$ is $\mathfrak{m}$-measurable and we have*

$$f_1{}' \sim_{\mathfrak{m}'} f_2{}' \quad \Leftrightarrow \quad \iota^* f_1{}' \sim_{\mathfrak{m}} \iota^* f_2{}' \tag{2.1.72}$$

(b) *The algebra $\mathscr{A}^* := \iota^*(\mathscr{A}')$ is a sub-algebra of $\mathscr{A}$ which is compatible with the extended metric-measure space $\mathbb{X}$.*
(c) *For every $p \in [1, +\infty]$ ι^* induces a linear isomorphism between $L^p(X', \mathfrak{m}')$ and $L^p(X, \mathfrak{m})$, whose inverse is denoted by ι_*. For every $f \in \mathscr{A}^*/\sim_{\mathfrak{m}}$ the class $\iota_* f$ contains all the elements $f' \in \mathscr{A}'$ satisfying $\iota^* f' = f$.*

Proof The proof of **(a)** is immediate: (2.1.72) is a consequence of the fact that the set $N' := \{f_1' \neq f_2'\}$ satisfies

$$\iota^{-1} N' = \{\iota^* f_1 \neq \iota^* f_2\}, \quad \mathfrak{m}(\iota^{-1} N') = \mathfrak{m}'(N).$$

(b) It is immediate to check that $\mathscr{A}^*$ is a unital algebra included in $\mathscr{A}$. It is not difficult to check that $\mathscr{A}^*$ satisfies (2.1.52): by (2.1.70) if $f' \in \mathscr{A}_1'$ then $\iota^* f' \in \mathscr{A}_1^*$ and since $\mathscr{A}'$ is compatible with $\mathbb{X}'$, for every $x, y \in X$

$$\begin{aligned}\mathsf{d}(x, y) &\overset{(2.1.70)}{=} \mathsf{d}'(\iota(x), \iota(y)) = \sup_{f' \in \mathscr{A}_1'} |f'(\iota(x)) - f'(\iota(y))| = \sup_{f' \in \mathscr{A}_1'} |\iota^* f'(x) - \iota^* f'(y)| \\ &= \sup_{f \in \mathscr{A}_1^*} |f(x) - f(y)|.\end{aligned}$$

(c) Thanks to property (E3) of Definition 2.1.28 we have

$$\int_{X'} \Phi(f') \,\mathrm{d}\mathfrak{m}' = \int_X \Phi(f' \circ \iota) \,\mathrm{d}\mathfrak{m} = \int_X \Phi(\iota^* f') \,\mathrm{d}\mathfrak{m} \tag{2.1.73}$$

for every nonnegative continuous function $\Phi : \mathbb{R} \to [0, +\infty)$, so that ι^* induces a linear isometry from each $L^p(X', \mathfrak{m}')$ into $L^p(X, \mathfrak{m})$. It is therefore sufficient to prove that ι^* is surjective; since ι^* is an isometry with respect to the L^1-norm, this is equivalent to the density of the image of ι^* in $L^1(X, \mathfrak{m})$. Since the image contains (the equivalence classes of elements in) $\mathscr{A}^*$, the density follows by (b) and Lemma 2.1.27. The last statement is a consequence of (2.1.72). □

On of the most useful application of the concept of measure-preserving embeddings is the possibility to construct a compactification $\mathbb{X}'$ of $\mathbb{X}$ starting from a compatible algebra $\mathscr{A}$. As a byproduct, we will obtain a compatible algebra $\mathscr{A}'$ in $\mathbb{X}'$ such that

$$f' \in \mathscr{A}' \quad \Leftrightarrow \quad f = f' \circ \iota \in \mathscr{A}. \tag{2.1.74}$$

As a general fact, every completely regular space (X, τ) can be homeomorphically imbedded as a dense subset of a compact Hausdorff space βX (called the Stone-Cech compactification, [53, § 38]), where every function $f \in \mathrm{C}_b(X)$ admits a unique continuous extension. The Gelfand theory of Banach algebras applied to $\mathrm{C}_b(X)$ provides one of the most effective construction of such a compactification and has the advantage to be well adapted to the setting of extended metric-topological spaces and compatible sub-algebras.

Let us briefly recall the construction. We consider $\mathscr{A}$ as a vector subspace of $\mathrm{C}_b(X, \tau)$ endowed with the sup norm $\|\cdot\|_\infty$ and we call $\overline{\mathscr{A}}$ the (strong) closure of $\mathscr{A}$ in $\mathrm{C}_b(X, \tau)$. Since $(\mathscr{A}, \|\cdot\|_\infty)$ is a normed space we can consider the dual Banach space $(\mathscr{A}^*, \|\cdot\|_{\mathscr{A}^*})$ endowed with the weak* topology $\hat{\tau}$ and the distinguished subset of characters.

Definition 2.1.32 (Characters) A character of $\mathscr{A}$ is an element φ of $\mathscr{A}^* \setminus \{0\}$ satisfying

$$\varphi(fg) = \varphi(f)\varphi(g) \quad \text{for every } f, g \in \mathscr{A}. \tag{2.1.75}$$

We will denote by $\hat{X}$ the subset of the characters of $\mathscr{A}$.

Let us first recall a preliminary list of useful properties of $\hat{X}$.

Proposition 2.1.33 *Let us consider the set*

$$\Sigma := \{\psi \in \mathscr{A}^* : \|\psi\|_{\mathscr{A}^*} \le 1,\ \psi(f) \ge 0 \textit{ for every } f \ge 0,\ \psi(\mathbb{1}) = 1\}. \tag{2.1.76}$$

(a) *Σ is a weakly* compact convex subset of $\mathscr{A}^*$ contained in $\{\psi \in \mathscr{A}^* : \|\psi\|_{\mathscr{A}^*} = 1\}$.*
(b) *$\hat{X}$ is a (weakly*) compact subset of Σ.*
(c) *Every point of $\hat{X}$ is an extremal point of Σ.*

Proof

(a) Is an immediate consequence of Banach-Alaouglu-Bourbaki theorem.

(b) It is not difficult to check that every element $f \in \mathscr{A}$ with $0 < m \le f \le M$ admits a nonnegative square root $g \in \overline{\mathscr{A}}$ such that $g^2 = f$: it is sufficient to define $h := 1 - f/M \in \mathscr{A}$ taking values in $[0, 1 - m/M]$ and use the power series expansion of the square root function in $]0, 2[$:

$$g = \sqrt{M}\sum_{n=0}^{\infty}\binom{1/2}{n}(-1)^n h^n = \sum_{n=0}^{\infty}\binom{1/2}{n} M^{1/2-n}(f - M)^n.$$

The relation $\varphi(f) = (\varphi(g))^2$ shows that $\varphi f \ge 0$ and since every nonnegative $f \in \mathscr{A}$ can be strongly approximated by uniformly positive elements we obtain that every $\varphi \in \hat{X}$ satisfies $\varphi(f) \ge 0$ for every nonnegative $f \in \mathscr{A}$. Moreover, since $\varphi \ne 0$, there exists an element $f \in \mathscr{A}$ such that $\varphi(f) \ne 0$; from $0 \ne \varphi(f) = \varphi(f\mathbb{1}) = \varphi(f)\varphi(\mathbb{1})$ we deduce that $\varphi(\mathbb{1}) = 1$. By comparison we obtain

$$\varphi(\gamma\mathbb{1}) = \gamma \quad \text{for every } \gamma \in \mathbb{R}; \quad -\alpha \le f \le \beta \quad \Rightarrow \quad -\alpha \le \varphi(f) \le \beta, \tag{2.1.77}$$

and in particular

$$-\|f\|_\infty \le \varphi(f) \le \|f\|_\infty, \tag{2.1.78}$$

so that every element of $\hat{X}$ is included in the weakly* compact set Σ. Since condition (2.1.75) characterizes a weakly* closed set, we conclude that $\hat{X}$ is a compact Hausdorff space endowed with the weak* topology of $\mathscr{A}^*$.

(c) Whenever $\psi \in \Sigma$ and $f \in \mathscr{A}$, the inequality

$$0 \le \psi((f + \kappa\mathbb{1})^2) = \psi(f^2) + 2\kappa\psi(f) + \kappa^2\psi(\mathbb{1}) = \psi(f^2) + 2\kappa\psi(f) + \kappa^2$$

for every $\kappa \in \mathbb{R}$ shows that

$$\big(\psi(f)\big)^2 \le \psi(f^2) \quad \text{for every } \psi \in \Sigma.$$

If $\hat{X} \ni \varphi = \frac{1}{2}\varphi_1 + \frac{1}{2}\varphi_2$ with $\varphi_i \in \Sigma$, we obtain

$$\begin{aligned}\frac{1}{2}\varphi_1(f^2) + \frac{1}{2}\varphi_2(f^2) = \varphi(f^2) = \big(\varphi(f)\big)^2 \\ &= \frac{1}{4}\big(\varphi_1(f)\big)^2 + \frac{1}{4}\big(\varphi_2(f)\big)^2 + \frac{1}{2}\varphi_1(f)\varphi_2(f) \\ &\le \frac{1}{4}\varphi_1(f^2) + \frac{1}{4}\varphi_2(f^2) + \frac{1}{2}\varphi_1(f)\varphi_2(f)\end{aligned}$$

thus showing that

$$\Big(\frac{1}{2}\varphi_1(f)-\frac{1}{2}\varphi_2(f)\Big)^2=\frac{1}{4}\varphi_1(f^2)+\frac{1}{4}\varphi_2(f^2)-\frac{1}{2}\varphi_1(f)\varphi_2(f)\le 0$$

for every $f\in\mathscr{A}$ and therefore $\varphi_1=\varphi_2$.

□

We can now define a canonical embedding $\iota: X\to\hat X$ by

$$\iota(x)=\hat x,\quad \hat x(f):=f(x)\quad \text{for every } x\in X,\ f\in\mathscr{A}, \tag{2.1.79}$$

and we call $\Gamma:\overline{\mathscr{A}}\to \mathrm{C}_b(\hat X,\hat\tau)$ the Gelfand transform

$$\hat f=\Gamma f,\quad \hat f(\varphi):=\varphi(f)\quad \text{for every } \varphi\in\hat X. \tag{2.1.80}$$

We will set $\hat{\mathscr{A}}:=\Gamma(\mathscr{A})$.

Theorem 2.1.34 (Gelfand Compactification of Extended Metric Topological Measure Spaces)

(a) *ι is a continuous and injective, $\iota(X)$ is a dense subset of the compact Hausdorff space $\hat X$ endowed with the weak* topology $\hat\tau$. If $\mathscr{A}$ is separable w.r.t. the uniform norm then $\hat X$ is metrizable.*

(b) *ι is a homeomorphism between X and $\iota(X)$ if and only if $\mathscr{A}$ is adapted, i.e. it generates the topology τ.*

(c) *Every function $f\in\overline{\mathscr{A}}$ admits a unique extension $\hat f=\Gamma f$ to $\hat X$ and the unital algebra $\hat{\mathscr{A}}:=\Gamma(\mathscr{A})$ is uniformly dense in $\mathrm{C}_b(\hat X,\hat\tau)$. The pull back algebra $\mathscr{A}^*$ coincides with $\mathscr{A}$.*

(d) *The measure $\hat{\mathfrak{m}}:=\iota_\sharp\mathfrak{m}$ is a Radon measure on $\hat X$ concentrated on the $\hat{\mathfrak{m}}$-measurable subset $\iota(X)$.*

(e) *If I is the directed set of all the finite collections of functions in $\mathscr{A}_1$, setting*

$$\hat{\mathsf{d}}_i(\varphi_1,\varphi_2):=\sup_{f\in i}|\varphi_1(f)-\varphi_2(f)|=\sup_{f\in i}|\hat f_i(\varphi_1)-\hat f_i(\varphi_2)|, \tag{2.1.81}$$

$$\hat{\mathsf{d}}(\varphi_1,\varphi_2):=\sup_{f\in\mathscr{A}_1}|\varphi_1(f)-\varphi_2(f)|=\sup_{i\in I}\hat{\mathsf{d}}_i(\varphi_1,\varphi_2), \tag{2.1.82}$$

$\hat{\mathsf{d}}_i$ are continuous and bounded semidistances on $\hat X$ and $\hat{\mathsf{d}}$ is an extended distance on $\hat X$ satisfying

$$\hat{\mathsf{d}}_i(\hat x,\hat y)=\mathsf{d}_i(x,y),\quad \hat{\mathsf{d}}(\hat x,\hat y)=\mathsf{d}(x,y)\quad \textit{for every } x,y\in X. \tag{2.1.83}$$

(f) *ι is a measure preserving embedding of $(\mathbb{X},\mathscr{A})$ into the compact extended metric measure space $\hat{\mathbb{X}}':=(\hat X,\hat\tau,\hat{\mathsf{d}},\hat{\mathfrak{m}})$ endowed with the compatible algebra $\hat{\mathscr{A}}=\Gamma(\mathscr{A})$.*

(g) *The map* $\iota_* : L^p(X, \mathfrak{m}) \to L^p(\hat{X}, \hat{\mathfrak{m}})$ *is the unique linear isometric extension of* $\Gamma : \mathscr{A} \to \hat{\mathscr{A}}$ *to* $L^p(X, \mathfrak{m})$ *for every* $p \in [1, \infty[$.

Proof

(a) Since the weak* topology is the coarsest topology that makes all the maps $\varphi \mapsto \varphi(f)$ continuous for every $f \in \mathscr{A}$, the continuity of ι is equivalent to check the continuity of $x \mapsto f(x)$ from (X, τ) to $\mathbb{R}$, which is guaranteed by the τ-continuity of f. ι is also injective, since $\mathscr{A}$ separates the points of X.

Let us now prove that $\iota(X)$ is dense in $\hat{X}$. Let us denote by Y the weak* closure of $\iota(X)$ and let us first consider the closed convex hull $K := \overline{\text{co}(Y)}$ of Y in the weak* topology. Since K is bounded and weakly* closed, K is compact. If a point $\psi \in \hat{X}$ does not belong to K we can apply the second geometric form of Hahn-Banach Theorem [55, Thm. 3.21] to find $f \in \mathscr{A}$ separating ψ from K: there exists $\alpha \in \mathbb{R}$ such that $\zeta(f) \geq \alpha > \psi(f)$ for every $\zeta \in K$. Choosing $\zeta = \hat{z} = \iota(z)$ for $z \in X$ we deduce that $f(z) \geq \alpha$ for every $z \in X$ and therefore $\psi(f) \geq \alpha$ since ψ is a nonnegative functional.

Thus $\psi \in \overline{\text{co}(Y)}$; since $Y \subset \Sigma$, also $\overline{\text{co}(Y)} \subset \Sigma$; since ψ is an extreme point of Σ, we deduce that ψ is an extreme point of $\overline{\text{co}(Y)}$; applying Milman's theorem [55, Thm. 3.25] we conclude that $\psi \in Y$.

Finally, if $\mathscr{A}$ is separable then the unit ball of $\mathscr{A}^*$ endowed with the weak* topology is metrizable so that $\hat{X}$ is metrizable as well.

(b) It is easy to check that $\hat{f} \in C_b(\hat{X})$ and that $\hat{\mathscr{A}}$ is an algebra. Clearly $\hat{\mathbb{1}}$ is the unit function in $C_b(\hat{X})$ and $\hat{\mathscr{A}}$ separates the points of $\hat{X}$: if $\varphi_i \in \hat{X}$ satisfies $\varphi_1(f) = \varphi_2(f)$ for every $f \in \mathscr{A}$, then $\varphi_1 = \varphi_2$ in $\mathscr{A}^*$. Applying Stone-Weierstrass Theorem, we conclude that $\hat{\mathscr{A}}$ is dense in $C_b(\hat{X})$. By construction Γ is a isomorphism between $\mathscr{A}$ and $\hat{\mathscr{A}}$, and ι^* is its inverse.

(c) Since $(\hat{X}, \hat{\tau})$ is compact and $\hat{\mathscr{A}}$ separates the points of $\hat{X}$, $\hat{\tau}$ is the initial topology of $\hat{\mathscr{A}}$. Thus a net $\hat{x}_i = \iota(x_i)$ in $\iota(X)$, $i \in I$, converges to $\hat{x} = \iota(x)$ if and only if $\lim_{i \in I} \hat{f}(\hat{x}_i) = \hat{f}(\hat{x})$ for every $\hat{f} \in \hat{\mathscr{A}}$. By (2.1.79) and (2.1.80) the latter property is equivalent to $\lim_{i \in I} f(x_i) = f(x)$ for every $f \in \mathscr{A}$, so ι is a homeomorphism between X endowed with the initial topology of $\mathscr{A}$ and $\iota(X)$.

(d) This is a general property for the push-forward of Radon measures through a continuous map: since $\mathfrak{m}$ is tight, we can find an increasing sequence of compact sets $K_n \subset X$ such that $\mathfrak{m}(X \setminus K_n) \leq 1/n$. Since ι is continuous, $\hat{K}_n = \iota(K_n) \subset \iota(X)$ are compact in $\hat{X}$ and $\hat{\mathfrak{m}}(\hat{X} \setminus \hat{K}_n) \leq 1/n$ so that $\hat{\mathfrak{m}}(\hat{X} \setminus \hat{K}) = 0$ where $\hat{K} = \cup_n \hat{K}_n \subset \iota(X)$.

(e) and **(f)** are immediate.

(g) Thanks to Lemma 2.1.31 the Gelfand isomorphism Γ preserves the equivalence classes in the sense that for every $f_1, f_2 \in \mathscr{A}$,

$$f_1 \sim_{\mathfrak{m}} f_2 \quad \Leftrightarrow \quad \hat{f}_1 \sim_{\hat{\mathfrak{m}}} \hat{f}_2$$

so that

$$\iota_*([f]_{\mathfrak{m}}) = [\hat{f}]_{\hat{\mathfrak{m}}} \quad \text{for every } f \in \mathscr{A}.$$

Since ι_* is a linear isometry and the equivalence classes of elements of $\mathscr{A}$ are dense in $L^p(X, \mathfrak{m})$ we conclude.

□

Remark 2.1.35 (A Universal Model) The compactification $\hat{\mathbb{X}} = (\hat{X}, \hat{\tau}, \hat{\mathsf{d}}, \hat{\mathfrak{m}})$ with the Gelfand algebra $\hat{\mathscr{A}}$ is a particular case of the setting considered within Example 2.1.11, where B is the space $\overline{\mathscr{A}}$, $L = \overline{\mathscr{A}_1}$ and $\hat{X}$ is a weakly* compact subset of the dual ball of B' (in fact concentrated on its extremal set). It follows that any e.m.t.m. space has a measure-preserving isometric immersion in a space with the characteristics of Example 2.1.11.

The previous construction is also useful to quickly get a completion of an e.m.t.m. space. We start from the compactification $\hat{\mathbb{X}} = (\hat{X}, \hat{\tau}, \hat{\mathsf{d}}, \hat{\mathfrak{m}})$ of $(\mathbb{X}, \mathscr{A}(\mathbb{X}))$ obtained by the canonical algebra $\mathrm{Lip}_b(X, \mathsf{d}, \mathfrak{m})$ and we set:

$$\bar{X} := \text{ the } \hat{\mathsf{d}}\text{-closure of } \iota(X) \text{ into } \hat{X}. \tag{2.1.84}$$

We obviously define $\bar{\tau}$, $\bar{\mathsf{d}}$ and $\bar{\mathfrak{m}}$ respectively as the restrictions of $\hat{\tau}$, $\hat{\mathsf{d}}$, and $\hat{\mathfrak{m}}$ to $\bar{X}$.

Corollary 2.1.36 (Completion) *The map $\iota : X \to \bar{X}$ is a measure preserving embedding of $\mathbb{X}$ into the e.m.t.m. space $\bar{\mathbb{X}} = (\bar{X}, \bar{\tau}, \bar{\mathsf{d}}, \bar{\mathfrak{m}})$ such that*

(C1) *$(\bar{X}, \bar{\mathsf{d}})$ is complete,*
(C2) *ι is a homeomorphism of X onto $\iota(X)$,*
(C3) *$\iota(X)$ is $\bar{\mathsf{d}}$-dense in $\bar{X}$,*
(C4) *every function $f \in \mathrm{Lip}_b(X, \tau, \mathsf{d})$ admits a unique extension to a function $\bar{f} \in \mathrm{Lip}_b(\bar{X}, \bar{\tau}, \bar{\mathsf{d}})$. The map $f \mapsto \bar{f}$ is an isomorphism of $\mathrm{Lip}_b(X, \tau, \mathsf{d})$ onto $\mathrm{Lip}_b(\bar{X}, \bar{\tau}, \bar{\mathsf{d}})$.*

The space $\bar{\mathbb{X}} := (\bar{X}, \hat{\tau}, \hat{\mathsf{d}}, \hat{\mathfrak{m}})$ is a completion of $\mathbb{X}$.

Notice that in the simple case when τ is the topology induced by the distance d, the previous construction coincides with the usual completion of a metric space.

Remark 2.1.37 It is not difficult to check that if another e.m.t.m. space $\mathbb{X}'$ satisfies the properties (C1)-(C4) of Corollary 2.1.36 then $\mathbb{X}'$ is isomorphic to $\bar{\mathbb{X}}$, so that the completion of an e.m.t.m. space is unique up to isomorphisms. We may identify X with the $\bar{\mathsf{d}}$-dense subset $\iota(X)$ in $\bar{X}$, so that for every function $f' : \bar{X} \to \mathbb{R}$, $\iota^* f'$ is just the restriction of f' to the $\bar{\mathfrak{m}}$-measurable set X and $\mathfrak{m}$ can be considered as the restriction of $\bar{\mathfrak{m}}$ to X. Since $\bar{\mathfrak{m}}(\bar{X} \setminus X) = 0$, we can identify $\bar{\mathfrak{m}}$ and $\mathfrak{m}$. Every function $f : X \to \mathbb{R}$ can be considered defined $\bar{\mathfrak{m}}$-a.e. and the trivial extension provides a realization of ι_* (which of course does not coincides with the extension $\bar{f}$ of $f \in \mathrm{Lip}_b(X, \tau, \mathsf{d})$ by continuity).

We conclude this section with an example of application of the compactification trick to prove a useful approximation result.

Corollary 2.1.38 *Let* $\mathbb{X} = (X, \tau, \mathsf{d}, \mathfrak{m})$ *be an e.m.t.m. space endowed with an* adapted *algebra* $\mathscr{A}$. *Then for every bounded upper semicontinuous function* $g : X \to \mathbb{R}$

$$g(x) = \inf\Big\{h(x) : h \in \mathscr{A},\ h \geq g\Big\}, \tag{2.1.85}$$

$$\int_X g \,\mathrm{d}\mathfrak{m} = \inf\Big\{\int_X h(x) \,\mathrm{d}\mathfrak{m} : h \in \mathscr{A},\ h \geq g\Big\}. \tag{2.1.86}$$

Proof Let $\iota : X \to \hat{X}$ be the compactification of $\mathbb{X}$ induced by $\mathscr{A}$ and let $\hat{g} : \hat{X} \to \mathbb{R}$ be the upper semicontinuous envelope of g to $\hat{X}$, i.e. the lowest upper semicontinuous function whose restriction to X is larger than or equal to g. We know that for every $\hat{x} = \iota(x) \in X$

$$\hat{g}(\hat{x}) = \inf_{U \in \mathscr{U}_{\hat{x}}} \sup\big\{g(y) : y \in X,\ \iota(y) \in U\big\} = g(x),$$

since g is upper semicontinuous; here we use the fact that ι is an homeomorphism between (X, τ) and $(\iota(X), \hat{\tau})$, since $\mathscr{A}$ is adapted. Thanks to (the u.s.c. version of) (2.1.4) we deduce that

$$\hat{g}(\hat{x}) = \inf\big\{h(\hat{x}) : h \in \mathrm{C}_b(\hat{X}, \hat{\tau}),\ h \geq \hat{g}\big\}$$

and since $\hat{\mathscr{A}} = \Gamma(\mathscr{A})$ is uniformly dense in $\mathrm{C}_b(\hat{X}, \hat{\tau})$ we deduce the formula

$$\hat{g}(\hat{x}) = \inf\big\{\hat{h}(\hat{x}) : h \in \mathscr{A},\ h \geq g \text{ in } X\big\},$$

which yields (2.1.85). Since $H := \{h \in \mathrm{C}_b(\hat{X}, \hat{\tau}),\ h \geq \hat{g}\big\}$ is a directed set by the order relation $h_0 \prec h_1 \Leftrightarrow h_0(x) \leq h_1(x)$ for every $x \in \hat{X}$, (2.1.8) yields

$$\int_{\hat{X}} \hat{g} \,\mathrm{d}\hat{\mathfrak{m}} = \lim_{h \in H} \int_{\hat{X}} h \,\mathrm{d}\hat{\mathfrak{m}} = \inf_{h \in H} \int_{\hat{X}} h \,\mathrm{d}\hat{\mathfrak{m}}. \tag{2.1.87}$$

Since $\Gamma(\mathscr{A})$ is uniformly dense in $\mathrm{C}_b(\hat{X}, \hat{\tau})$, every function $h \in H$ can be uniformly approximated from above by functions of the forms $\hat{f} = \Gamma(f)$ for $f \in \mathscr{A}$, $\Gamma(f) \geq h \geq \hat{g}$; we deduce that

$$\int_{\hat{X}} \hat{g} \,\mathrm{d}\hat{\mathfrak{m}} = \inf_{f \in \mathscr{A},\ f \geq g} \int_{\hat{X}} \hat{f} \,\mathrm{d}\hat{\mathfrak{m}} = \inf_{f \in \mathscr{A},\ f \geq g} \int_X f \,\mathrm{d}\mathfrak{m}.$$

□

2.1.8 Notes

Section 2.1.1: general references for measure theory are [18, 56]; here we mainly followed the approach to Radon measures given by [56], trying to minimize the topological assumptions. The main points are the *complete regularity* of (X, τ) (which is almost needed for the standard formulation of the weak convergence of probability measures, see the Appendix of [56]) and the Radon property of the reference measure $\mathfrak{m}$. Complete regularity is in fact equivalent to the fact that continuous functions characterizes the topology of X (see Sect. A.2 in the Appendix) and it is also important for the formulation of the extended metric-topological spaces and for the compactification argument of Sect. 2.1.7.

Section 2.1.2: here we followed very closely the presentation of [3, Section 3]. Extended metrics and the use of an auxiliary topology have already been considered in [8], under a slightly different set of compatibility conditions.

Section 2.1.4: the section just recalls the basic properties of the Kantorovich-Rubinstein distance, adapted to the extended setting, General references are the books [6, 63, 64]; notice that K_δ is sometimes called Kantorovich-Wasserstein distance of order 1 and denoted by W_1. The approximation result is quite similar to [3], where the distance of order 2 has been considered.

Section 2.1.5: most of the results are classic for the local slope (2.1.50). We adapted the same approach of [8] to the more refined (and stronger) local Lipschitz constant, which in the present setting also depends on the topology τ. Equation (2.1.51c) plays a crucial role in the locality property of the Cheeger energy and (2.1.51e) is quite useful to derive contraction estimate for its L^2 gradient flow. It is in fact possible to prove a more refined property, see [51].

Section 2.1.6 contains the main definition of algebras of functions compatible with the extended metric-topological setting. The basic requirements is that the algebra is sufficiently rich to recover the extended distance. We also collected a few results, mainly based on Bernstein polynomials [49], which will be quite useful to replace Lipschitz truncations with smoother polynomial maps preserving the algebraic structure.

Section 2.1.7: measure-preserving isometric embeddings play an important role in the theory of metric-measure spaces, in particular when one studies their convergence (see [64, Chap. 27], [40]). Here we adapted this notion to the presence of the auxiliary topology τ and of the compatible algebra $\mathscr{A}$. Another typical application arises in regular representation of Dirichlet spaces (the so-called transfer method, [52, Chap. VI]); the idea of using the Gelfand transform to construct a suitable compactification is taken from [31], [32, Appendix A.4] and it is based on one of the possible construction of the Stone-Cech compactification of a completely regular topological space.

2.2 *Continuous Curves and Nonparametric Arcs*

This section mostly contains classic material on the topology of space of curves, adapted to the extended metric-topological setting. Its main goal is to construct a useful setting to deal with Radon measures on (non parametric) rectifiable curves. Differently from other approaches (see e.g. [2, 45, 54]) we first study class of equivalent curves (up to reparametrizations) without assuming their rectifiability. In Sect. 2.2.1 we study the natural e.m.t. structure on $\mathrm{C}([a,b];(X,\tau))$, and in Sect. 2.2.2 we consider the natural quotient space of (continuous) arcs $\mathrm{A}(X,\tau)$, which behaves quite well with respect to topology and distance. The last part Sect. 2.2.3 focuses on d-rectifiable arcs $\mathrm{RA}(X,\mathsf{d})$, considered as a natural Borel subset of $\mathrm{A}(X,\tau)$. Here we have the arc-length reparametrization at our disposal, and we study measurability properties of important operations, like evaluations, integrals, length, and reparametrizations. We will also state a useful compactness result, which is a natural generalization of Arzelà-Ascoli Theorem. Theorem 2.2.13 will collect most of the main properties we will use in the next chapters.

2.2.1 Continuous Curves

Let (X,τ) be a completely regular Hausdorff space. We will denote by $\mathrm{C}([a,b];(X,\tau))$ the set of τ-continuous curves $\gamma:[a,b]\to X$ endowed with the compact-open topology τ_{C} (we will simply write $\mathrm{C}([a,b];X)$, when the topology τ will be clear from the context). By definition, a subbasis generating τ_C is given by the collection of sets

$$S(K,V):=\Big\{\gamma\in\mathrm{C}([a,b];X):\gamma(K)\subset V\Big\},\qquad K\subset[a,b]\text{ compact, }V\text{ open in }X. \tag{2.2.1}$$

Remark 2.2.1 Thanks to the particular structure of the domain $[a,b]$, we can also consider an equivalent basis associated to partitions $\mathcal{P}=\{a=t_0<t_1<\cdots<t_J=b\}$:

$$\cap_{j=1}^{J}\Big\{\gamma\in\mathrm{C}([a,b];X):\gamma([t_{j-1},t_j])\subset W_j\Big\},\qquad t_j\in\mathcal{P},\ W_j\text{ open in }X. \tag{2.2.2}$$

It is also not restrictive to consider partitions $\mathcal{P}$ induced by rational points $t_j\in\mathbb{Q}\cap\,]a,b[$, $j=1,\cdots,J$. It is sufficient to show that if γ_0 belongs to an open set $U=\cap_{h=1}^{H}S(K_h,V_h)$ arising from the finite intersection of elements of the subbasis (2.2.1), we can also find a set U' of the form (2.2.2) such that $\gamma_0\in U'\subset U$.

To this aim, it is not restrictive to add to the collection $\{K_h, V_h\}_{h=1}^H$ the pair $K_0 = [a, b]$, $V_0 = X$; we can cover each K_h with a finite number of intervals $I_{h,k} = [\alpha_{h,k}, \beta_{h,k}]$, $1 \le k \le k(h)$, such that $\gamma_0(I_{h,k}) \subset V_h$. We can then take the partition $\mathcal{P}$ of $[a, b]$ containing all the extrema of $I_{h,k}$ (notice that a and b are included). If t_j, $j \ge 1$, is a point of the partition $\mathcal{P}$, we set $W_j := \cap\big\{V_h : \exists k \in \{1, \cdots, k(h)\},\ [t_{j-1}, t_j] \subset I_{h,k}\big\}$.

Let us recall a few simple and useful properties of the compact-open topology [53, § 46]:

(CO1) If the topology τ is induced by a distance δ, then the topology τ_{C} is induced by the uniform distance $\mathsf{d}_\delta(\gamma, \gamma') := \sup_{t \in [a,b]} \delta(\gamma(t), \gamma'(t))$ and convergence w.r.t. the compact-open topology coincides with the uniform convergence w.r.t. δ. If moreover τ is separable then also τ_{C} is separable.

(CO2) The evaluation map $\mathsf{e} : [a, b] \times \mathrm{C}([a, b]; X) \to X$, $\mathsf{e}(t, \gamma) := \gamma(t)$, is continuous.

(CO3) If $f : X \to Y$ is a continuous function with values in a Hausdorff space (Y, τ_Y), the composition map $\gamma \mapsto f \circ \gamma$ is continuous from $\mathrm{C}([a, b]; X)$ to $\mathrm{C}([a, b]; Y)$.

(CO4) If $\sigma : [0, 1] \to [0, 1]$ is nondecreasing and continuous the map $\gamma \mapsto \gamma \circ \sigma$ is continuous from $\mathrm{C}([a, b]; X)$ to $\mathrm{C}([a, b]; X)$.

(CO5) If τ is Polish, then τ_{C} is Polish.

The proof of the first two properties can be found, e.g., in [53, Thm. 46.8, 46.10]; in the case when τ is metrizable and separable, the separability of τ_{C} follows by [28, 4.2.18].

In order to prove (CO3) it is sufficient to show that the inverse image of an arbitrary element $S(K, W)$ (as in (2.2.1)) of the subbasis generating the compact-open topology of $\mathrm{C}([a, b]; Y)$ is an element of the corresponding subbasis of the topology of $\mathrm{C}([a, b]; X)$. In fact, $f \circ \gamma \in S(K, W)$ if and only if $\gamma \in S(K, f^{-1}(W))$. (CO4) can be proved by a similar argument: if $\gamma \circ \sigma \in S(K, V)$ if and only if $\gamma \in S(\sigma(K), V)$. Finally (CO5) is a consequence of (CO1).

We will denote by $\mathsf{e}(\gamma)$ the image $\gamma([a, b])$ in X.

(Semi)distances on $\mathrm{C}([a, b]; X)$

An extended semidistance $\delta : X \times X \to [0, \infty]$ induces an extended semidistance δ_{C} in $\mathrm{C}([a, b]; X)$ by

$$\delta_{\mathrm{C}}(\gamma_1, \gamma_2) := \sup_{t \in [a,b]} \delta(\gamma_1(t), \gamma_2(t)). \tag{2.2.3}$$

We have

Proposition 2.2.2 *If* (X, τ, d) *is an extended metric-topological space, than also* $(\mathrm{C}([a, b]; (X, \tau)), \tau_{\mathrm{C}}, \mathsf{d}_{\mathrm{C}})$ *is an extended metric-topological space.*

If moreover τ' *is an auxiliary topology for* (X, τ, d) *according to Definition 2.1.6, then* τ'_{C} *is an auxiliary topology for* $(\mathrm{C}([a, b]; (X, \tau)), \tau_{\mathrm{C}}, \mathsf{d}_{\mathrm{C}})$ *as well.*

Proof Notice that for every $f \in \mathrm{Lip}_{b,1}(X, \tau, \mathsf{d})$ and $t \in [a, b]$ the map $\mathsf{f}_t := f \circ \mathsf{e}_t$ belongs to $\mathrm{Lip}_{b,1}(\mathrm{C}([a, b]; X), \tau_{\mathrm{C}}, \mathsf{d}_{\mathrm{C}})$ and it is easy to check by (2.1.17) that

$$\mathsf{d}_{\mathrm{C}}(\gamma_1, \gamma_2) = \sup \Big\{ |\mathsf{f}_t(\gamma_1) - \mathsf{f}_t(\gamma_2)| : f \in \mathrm{Lip}_{b,1}(X, \tau, \mathsf{d}),\ t \in [a, b] \Big\}. \tag{2.2.4}$$

For every finite collection $\mathcal{K} := \{K_h\}_{h=1}^H$ of compact subsets of $[a, b]$ and for every finite collection $\mathcal{F} := (f_h)_{h=1}^H$ in $\mathrm{Lip}_b(X, \tau, \mathsf{d})$ let us consider the function $F = F_{\mathcal{K},\mathcal{F}}$ defined by

$$F(\gamma) := \max_{1 \le h \le H} \max_{t \in K_h} f_h(\gamma(t)). \tag{2.2.5}$$

It is easy to check that $F \in \mathrm{Lip}_b(\mathrm{C}([a, b]; X), \tau_{\mathrm{C}}, \mathsf{d}_{\mathrm{C}})$; we want to show that this family of functions separates points from closed sets. To this aim, we fix a closed set $\mathrm{F} \subset \mathrm{C}([a, b]; X)$ and a curve $\gamma_0 \in \mathrm{C}([a, b]; X) \setminus \mathrm{F}$.

By the definition of compact-open topology, we can find a collection $\mathcal{K} = \{K_h\}_{h=1}^H$ of compact sets of $[a, b]$ and open sets $U_h \subset X$ such that the open set $\mathrm{U} = \{\gamma : \gamma(K_h) \subset U_h,\ 1 \le h \le H\}$ is included in $\mathrm{C}([a, b]; X) \setminus \mathrm{F}$ and contains γ_0. By (2.1.2) and the compactness we can find nonnegative functions $\mathcal{F} = (f_h)_{h=1}^H \subset \mathrm{Lip}_b(X, \tau, \mathsf{d})$ such that $f_h \equiv 1$ on $X \setminus U_h$ and $f_h|_{\gamma_0(K_h)} \equiv 0$. It follows that $F = F_{\mathcal{K},\mathcal{F}}$ satisfies $F(\gamma_0) = 0$ and $F(\gamma) \ge 1$ for every γ in the complement of U, in particular in F.

Let us eventually check the last statement, by checking that τ'_{C} satisfies properties (A1,2,3) of Definition 2.1.6. (A1,2) are obvious. Concerning (A3) we select a sequence $f_n \in \mathrm{Lip}_b(X, \tau', \mathsf{d})$ satisfying (2.1.28) and the countable collection of maps $\mathcal{F} := (\mathsf{f}_{n,t})_{n \in \mathbb{N},\ t \in [a,b] \cap \mathbb{Q}}$, $\mathsf{f}_{n,t} := f_n \circ \mathsf{e}_t$. It is clear that for every $\gamma_1, \gamma_2 \in \mathrm{C}([a, b]; X)$ and every $n \in \mathbb{N}$

$$\sup_{t \in [a,b]} |\mathsf{f}_{n,t}(\gamma_1) - \mathsf{f}_{n,t}(\gamma_2)| = \sup_{t \in [a,b] \cap \mathbb{Q}} |\mathsf{f}_{n,t}(\gamma_1) - \mathsf{f}_{n,t}(\gamma_2)|$$

(2.1.28) then yields

$$\mathsf{d}_{\mathrm{C}}(\gamma_1, \gamma_2) = \sup_{\mathsf{f} \in \mathcal{F}} |\mathsf{f}(\gamma_1) - \mathsf{f}(\gamma_2)|.$$

□

We can state a useful compactness criterium, which is the natural extension of Arzelà-Ascoli Theorem to extended metric-topological structures.

Proposition 2.2.3 *Let us suppose that* (X, τ) *is compact and let* $\Gamma \subset \mathrm{C}([a, b]; X)$ *be* d*-equicontinuous, i.e. there exists* $\omega : [0, \infty) \to [0, \infty)$ *concave, nondecreasing and continuous with* $\omega(0) = 0$ *such that*

$$\mathsf{d}(\gamma(r), \gamma(s)) \le \omega(|r - s|) \quad \textit{for every } r, s \in [a, b],\ \gamma \in \Gamma. \tag{2.2.6}$$

Then Γ *is relatively compact with respect to* τ_{C}.

Proof Let γ_i, $i \in I$, be a net in Γ. Since X is compact, we can find a subnet $h : J \to I$ and a limit curve $\gamma : [a, b] \to X$ such that $\lim_{j\in J} \gamma_{h(j)} = \gamma$ with respect to the topology of pointwise convergence. Passing to the limit in (2.2.6) we immediately see that

$$\mathsf{d}(\gamma(r), \gamma(s)) \le \liminf_{j\in J} \mathsf{d}(\gamma_{h(j)}(r), \gamma_{h(j)}(s)) \le \omega(|r - s|) \quad \text{for every } r, s \in [a, b],$$

so that $\gamma \in \mathrm{C}([a, b]; (X, \mathsf{d}))$.

Let us now consider a neighborhood $\mathcal{U} = \cap_{h=1}^{H} S(K_h, V_h)$ of γ: we will show that there exists $j_0 \in J$ such that $\gamma_{h(j)} \in \mathcal{U}$ for every $j \succeq j_0$. Let us consider functions $f_h \in \mathrm{Lip}_b(X, \tau, \mathsf{d})$ satisfying $f_h \equiv 0$ on $\gamma(K_h)$ and $f_h \equiv 1$ on $X \setminus V_h$ as in the previous Lemma and let $L > 0$ be the maximum of the Lipschitz constants of f_h. We call $\boldsymbol{f}_j : [a, b] \to \mathbb{R}^H$ the family of curves $\boldsymbol{f}_j(t) := (f_1(\gamma_{h(j)}(t)), f_2(\gamma_{h(j)}(t)), \cdots, f_H(\gamma_{h(j)}(t))$ indexed by $j \in J$. Since f_h are continuous, we have

$$\lim_{j\in J} \boldsymbol{f}_j(t) = \boldsymbol{f}(t) \quad \text{pointwise}, \quad \boldsymbol{f}(t) := (f_1(\gamma(t)), f_2(\gamma(t)), \cdots, f_H(\gamma(t)). \tag{2.2.7}$$

On the other hand, we have

$$|\boldsymbol{f}_j(r) - \boldsymbol{f}_j(s)| \le HL\omega(|r - s|) \tag{2.2.8}$$

so that Ascoli-Arzelà theorem (in $\mathrm{C}([a, b]; \mathbb{R}^H)$) yields

$$\lim_{j\in J} \sup_{t\in[a,b]} |\boldsymbol{f}_j(t) - \boldsymbol{f}(t)| = 0. \tag{2.2.9}$$

Choosing $j_0 \in J$ so that $\sup_{t\in[a,b]} |\boldsymbol{f}_j(t) - \boldsymbol{f}(t)| \le 1/2$ for every $j \succeq j_0$, since $f_h(\gamma(t)) \equiv 0$ whenever $t \in K_h$ we deduce that $f_h(\gamma_j(t)) \le 1/2$ whenever $t \in K_h$ so that $\gamma_j(K_h) \subset V_h$. □

2.2.2 Arcs

Let us denote by Σ the set of continuous, nondecreasing and surjective maps $\sigma : [0, 1] \to [0, 1]$ and by Σ' the subset of Σ of the invertible maps, thus increasing homeomorphisms of $[0, 1]$. We also set

$$\Sigma_2 := \Big\{\sigma \in \Sigma : |\sigma(r) - \sigma(s)| \le 2|r - s|\Big\}. \tag{2.2.10}$$

On $\mathrm{C}([0, 1]; (X, \tau))$ we introduce the symmetric and reflexive relation

$$\gamma_1 \sim \gamma_2 \quad \text{if} \quad \exists \sigma_i \in \Sigma :\ \gamma_1 \circ \sigma_1 = \gamma_2 \circ \sigma_2. \tag{2.2.11}$$

Notice that $\gamma \sim \gamma \circ \sigma$ for every $\sigma \in \Sigma$. It is also not difficult to check that if a map $\gamma : [0, 1] \to X$ satisfies $\gamma \circ \sigma \in \mathrm{C}([0, 1]; (X, \tau))$ for some $\sigma \in \Sigma$ then $\gamma \in \mathrm{C}([0, 1]; (X, \tau))$. In fact, if $A \subset X$ is a closed set, then $B := (\gamma \circ \sigma)^{-1}(A) = \sigma^{-1}(\gamma^{-1}(A))$ is closed in $[0, 1]$ and therefore compact. Since σ is surjective, for every $Z \subset [0, 1]$ we have $\sigma(\sigma^{-1}(Z)) = Z$ so that $\gamma^{-1}(A) = \sigma(B)$ is closed since it is the continuous image of a compact set.

We want to show that $\sim$ also satisfies the transitive property, so that it is an equivalence relation.

If $\delta : X \times X \to [0, +\infty]$ is a τ-l.s.c. extended semidistance we also introduce

$$\delta_{\mathrm{A}}(\gamma_1, \gamma_2) := \inf_{\sigma_i \in \Sigma} \delta_{\mathrm{C}}(\gamma_1 \circ \sigma_1, \gamma_2 \circ \sigma_2) \quad \text{for every } \gamma_i \in \mathrm{C}([0, 1]; (X, \tau)) \tag{2.2.12}$$

and the set

$$\mathrm{C}([0, 1]; (X, \tau, \delta)) := \big\{\gamma \in \mathrm{C}([0, 1]; (X, \tau)) : \lim_{s \to t} \delta(\gamma(s), \gamma(t)) = 0 \quad \text{for every } t \in [0, 1]\big\}. \tag{2.2.13}$$

It is not difficult to check that

$$\gamma \in \mathrm{C}([0, 1]; (X, \tau, \delta)) \quad \Rightarrow \quad \lim_{r \downarrow 0} \sup_{|t-s| \le r} \delta(\gamma(s), \gamma(t)) = 0; \tag{2.2.14}$$

in the case $\delta = \mathsf{d}$ we have

$$\mathrm{C}([0, 1]; (X, \tau, \mathsf{d})) = \mathrm{C}([0, 1]; (X, \mathsf{d})). \tag{2.2.15}$$

Theorem 2.2.4 (Reparametrizations and Semidistances) *Let $\delta : X \times X \to [0, +\infty]$ be a τ-l.s.c. extended semidistance and let $\gamma_i \in \mathrm{C}([0, 1]; (X, \tau, \delta))$, $i = 1, 2, 3$. We have*

$$\delta_{\mathrm{A}}(\gamma_1, \gamma_2) = \inf_{\sigma \in \Sigma'} \delta_{\mathrm{C}}(\gamma_1, \gamma_2 \circ \sigma) \tag{2.2.16a}$$

$$= \min_{\sigma_i \in \Sigma_2} \delta_C(\gamma_1 \circ \sigma_1, \gamma_2 \circ \sigma_2) \tag{2.2.16b}$$

$$= \inf_{\gamma_i' \sim \gamma_i} \delta_C(\gamma_1', \gamma_2'). \tag{2.2.16c}$$

In particular δ_A satisfies the triangle inequality

$$\delta_A(\gamma_1, \gamma_3) \le \delta_A(\gamma_1, \gamma_2) + \delta_A(\gamma_2, \gamma_3). \tag{2.2.17}$$

Proof If $\sigma \in \Sigma$ we will still denote by σ its extension to $\mathbb{R}$ defined by the map $r \mapsto \sigma(0 \vee r \wedge 1)$. We introduce the $(1+\varepsilon)$-Lipschitz map $j_{\sigma,\varepsilon} : \mathbb{R} \to \mathbb{R}$ defined by

$$s = j_{\sigma,\varepsilon}(r) \quad \Leftrightarrow \quad s + \varepsilon\sigma(s) = (1+\varepsilon)r \tag{2.2.18}$$

and the maps $\hat{\sigma}_\varepsilon, \sigma_\varepsilon : \mathbb{R} \to \mathbb{R}$

$$\hat{\sigma}_\varepsilon(r) := (1+\varepsilon)^{-1}(\varepsilon r + \sigma(r)), \quad |\hat{\sigma}_\varepsilon(r) - \sigma(r)| \le \frac{\varepsilon}{1+\varepsilon}|r - \sigma(r)| \le \varepsilon \quad \text{if } r \in [0,1], \tag{2.2.19}$$

$$\sigma_\varepsilon(r) := \varepsilon^{-1}((1+\varepsilon)r - j_{\sigma,\varepsilon}(r)) = \sigma(j_{\sigma,\varepsilon}(r)). \tag{2.2.20}$$

Notice that the restrictions to $[0, 1]$ of the maps $\hat{\sigma}_\varepsilon, j_{\sigma,\varepsilon}, \sigma_\varepsilon$ operate in $[0, 1]$ and $\hat{\sigma}_\varepsilon \in \Sigma'$, $j_{\sigma,\varepsilon}, \sigma_\varepsilon \in \Sigma$, σ_ε is $(1 + 1/\varepsilon)$-Lipschitz. We also denote by ω_γ the δ-modulus of continuity of $\gamma \in C([0, 1]; (X, \tau, \delta))$, i.e.

$$\omega_\gamma(r) := \sup \Big\{\delta(\gamma(s), \gamma(t)) : s, t \in [0,1],\ |s - t| \le r\Big\}, \tag{2.2.21}$$

observing that $\lim_{r\downarrow 0} \omega_{\gamma_i}(r) = 0$ in the case of the curves γ_1, γ_2 considered by the lemma.

In order to prove (2.2.16a), we observe that that δ_C is invariant w.r.t. composition with arbitrary $\sigma \in \Sigma$

$$\delta_C(\gamma_1, \gamma_2) = \delta_C(\gamma_1 \circ \sigma, \gamma_2 \circ \sigma) \quad \text{for every } \gamma_i \in C([0,1]; X),\ \sigma \in \Sigma, \tag{2.2.22}$$

and every $\sigma \in \Sigma$ can be uniformly approximated by the increasing homeomorphisms $\hat{\sigma}_\varepsilon \in \Sigma'$; we easily get

$$\delta_C(\gamma \circ \sigma, \gamma \circ \hat{\sigma}_\varepsilon) \le \omega_\gamma\big(\sup_{r\in[0,1]} |\sigma(r) - \hat{\sigma}_\varepsilon(r)|\big) \overset{(2.2.19)}{\le} \omega_\gamma(\varepsilon), \tag{2.2.23}$$

so that the triangle inequality for δ_C yields

$$\begin{aligned}\delta_C(\gamma_1 \circ \sigma_1, \gamma_2 \circ \sigma_2) &\geq \delta_C(\gamma_1 \circ \hat{\sigma}_{1,\varepsilon}, \gamma_2 \circ \sigma_2) - \omega_{\gamma_1}(\varepsilon) \geq \delta_C(\gamma_1, \gamma_2 \circ \sigma_2') - \omega_{\gamma_1}(\varepsilon) \\ &\geq \delta_C(\gamma_1, \gamma_2 \circ \hat{\sigma}_{2,\varepsilon}') - \omega_{\gamma_1}(\varepsilon) - \omega_{\gamma_2}(\varepsilon) \\ &\geq \inf_{\sigma \in \Sigma'} \delta_C(\gamma_1, \gamma_2 \circ \sigma) - \omega_{\gamma_1}(\varepsilon) - \omega_{\gamma_2}(\varepsilon)\end{aligned}$$

where $\sigma_2' := \sigma_2 \circ (\hat{\sigma}_{1,\varepsilon})^{-1}$ and $\hat{\sigma}_{2,\varepsilon}'$ is obtained by $(\sigma_2')_\varepsilon$ as in (2.2.19). Taking the infimum w.r.t. $\sigma_1, \sigma_2 \in \Sigma$ and passing to the limit as $\varepsilon \downarrow 0$ we obtain

$$\delta_A(\gamma_1, \gamma_2) \geq \inf_{\sigma \in \Sigma'} \delta_C(\gamma_1, \gamma_2 \circ \sigma).$$

Since the opposite inequality is obvious, we get (2.2.16a).

The triangle inequality is an immediate consequence of (2.2.16a): if $\gamma_1, \gamma_2, \gamma_3 \in \mathrm{C}([0,1]; (X, \tau, \delta))$ and $\sigma_1, \sigma_3 \in \Sigma'$ we have

$$\delta_A(\gamma_1, \gamma_3) \leq \delta_C(\gamma_1 \circ \sigma_1, \gamma_3 \circ \sigma_3) \leq \delta_C(\gamma_1 \circ \sigma_1, \gamma_2) + \delta_C(\gamma_2, \gamma_3 \circ \sigma_3);$$

taking the infimum w.r.t. σ_1, σ_3 we obtain (2.2.17).

Let us now prove (2.2.16b), i.e. the infimum in (2.2.12) is attained at a pair $\varrho_i \in \Sigma$, $i = 1, 2$, given by 2-Lipschitz maps. We observe that

$$\delta_C(\gamma_1, \gamma_2 \circ \varrho) \overset{(2.2.22)}{=} \delta_C(\gamma_1 \circ j_{\varrho,\varepsilon}, \gamma_2 \circ \varrho \circ j_{\varrho,\varepsilon}) \overset{(2.2.20)}{=} \delta_C(\gamma_1 \circ j_{\varrho,\varepsilon}, \gamma_2 \circ \varrho_\varepsilon) \tag{2.2.24}$$

and $j_{\varrho,\varepsilon}$ is $(1+\varepsilon)$-Lipschitz, ϱ_ε is $(1+\varepsilon^{-1})$-Lipschitz. Choosing $\varepsilon = 1$ we deduce that the infimum in (2.2.12) can be restricted to Σ_2. Since Σ_2 is compact w.r.t. uniform convergence, γ_i are continuous and δ_C is lower semicontinuous w.r.t. τ_C, we obtain (2.2.16b).

Let us eventually consider (2.2.16c); since $\inf_{\gamma_i' \sim \gamma_i} \delta_C(\gamma_1', \gamma_2') \leq \delta_A(\gamma_1, \gamma_2)$ it is sufficient to prove the opposite inequality. By (2.2.17) we have

$$\delta_C(\gamma_1', \gamma_2') \geq \delta_A(\gamma_1', \gamma_2') \geq -\delta_A(\gamma_1', \gamma_1) + \delta_A(\gamma_1, \gamma_2) - \delta_A(\gamma_2, \gamma_2') = \delta_A(\gamma_1, \gamma_2).$$

□

Corollary 2.2.5 *The relation $\sim$ satisfies the transitive property and it is an equivalent relation. Moreover*

(a) *The space $\mathrm{A}(X, \tau) := \mathrm{C}([0,1]; (X, \tau))/\sim$ endowed with the quotient topology τ_A is an Hausdorff space. We will denote by $[\gamma]$ the corresponding equivalence class associated to $\gamma \in \mathrm{C}([0,1]; (X, \tau))$ and by $\mathsf{q} : \mathrm{C}([0,1]; (X, \tau)) \to \mathrm{A}(X, \tau)$ the quotient map $\gamma \mapsto [\gamma]$.*
(b) *If δ is a τ-continuous semidistance, then δ_A is a τ_A continuous semidistance (considered as a function between equivalence classes of curves).*

(c) *If the topology τ is induced by the distance δ then the quotient topology τ_{A} is induced by δ_{A} (considered as a distance between equivalence classes of curves).*
(d) *If (X, τ) is a Polish space, then $(\mathrm{A}(X, \tau), \tau_{\mathrm{A}})$ is a Souslin metrizable space.*

Proof

(a) Let us first prove the transitivity of $\sim$. Let $\gamma_i \in \mathrm{C}([0, 1]; X)$, $i = 1, 2, 3$, such that $\gamma_1 \sim \gamma_2$ and $\gamma_2 \sim \gamma_3$ and let $K := \cup_{i=1}^{3} \gamma_i([0, 1])$. K is a compact and separable set; applying Remark 2.1.5 we can find a bounded and continuous semidistance δ whose restriction to $K \times K$ induces the τ-topology.

By the very definition (2.2.12) of δ_{A} we get $\delta_{\mathrm{A}}(\gamma_1, \gamma_2) = 0$ and $\delta_{\mathrm{A}}(\gamma_2, \gamma_3) = 0$, so that (2.2.17) yields $\delta_{\mathrm{A}}(\gamma_1, \gamma_3) = 0$ and (2.2.16b) yields $\gamma_1 \sim \gamma_3$.

Let us now show that $(\mathrm{A}(X), \tau_{\mathrm{A}})$ is Hausdorff. We fix two curves $\gamma_1, \gamma_2 \in \mathrm{C}([0, 1]; X)$ such that $[\gamma_1] \neq [\gamma_2]$, we consider the compact and separable subspace $K := \gamma_1([0, 1]) \cup \gamma_2([0, 1])$, and a bounded τ-continuous semidistance δ, generated as in Remark 2.1.5, whose restriction to K induces the τ-topology.

We notice that the maps $\gamma \mapsto \delta_{\mathrm{C}}(\gamma_i, \gamma) = \sup_{t\in[0,1]} \delta(\gamma_i(t), \gamma(t))$ are continuous in $\mathrm{C}([0, 1]; X)$; since the composition maps $\gamma \mapsto \gamma \circ \varrho$, $\varrho \in \Sigma'$, are continuous, we deduce that the maps $\gamma \mapsto \delta_{\mathrm{A}}([\gamma_i], [\gamma]) = \inf_{\varrho\in\Sigma'} \delta_{\mathrm{C}}(\gamma_i, \gamma \circ \varrho)$ are upper semicontinuous from $\mathrm{C}([0, 1]; X)$ to $\mathbb{R}$. By the above discussion, if $[\gamma_1] \neq [\gamma_2]$ we get $\delta_{\mathrm{A}}(\gamma_1, \gamma_2) = \delta > 0$ so that the open sets $U_i := \big\{\gamma \in \mathrm{C}([0, 1]; X) : \delta_{\mathrm{A}}(\gamma_i, \gamma) < \delta/2\big\}$ are disjoint (by the triangle inequality) saturated open neighborhoods of γ_i.

(b) Since δ_{A} is continuous w.r.t. τ_{C} and it is invariant w.r.t. the equivalence relation, it is clear that δ_{A} is continuous w.r.t. τ_{A}.

(c) If $\gamma_0 \in \mathrm{C}([0, 1]; X)$ every τ_{A} open neighborhood U_0 of $[\gamma_0]$ in $\mathrm{A}(X)$ corresponds to a saturated open set U of $\mathrm{C}([0, 1]; X)$ containing γ_0. In particular, there exists $\varepsilon > 0$ such that U contains all the curves $\gamma \in \mathrm{C}([0, 1]; X)$ with $\delta_{\mathrm{C}}(\gamma, \gamma_0) < \varepsilon$ and all the curves of the form $\gamma\circ\sigma$, $\gamma_0\circ\sigma_0$ for arbitrary $\sigma, \sigma_0 \in \Sigma$. It follows that U_0 contains $\{[\gamma] : \delta_{\mathrm{A}}([\gamma], [\gamma_0]) < \varepsilon\}$.

(d) The Souslin property is an immediate consequence of the fact that $(\mathrm{A}(X, \tau), \tau_{\mathrm{A}})$ is obtained as a quotient space of the Polish space $(\mathrm{C}([0, 1]; (X, \tau)), \tau_{\mathrm{C}})$. The metrizability follows by the previous claim.

□

The image of an arc $[\gamma]$ is independent of the parametrization and it will still be denoted by $\mathsf{e}([\gamma])$; we will also set $\mathsf{e}_i([\gamma]) := \mathsf{e}_i(\gamma)$, $i = 0, 1$. Every function $f : \mathrm{A}(X, \tau) \to Y$ is associated to a "lifted" function $\tilde{f} : \mathrm{C}([0, 1]; (X, \tau)) \to Y$, $\tilde{f}(\gamma) := f([\gamma])$, whose values are invariant by reparametrizations of γ. f is continuous (w.r.t. the quotient topology τ_{A}) if and only if $\tilde{f}$ is continuous (w.r.t. the compact-open topology τ_{C}).

Let us now consider the case of an extended metric-topological space. We denote by $\mathrm{A}(X, \mathsf{d})$ the subset of arcs admitting an equivalent d-continuous parametrization (or, equivalently, whose equivalent parametrizations are d-continuous). Notice that $\mathrm{A}(X, \mathsf{d}) \subset \mathrm{A}(X, \tau)$.

Proposition 2.2.6 *Let* (X, τ, d) *be an extended metric-topological space.* $(\mathrm{A}(X, \mathsf{d}), \tau_{\mathrm{A}}, \mathsf{d}_{\mathrm{A}})$ *is an extended metric-topological space.*

Proof The fact that d_{A} is an extended distance on $\mathrm{A}(X, \mathsf{d})$ is a consequence of the previous results (applied to the topology induced by d). Let us now consider the two properties (X1) and (X2) of Definition 2.1.3 separately.

Proof of Property (X1) For every $J \in \mathbb{N}$, let us denote by $\mathcal{P}_J \subset [0, 1]^{J+1}$ the collection of all partitions $P = (t_0, t_1, \cdots t_J)$ with $0 = t_0 \le t_1 \le t_2 \le \cdots \le t_J = 1$. For every $P \in \mathcal{P}_J$ and every finite collection of functions $\mathcal{F} = (f_j)_{j=1}^J \subset \mathrm{Lip}_b(X, \tau, \mathsf{d})$ we consider the functions $F_{P,\mathcal{F}}$, $F_{\mathcal{F}}$ defined by

$$F_{(t_0,t_1,\cdots,t_J),\mathcal{F}}(\gamma) := \max_{1\le j\le J} \max_{t\in[t_{j-1},t_j]} f_j(\gamma(t)), \quad F_{\mathcal{F}}(\gamma) = \min_{P\in\mathcal{P}_J} F_{P,\mathcal{F}}(\gamma). \tag{2.2.25}$$

By fixing the uniform partition $P_J = (0, 1/J, 2/J, \cdots, j/J, \cdots, 1)$ as a reference, it is not difficult to check that

$$F_{\mathcal{F}}(\gamma) = \min_{\gamma'\sim\gamma} F_{P_J,\mathcal{F}}(\gamma'). \tag{2.2.26}$$

Since $F_{\mathcal{F}}$ only depends on the equivalence class of γ, it induces a map (denoted by $\bar{F}_{\mathcal{F}}$) on $\mathrm{A}(X, \mathsf{d})$.

Claim: $\bar{F}_{\mathcal{F}}$ *Is* τ_{A}*-continuous* It is sufficient to show that $F_{\mathcal{F}}$ is τ_{C} continuous. Let us consider the corresponding map $H : \mathcal{P}_J \times \mathrm{C}([0, 1]; \mathbb{R}^J) \to \mathbb{R}$,

$$H((t_0, t_1, \cdots, t_N), \boldsymbol{f}) := \max_{1\le j\le J} \max_{t\in[t_{j-1},t_j]} f_j(t), \quad \boldsymbol{f} := (f_1, \cdots, f_J) \in \mathrm{C}([0, 1]; \mathbb{R}^J). \tag{2.2.27}$$

H is continuous map (where P_J is endowed with the product topology of $[0, 1]^{J+1}$); since P_J is compact, also $\tilde{H}(\boldsymbol{f}) := \min_{(t_0,t_1,\cdots,t_N)\in P_J} H((t_0, t_1, \cdots, t_N), \boldsymbol{f})$ is continuous. Since $F_{\mathcal{F}}(\gamma) = \tilde{H}(\boldsymbol{f} \circ \gamma)$ we conclude that $F_{\mathcal{F}}$ is τ_{C}-continuous.

Claim: $\bar{F}_{\mathcal{F}}$ *Is* d_{A}*-Lipschitz* Let $[\gamma_i] \in \mathrm{A}(X, \mathsf{d})$, $\varepsilon > 0$, and $\varrho \in \Sigma'$ such that $\mathsf{d}_{\mathrm{A}}([\gamma_1], [\gamma_2]) \ge \mathsf{d}_{\mathrm{C}}(\gamma_1 \circ \varrho, \gamma_2) - \varepsilon$. We select a sequence $P \in \mathcal{P}_J$ such that $\bar{F}([\gamma_2]) = F_{P,\mathcal{F}}(\gamma_2)$ and we observe that

$$\begin{aligned}\bar{F}_{\mathcal{F}}([\gamma_1]) - \bar{F}_{\mathcal{F}}([\gamma_2]) &\le F_{\varrho(P),\mathcal{F}}(\gamma_1) - F_{P,\mathcal{F}}(\gamma_2) \le F_{P,\mathcal{F}}(\gamma_1 \circ \varrho) - F_{P,\mathcal{F}}(\gamma_2)\\ &\le L\mathsf{d}_{\mathrm{C}}(\gamma_1 \circ \varrho, \gamma_2) \le L\big(\mathsf{d}_{\mathrm{A}}([\gamma_1], [\gamma_2]) + \varepsilon\big),\end{aligned}$$

where L is the greatest Lipschitz constant of the functions f_j. Since $\varepsilon > 0$ is arbitrary and we can invert the order of γ_1 and γ_2 we conclude that $\bar{F}_{\mathcal{F}}$ is d_{A}-Lipschitz.

Conclusion Now, if U is a saturated open set containing γ_0, thanks to Remark 2.2.1 we can find a collection of open sets $U_j \subset X$ such that

$$\mathrm{U} \supset \mathrm{U}' = \{\gamma \in \mathrm{C}([a,b];X) : \text{there exists } (t_0', t_1', \cdots, t_J') \in \mathcal{P}_J : \gamma([t_{j-1}', t_j']) \subset U_j\},$$

$$\gamma_0 \in \mathrm{U}';$$

in particular, there exist $(t_0, t_1, \cdots, t_J) \in \mathcal{P}_J$ such that $\gamma_0([t_{j-1}, t_j]) \subset U_j$.

Selecting $f_j \in \mathrm{Lip}_b(X, \tau, \mathsf{d})$ so that $f_j|_{X\setminus U_j} \equiv 1$, $f_j|_{\gamma_0([t_{j-1},t_j])} \equiv 0$, as in the proof of Proposition 2.2.2, we conclude that $\bar{F}_{\mathcal{F}}([\gamma_0]) = F_{\mathcal{F}}(\gamma_0) = 0$, $\bar{F}_{\mathcal{F}}|_{\mathrm{A}(X,\mathsf{d})\setminus \mathrm{U}} \geq 1$.

Proof of Property (X2) By Lemma 2.1.4 we can find a directed set I and a monotone family $(\mathsf{d}_i)_{i\in I}$ of τ-continuous, bounded semidistances such that $\mathsf{d} = \sup_I \mathsf{d}_i$. We can therefore introduce the corresponding continuous and bounded semidistances $\mathsf{d}_{\mathrm{C},i}$ on $\mathrm{C}([0,1];X)$ and $\mathsf{d}_{\mathrm{A},i}$ on $\mathrm{A}(X)$ by

$$\mathsf{d}_{\mathrm{C},i}(\gamma_1,\gamma_2) := \sup_{t\in[0,1]} \mathsf{d}_i(\gamma_1(t),\gamma_2(t)), \quad \mathsf{d}_{\mathrm{A},i}([\gamma_1],[\gamma_2]) := \inf\Big\{\mathsf{d}_{\mathrm{C},i}(\gamma_1',\gamma_2') : \gamma_j' \sim \gamma_j\Big\}. \tag{2.2.28}$$

Let us first notice that for every $\gamma_1, \gamma_2 \in \mathrm{C}([0,1];X)$ we have

$$\mathsf{d}_{\mathrm{C}}(\gamma_1,\gamma_2) = \lim_{i\in I} \mathsf{d}_{\mathrm{C},i}(\gamma_1,\gamma_2) = \sup_{i\in I} \mathsf{d}_{\mathrm{C},i}(\gamma_1,\gamma_2). \tag{2.2.29}$$

By claim (b) of Corollary 2.2.5 the semidistances $\mathsf{d}_{\mathrm{A},i}$ are τ_{A}-continuous. Let us now fix $\gamma_1, \gamma_2 \in \mathrm{C}([0,1];(X,\mathsf{d}))$ with d-modulus of continuity $\omega_{\gamma_1}, \omega_{\gamma_2}$ as in (2.2.21); by the reparametrization Theorem 2.2.4 we can find $\varrho_{1,i}, \varrho_{2,i} \in \Sigma_2$ such that

$$\mathsf{d}_{\mathrm{A},i}([\gamma_1],[\gamma_2]) = \mathsf{d}_{\mathrm{C},i}(\gamma_1\circ\varrho_{1,i}, \gamma_2\circ\varrho_{2,i}) \quad \text{for every } i \in I.$$

Since Σ_2 is compact, we can find a directed subset J and a monotone final map $h : J \to I$ such that the subnets $j \mapsto \varrho_{1,h(j)}$ and $j \mapsto \varrho_{2,h(j)}$ are convergent to elements $\varrho_1, \varrho_2 \in \Sigma_2$. By (2.2.29), for every $\varepsilon > 0$ we may find $i_0 \in I$ such that

$$\mathsf{d}_{\mathrm{A}}([\gamma_1],[\gamma_2]) \leq \mathsf{d}_{\mathrm{C}}(\gamma_1\circ\varrho_1, \gamma_2\circ\varrho_2) \leq \mathsf{d}_{\mathrm{C},i}(\gamma_1\circ\varrho_1, \gamma_2\circ\varrho_2) + \varepsilon \quad \text{for every } i \succeq i_0. \tag{2.2.30}$$

On the other hand, there exists $j_0 \in J$ such that $h(j_0) \succeq i_0$ and for every $j \succeq j_0$

$$\sup_{t\in[0,1]} |\varrho_{1,h(j)}(t) - \varrho_1(t)| \leq \varepsilon, \qquad \sup_{t\in[0,1]} |\varrho_{2,h(j)}(t) - \varrho_2(t)| \leq \varepsilon,$$

so that for every $j \succeq j_0$

$$\mathsf{d}_{\mathrm{C},h(j)}(\gamma_1 \circ \varrho_{1,h(j)}, \gamma_2 \circ \varrho_{2,h(j)}) \geq \mathsf{d}_{\mathrm{C},h(j)}(\gamma_1 \circ \varrho_1, \gamma_2 \circ \varrho_2) - \omega_{\gamma_1}(\varepsilon) - \omega_{\gamma_2}(\varepsilon). \tag{2.2.31}$$

Combining (2.2.30) with (2.2.31) we thus obtain

$$\mathsf{d}_{\mathrm{A}}([\gamma_1], [\gamma_2]) \leq \mathsf{d}_{\mathrm{A},h(j)}([\gamma_1], [\gamma_2]) + \varepsilon + \omega_{\gamma_1}(\varepsilon) + \omega_{\gamma_2}(\varepsilon) \tag{2.2.32}$$

for every $j \succeq j_0$. Passing to the limit w.r.t. $j \in J$ we get

$$\begin{aligned}\mathsf{d}_{\mathrm{A}}([\gamma_1], [\gamma_2]) &\leq \lim_{j\in J} \mathsf{d}_{\mathrm{A},h(j)}([\gamma_1], [\gamma_2]) + \varepsilon + \omega_{\gamma_1}(\varepsilon) + \omega_{\gamma_2}(\varepsilon)\\ &= \lim_{i\in I} \mathsf{d}_{\mathrm{A},i}([\gamma_1], [\gamma_2]) + \varepsilon + \omega_{\gamma_1}(\varepsilon) + \omega_{\gamma_2}(\varepsilon),\end{aligned}$$

which yields $\mathsf{d}_{\mathrm{A}}([\gamma_1], [\gamma_2]) = \lim_{i\in I} \mathsf{d}_{\mathrm{A},i}([\gamma_1], [\gamma_2])$ since $\varepsilon > 0$ is arbitrary. □

Corollary 2.2.7 *If τ' is an auxiliary topology for (X, τ, d) according to Definition 2.1.6 then τ'_{A} is an auxiliary topology for $(\mathrm{A}(X, \mathsf{d}), \tau_{\mathrm{A}}, \mathsf{d}_{\mathrm{A}})$.*

Proof Properties (A1,2) of Definition 2.1.6 are immediate. We now select an increasing sequence d_n of τ'-continuous distances satisfying (2.28'); arguing as in the previous proof we obtain that $\mathsf{d}_{\mathrm{A}}([\gamma_1], [\gamma_2]) = \sup_{n\in\mathbb{N}} \mathsf{d}_{\mathrm{A},n}([\gamma_1], [\gamma_2])$ for every $\gamma_1, \gamma_2 \in \mathrm{C}([0, 1]; (X, \mathsf{d}))$, which precisely yields (A3') for τ'_{A}. □

2.2.3 Rectifiable Arcs

If δ is an extended semidistance in X, $\gamma : [a, b] \to X$ and $J \subset [a, b]$ we set

$$\mathrm{Var}_\delta(\gamma; J) := \sup\Big\{\sum_{j=1}^{N} \delta(\gamma(t_j), \gamma(t_{j-1})) : \{t_j\}_{j=0}^{N} \subset J, \quad t_0 < t_1 < \cdots < t_N\Big\}. \tag{2.2.33}$$

$\mathrm{BV}([a, b]; (X, \delta))$ will denote the space of maps $\gamma : [a, b] \to X$ such that $\mathrm{Var}_\delta(\gamma; [a, b]) < \infty$ (we will sometimes omit δ in the case of the distance d). We will set

$$\ell(\gamma) := \mathrm{Var}_{\mathsf{d}}(\gamma; [a, b]), \quad V_\gamma(t) := \mathrm{Var}_{\mathsf{d}}(\gamma; [a, t]) \quad t \in [a, b], \tag{2.2.34}$$

and

$$V_{\gamma,\ell}(t) := \ell(\gamma)^{-1} V_\gamma(t) \quad \text{whenever } \ell(\gamma) > 0, \qquad V_{\gamma,\ell}(t) := 0 \quad \text{if } \ell(\gamma) = 0; \tag{2.2.35}$$

notice that if $\ell(\gamma) = 0$ then γ is constant.

Lemma 2.2.8 *Let (X, τ, d) be an extended metric-topological space. If $\gamma \in \mathrm{C}([a, b]; (X, \tau))$ satisfies $\mathrm{Var}_{\mathsf{d}}(\gamma; [a, b]) < \infty$ then $\gamma \in \mathrm{C}([a, b]; (X, \mathsf{d}))$, and its variation map V_γ is continuous in $[a, b]$. We will set*

$$\begin{aligned} \mathrm{BVC}([a, b]; X) &:= \mathrm{BV}([a, b]; (X, \mathsf{d})) \cap \mathrm{C}([a, b]; (X, \tau)) \\ &= \mathrm{BV}([a, b]; (X, \mathsf{d})) \cap \mathrm{C}([a, b]; (X, \mathsf{d})). \end{aligned}$$

Proof Thanks to the obvious estimate

$$\mathsf{d}(\gamma(r), \gamma(s)) \le V_\gamma(s) - V_\gamma(r) \quad \text{for every } r, s \in [a, b],\ r \le s, \tag{2.2.36}$$

it is easy to check γ is d-continuous at every continuity point of V_γ. Let us fix $r, s, t \in (a, b]$ with $r < s < t$; passing to the limit in the inequality (2.2.36) keeping r fixed and letting $s \uparrow t$ we get by the τ-continuity of γ, the lower semicontinuity of d and the monotonicity of V_γ

$$\mathsf{d}(\gamma(t), \gamma(r)) \le V_\gamma(t-) - V_\gamma(r) \quad V_\gamma(t-) := \lim_{s \uparrow t} V_\gamma(s).$$

We thus obtain $\lim_{r \uparrow t} \mathsf{d}(\gamma(t), \gamma(r)) = 0$. A similar argument yields the right continuity of γ and therefore the continuity of V_γ. □

Since the length functional $\gamma \mapsto \ell(\gamma)$ is lower semicontinuous with respect to the compact-open topology of $\mathrm{C}([a, b]; (X, \tau))$, the set $\mathrm{BVC}([a, b]; X)$ is an F_σ in $\mathrm{C}([a, b]; (X, \tau))$. We consider two other subsets:

$$\mathrm{BVC}_c([a, b]; X) := \big\{\gamma \in \mathrm{BVC}([a, b]; X) : V_\gamma(t) = \ell(\gamma)(t - a)\big\} \tag{2.2.37}$$

whose curves have constant velocity, and

$$\mathrm{BVC}_k([a, b]; X) := \big\{\gamma \in \mathrm{BVC}_c([a, b]; X) : \ell(\gamma) > k\big\} \quad k \in [0, +\infty[\,, \tag{2.2.38}$$

whose curves have length strictly larger than k (and in particular are not constant). $\mathrm{BVC}_c([a, b]; X)$ is a Borel subset of $\mathrm{BVC}([a, b]; X)$ since it can be equivalently characterized by the condition

$$\gamma \in \mathrm{BVC}_c([a, b]; X) \quad \Leftrightarrow \quad \mathrm{Lip}(\gamma; [a, b]) \le \ell(\gamma), \tag{2.2.39}$$

and the maps Lip and ℓ are lower semicontinuous. $\mathrm{BVC}_0([a,b];X)$ is open in $\mathrm{BVC}_c([a,b];X)$.

Corollary 2.2.9 *If* (X,τ,d) *is Polish then* $\mathrm{BVC}([a,b];X)$, $\mathrm{BVC}_c([a,b];X)$, $\mathrm{BVC}_k([a,b];X)$, $k\ge 0$, *are Lusin spaces. If* (X,τ,d) *admits an auxiliary topology* τ' *(in particular if* (X,τ) *is Souslin) then they are* $\mathscr{F}(\mathrm{C}([a,b];X),\tau_{\mathrm{C}})$ *analytic sets.*

Proof The first statement follows by the fact that Borel sets in a Polish space (in this case the space $(\mathrm{C}([a,b];(X,\tau),\tau_{\mathrm{C}}))$ are Lusin.

The second claim is obvious for the F_σ set $\mathrm{BVC}([a,b];X)$; in the case of $\mathrm{BVC}_c([a,b];X)$ and $\mathrm{BVC}_k([a,b];X)$, it follows by the fact that they are Borel sets in the metrizable and separable space $(\mathrm{C}([a,b];(X,\tau'),\tau'_{\mathrm{C}})$, thus are $\mathscr{F}$-analytic (see (A3) in Sect. A.3) for the coarser topology τ'_{C}, and thus $\mathscr{F}$-analytic also with respect to τ_{C}. □

If $f\in \mathrm{C}(X)$ and $\gamma\in\mathrm{BVC}([a,b];X)$ then the integral $\int_\gamma f$ is well defined by Riemann-Stieltjes integration of $f\circ\gamma$ with respect to $\mathrm{d}V_\gamma$; it can also be obtained as the limit of the Riemann sums

$$\int_\gamma f=\lim_{\tau(P)\downarrow 0}\sum_{j=1}^N f(\gamma(\xi_j))\mathsf{d}(\gamma(t_j),\gamma(t_{j-1})):$$
$$P=\{t_0=a\le\xi_1\le t_1\le\xi_2\le\cdots\le t_{N-1}\le\xi_N\le t_N=b\},\qquad \tau(P):=\sup_j|t_j-t_{j-1}| \tag{2.2.40}$$

since in (2.2.40) is equivalent to use $\mathsf{d}(\gamma(t_j),\gamma(t_{j-1}))$ or $V_\gamma(t_j)-V_\gamma(t_{j-1})$.

Notice that for every $\gamma\in\mathrm{BVC}([a,b];X)$ the map $V_\gamma:[a,b]\to[0,\ell(\gamma)]$ is continuous and surjective and

$$\text{there exists a unique }\ell(\gamma)\text{-Lipschitz map }R_\gamma\in\mathrm{BVC}_c([0,1];X)\text{ such that }\gamma=R_\gamma\circ(V_{\gamma,\ell}), \tag{2.2.41}$$

with $|R'_\gamma|(s)=\ell(\gamma)$ a.e. and

$$\int_\gamma f=\int_0^1 f(R_\gamma(s))|R'_\gamma|(s)\,\mathrm{d}s=\ell(\gamma)\int_0^1 f(R_\gamma(s))\,\mathrm{d}s. \tag{2.2.42}$$

Notice that if $\gamma\in\mathrm{BVC}_c([0,1];X)$ and $\ell(\gamma)>0$, then $V_{\gamma,\ell}\in\Sigma$ so that (2.2.41) yields $\gamma\sim R_\gamma$. Denoting by $\vartheta_\gamma:[0,1]\to[a,b]$ the right-continuous pseudo inverse of $V_{\gamma,\ell}$ (when $\ell(\gamma)>0$)

$$\vartheta_\gamma(s):=\max\Big\{t\in[a,b]:V_{\gamma,\ell}(t)=s\Big\}\quad s\in[0,1],\qquad\text{so that }V_{\gamma,\ell}(\vartheta_\gamma(s))=s\quad\text{in }[0,1], \tag{2.2.43}$$

we have $R_\gamma = \gamma \circ \vartheta_\gamma$. When $\ell(\gamma) = 0$ we set $\vartheta_\gamma(s) \equiv b$, and we still have $R_\gamma = \gamma \circ \vartheta_\gamma$. We also notice that

$$\int_\gamma f = \int f \, \mathrm{d}\nu_\gamma \quad \text{where} \quad \nu_\gamma := \ell(\gamma)(R_\gamma)_\sharp(\mathscr{L}^1 \llcorner [0,1]); \tag{2.2.44}$$

by (2.2.44) it is possible to extend the integral to every bounded or nonnegative Borel map $f : X \to \mathbb{R}$.

Lemma 2.2.10 *Let* $\gamma \in \mathrm{BVC}([0,1]; X)$ *and let* $\vartheta : [0,1] \to [0,1]$ *be an increasing map.*

(a) The map $\tilde{\gamma} := \gamma \circ \vartheta$ *belongs to* $\mathrm{BV}([0,1]; (X, \mathsf{d}))$ *and*

$$V_{\gamma\circ\vartheta,\ell}(t) \le V_{\gamma,\ell}(\vartheta(t)) \quad \textit{for every } t \in [0,1].$$

(b) If $\tilde{\gamma} \in \mathrm{C}([0,1]; (X, \tau))$, $\ell(\tilde{\gamma}) = \ell(\gamma)$, *and* $V_{\gamma,\ell}$ *is strictly increasing, then* $\vartheta \in \Sigma$.

(c) If $\vartheta \in \Sigma$ *then* $\tilde{\gamma}$ *still belongs to* $\mathrm{BVC}([0,1]; X)$ *and*

$$V_{\tilde{\gamma},\ell} = V_{\gamma\circ\vartheta,\ell} = V_{\gamma,\ell} \circ \vartheta \tag{2.2.45}$$

and

$$\ell(\tilde{\gamma}) = \ell(\gamma), \quad R_{\tilde{\gamma}} = R_\gamma, \quad \int_\gamma f = \int_{\tilde{\gamma}} f. \tag{2.2.46}$$

Proof Claims (a) and (c) follow easily by the definition (2.2.33), the characterization of R_γ and (2.2.42).

In order to check Claim (b), we choose a point $t \in (0,1)$ (the argument for the case $t = 0$ or $t = 1$ can be easily adapted) and we set $r_- = \lim_{s\uparrow t} \vartheta(s)$, $r_+ = \lim_{s\downarrow t} \vartheta(s)$. The identity $\tilde{\gamma} = \gamma \circ \vartheta$ and the continuity of $\tilde{\gamma}$ yield $\gamma(r_-) = \gamma(r_+)$ and

$$\begin{aligned}\ell(\tilde{\gamma}) &= \mathrm{Var}_{\mathsf{d}}(\tilde{\gamma}; [0,t]) + \mathrm{Var}_{\mathsf{d}}(\tilde{\gamma}; [t,1]) \le \mathrm{Var}_{\mathsf{d}}(\gamma; [0,r_-]) + \mathrm{Var}_{\mathsf{d}}(\gamma; [r_+,1]) \\ &= \ell(\gamma) - \mathrm{Var}_{\mathsf{d}}(\gamma; [r_-,r_+]) = \ell(\gamma)\big(1 + V_{\gamma,\ell}(r_-) - V_{\gamma,\ell}(r_+)\big).\end{aligned}$$

We deduce that $V_{\gamma,\ell}(r_-) = V_{\gamma,\ell}(r_+)$ so that $r_- = r_+$. As similar argument shows that $\vartheta(i) = i$, $i = 0, 1$, so that ϑ is surjective. □

On $\mathrm{BVC}([0,1]; X)$ we introduce the equivalence relation (2.2.11) and we will denote by $\mathrm{RA}(X, \mathsf{d})$ (or simply $\mathrm{RA}(X)$) the quotient space $\mathrm{BVC}([0,1]; X)/\sim$

endowed with the quotient topology τ_{A} induced by $\mathrm{C}([0,1];(X,\tau))$ and with the extended distance

$$\mathsf{d}_{\mathrm{A}}([\gamma_1],[\gamma_2]) := \inf\left\{\mathsf{d}_{\mathrm{C}}(\gamma_1',\gamma_2') : \gamma_i' \sim \gamma_i\right\} = \inf_{\varrho_i\in\Sigma} \mathsf{d}_{\mathrm{C}}(\gamma_1\circ\varrho_1,\gamma_2\circ\varrho_2) \tag{2.2.47}$$

as in (2.2.12). By Proposition 2.2.6 the space $(\mathrm{RA}(X,\mathsf{d}),\tau_{\mathrm{A}},\mathsf{d}_{\mathrm{A}})$ is an extended metric-topological space.

Lemma 2.2.11 (Reparametrizations of Rectifiable Arcs) *Let (X,τ,d) be an extended metric-topological space. We have:*

(a) *If $\gamma \in \mathrm{BVC}([0,1];X)$, $\gamma' \in \mathrm{C}([0,1];X)$ and $\gamma' \sim \gamma$ then $\gamma' \in \mathrm{BVC}([0,1];X)$.*
(b) *For every $\gamma,\gamma' \in \mathrm{BVC}([0,1];X)$ we have*

$$\gamma\sim\gamma' \quad\Leftrightarrow\quad R_\gamma = R_{\gamma'}, \tag{2.2.48}$$

and all the curves γ' equivalent to γ can be described as $\gamma' = R_\gamma\circ\sigma$ for some $\sigma\in\Sigma$.
(c) *For every $\gamma_i \in \mathrm{BVC}([0,1];X)$ the distance d_{A} satisfies* (2.2.16a,b,c) *and we have*

$$\mathsf{d}_{\mathrm{A}}(\gamma_1,\gamma_2) = \inf_{\sigma\in\Sigma'} \mathsf{d}_{\mathrm{C}}(R_{\gamma_1}, R_{\gamma_2}\circ\sigma) \tag{2.2.49a}$$

$$= \min_{\varrho_i\in\Sigma_2} \mathsf{d}_{\mathrm{C}}(R_{\gamma_1}\circ\varrho_1, R_{\gamma_2}\circ\varrho_2). \tag{2.2.49b}$$

(d) *The function ℓ and the evaluation maps $\mathsf{e}_0,\mathsf{e}_1$ are invariant w.r.t. parametrizations, so that we will still denote by ℓ and $\mathsf{e}_0,\mathsf{e}_1$ the corresponding quotient maps. $\ell : \mathrm{A}(X,\tau)\to[0,+\infty]$ is τ_{A}-lower semicontinuous and $\mathsf{e}_0,\mathsf{e}_1 : \mathrm{A}(X,\tau)\to X$ are continuous.*
(e) *If $f : X\to[0,+\infty]$ is lower semicontinuous then the map $\gamma\mapsto\int_\gamma f$ only depends on $[\gamma]$ and it is lower semicontinuous w.r.t. τ_{A} in* $\mathrm{RA}(X,\mathsf{d})$.

Proof

(a) Since $\gamma'\circ\sigma' = \gamma\circ\sigma$ for some $\sigma,\sigma'\in\Sigma$, we have $\ell(\gamma') = \ell(\gamma) < \infty$ by (2.2.46).
(b) The right implication $\Rightarrow$ in (2.2.48) follows by (2.2.46). The converse implication and the representation property follow by (2.2.41) and the fact that $V_{\gamma,\ell}\in\Sigma$, which in particular yield that $\gamma\sim R_\gamma$.

(c) is an immediate consequence of Theorem 2.2.4 for $\delta := \mathsf{d}$.
(d) The function ℓ is lower semicontinuous w.r.t. the τ_{C}-topology being the supremum of lower semicontinuous functions by (2.2.33). Since $\ell(\gamma)$ is independent on the choice of a representative in $[\gamma]$, it is also lower semicontinuous w.r.t. the

τ_{A} topology. A similar argument holds for the initial and final evaluation maps $\mathsf{e}_0, \mathsf{e}_1$.

(e) If f is continuous, we use the representation of the integral by Riemann sums

$$\int_\gamma f = \sup \sum_{j=1}^{N} \Big(\inf_{t\in[t_{j-1},t_j]} f(\gamma(t)) \Big) \mathsf{d}(\gamma(t_j), \gamma(t_{j-1})) : t_0 = 0 < t_1 < \cdots < t_{N-1} < t_N = 1 \tag{2.2.50}$$

which exhibits $\int_\gamma f$ as the supremum of τ_{C} lower semicontinuous functions. The invariance of the integral w.r.t. reparametrization yields the τ_{A} lower semicontinuity.

When f is τ-lower semicontinuous, we can represent it as the supremum of the (directed) set

$$f(x) = \sup_{g\in F} g(x), \quad F := \Big\{ g \in \mathrm{C}_b(X),\ 0 \le g \le f \Big\}.$$

Since ν_γ is a Radon measure, we have

$$\int_\gamma f = \int_X f\, \mathrm{d}\nu_\gamma = \sup_{g\in F} \int_X g\, \mathrm{d}\nu_\gamma = \sup_{g\in F} \int_\gamma g.$$

□

Lemma 2.2.12 $\mathrm{RA}(X, \mathsf{d})$ *is an* F_σ*-subset of* $(\mathrm{A}(X, \tau), \tau_{\mathrm{A}})$.

If (X, τ) *is a Polish space then* $(\mathrm{RA}(X, \mathsf{d}), \tau_{\mathrm{A}})$ *is a Lusin space. If* (X, τ, d) *admits an auxiliary topology* τ' *(in particular if* (X, τ) *is Souslin) then for every* $k \ge 0$ *the (relatively) open subsets*

$$\mathrm{RA}_k(X, \mathsf{d}) := \big\{ \gamma \in \mathrm{RA}(X, \mathsf{d}) : \ell(\gamma) > k \big\} \tag{2.2.51}$$

are $\mathscr{F}(\mathrm{RA}(X, \mathsf{d}))$*-analytic set for the* τ'_{A} *and the* τ_{A}*-topology.*

Proof Notice that $\mathrm{RA}(X, \mathsf{d})$ can be equivalently identified with the F_σ-subset of $\mathrm{A}(X, \tau)$ and of $\mathrm{A}(X, \mathsf{d})$ defined by $\{\gamma \in \mathrm{A}(X, \tau) : \ell(\gamma) < \infty\}$ with the induced topology τ_{A} and the extended distance d_{A} of $\mathrm{A}(X, \mathsf{d})$. From this point of view, $\mathrm{RA}(X, \mathsf{d})$ is a F_σ subset (i.e. it is the countable union of closed sets and therefore it is also a Borel set) since $\mathrm{RA}(X, \mathsf{d}) = \cup_{k\in\mathbb{N}}\{\gamma \in \mathrm{A}(X, \tau) : \ell(\gamma) \le k\}$ and the map $\ell : \mathrm{A}(X, \tau) \to [0, +\infty]$ is lower semicontinuous with respect to τ_{A} thanks to (d) of the previous Lemma 2.2.11. Since ℓ is also τ'_{A}-l.s.c. and τ'_{A} is metrizable, all the sets $\mathrm{RA}_k(X, \mathsf{d})$ are F_σ and thus $\mathscr{F}$-analytic.

Finally, if (X, τ) is Polish, Corollary 2.2.9 shows that $(\mathrm{BVC}_c([0, 1]; X), \tau_{\mathrm{C}})$ is a Lusin space. Lemma 2.2.11 shows that the quotient map q is a continuous bijection of $(\mathrm{BVC}_c([0, 1]; X), \tau_{\mathrm{C}})$ onto $(\mathrm{RA}(X, \mathsf{d}), \tau_{\mathrm{A}})$, so that the latter is Lusin as well. □

We conclude this section with a list of useful properties concerning the compactness in $\mathrm{RA}(X, \mathsf{d})$ and the continuity of the map $\gamma \mapsto \nu_\gamma$ defined by (2.2.44). For every $t \in [0, 1]$ we also introduce the arc-length evaluation maps

$$\hat{\mathsf{e}}_t : \mathrm{RA}(X, \mathsf{d}) \to X, \quad \hat{\mathsf{e}}_t := \hat{\mathsf{e}}_t \circ R, \quad \hat{\mathsf{e}}_t(\gamma) = R_\gamma(t) \quad \text{for every } \gamma \in \mathrm{RA}(X, \mathsf{d}). \tag{2.2.52}$$

When $t = 0, 1$ we still keep the notation γ_t for the initial and final points $\mathsf{e}_t(\gamma) = \hat{\mathsf{e}}_t(\gamma)$.

Theorem 2.2.13

(a) *If γ_i, $i \in I$, is a converging net in* $\mathrm{RA}(X, \mathsf{d})$ *with $\gamma = \lim_{i\in I} \gamma_i$ and $\lim_{i\in I} \ell(\gamma_i) = \ell(\gamma)$ then*

$$\lim_{i\in I} R_{\gamma_i} = R_\gamma \quad \text{in } \mathrm{C}([0, 1]; X), \quad \lim_{i\in I} \hat{\mathsf{e}}_t(\gamma_i) = \hat{\mathsf{e}}_t(\gamma) \quad \text{for every } t \in [0, 1], \tag{2.2.53}$$

and for every bounded and continuous function $f \in \mathrm{C}_b(X, \tau)$ we have

$$\lim_{i\in I} \int_{\gamma_i} f = \int_\gamma f. \tag{2.2.54}$$

In particular, we have

$$\lim_{i\in I} \nu_{\gamma_i} = \nu_\gamma \quad \text{weakly in } \mathcal{M}_+(X). \tag{2.2.55}$$

(b) *The map $\gamma \mapsto \nu_\gamma$ from* $\mathrm{RA}(X, \mathsf{d})$ *to $\mathcal{M}_+(X)$ is universally Lusin measurable.*
(c) *If $i \mapsto \gamma_i$ converges to γ in* $\mathrm{RA}(X)$, $\sup \ell(\gamma_i) < \infty$ *and $\nu_{\gamma_i} \rightharpoonup \mu$ in $\mathcal{M}_+(X)$ with $\mu(X) > 0$, then* $\mathrm{supp}(\mu) = \gamma([0, 1])$.
(d) *The map $[\gamma] \mapsto R_\gamma$ is universally Lusin measurable from* $\mathrm{RA}(X, \mathsf{d})$ *to* $\mathrm{BVC}_c([0, 1]; (X, \mathsf{d}))$ *endowed with the topology τ_{C} and it is also Borel if $\mathbb{X}$ has an auxiliary topology (in particular if (X, τ) is Souslin). For every $t \in [0, 1]$ the maps $\hat{\mathsf{e}}_t$:* $\mathrm{RA}(X, \mathsf{d}) \to X$ *are universally Lusin measurable (and Borel if $\mathbb{X}$ has an auxiliary topology).*
(e) *If $f \in \mathrm{B}_b(X)$ (or $f : X \to [0, +\infty]$ Borel) the map $\gamma \mapsto \int_\gamma f$ is Borel. In particular the family of measures $\{\nu_\gamma\}_{\gamma\in\mathrm{RA}(X)}$ is Borel.*
(f) *If (X, τ) is compact and $\Gamma \subset$* $\mathrm{RA}(X, \mathsf{d})$ *satisfies $\sup_{\gamma\in\Gamma} \ell(\gamma) < +\infty$ then Γ is relatively compact in* $\mathrm{RA}(X, \mathsf{d})$ *w.r.t. the τ_{A} topology.*
(g) *If (X, d) is complete and $\Gamma \subset$* $\mathrm{RA}(X, \mathsf{d})$ *satisfies the following conditions:*

1. $\sup_{\gamma\in\Gamma} \ell(\gamma) < +\infty$;
2. *there exists a τ-compact set $K \subset X$ such that $\mathsf{e}(\gamma) \cap K \neq \emptyset$ for every $\gamma \in \Gamma$;*
3. *$\{\nu_\gamma : \gamma \in \Gamma\}$ is equally tight, i.e. for every $\varepsilon > 0$ there exists a τ-compact set $K_\varepsilon \subset X$ such that $\nu_\gamma(X \setminus K_\varepsilon) \le \varepsilon$ for every $\gamma \in \Gamma$,*

then Γ is relatively compact in $\mathrm{RA}(X, \mathsf{d})$ *w.r.t. the τ_{A} topology.*

Proof

(a) In order to prove (2.2.53) we consider the compactification $(\hat{X}, \hat{\tau}, \hat{\mathsf{d}})$ given by Theorem 2.1.34 (here we can choose, e.g., $\mathscr{A} = \mathrm{Lip}_b(X, \tau, \mathsf{d})$; the measure $\mathfrak{m}$ does not play any role). Clearly the imbedding $\iota : X \to \hat{X}$ extends to a corresponding embedding of $\mathrm{RA}(X, \mathsf{d})$ in $\mathrm{RA}(\hat{X}, \hat{\mathsf{d}})$, simply by setting $\hat{\gamma}(t) := \iota \circ \gamma(t)$ and considering the corresponding equivalence class. We can apply Proposition 2.2.3 to the net $i \mapsto R_{\hat{\gamma}_i} = \hat{R}_{\gamma_i}$ and we find a limit curve $\hat{R}_* \in \mathrm{Lip}([0, 1]; (\hat{X}, \hat{\mathsf{d}}))$ with respect to the topology $\hat{\tau}_{\mathrm{C}}$. Since the projection from $\mathrm{C}([0, 1]; \hat{X})$ to $\mathrm{A}(\hat{X})$ is continuous, we deduce that $[R_*] = \hat{\gamma}$ so that $\hat{R}_*$ takes values in $\iota(X)$ and therefore can be written as $\iota \circ R_*$ for a curve $R_* \in \mathrm{Lip}([0, 1]; (X, \mathsf{d}))$ which is the limit of R_{γ_i} in $\mathrm{C}([0, 1]; X)$. Passing to the limit in the identities

$$\mathsf{d}(R_{\gamma_i}(r), R_{\gamma_i}(s)) \le \ell(\gamma_i)|r - s| \quad \text{we get} \quad \mathsf{d}(R_*(r), R_*(s)) \le \ell(\gamma)|r - s|. \tag{2.2.56}$$

so that $R_* = R_\gamma$.

Let us prove (2.2.54). We set $m := \inf f$ and $M = \sup f$ and we observe that $\ell(\gamma) = \int_\gamma \mathbb{1}$ so that the thesis follows by applying the lower semicontinuity property of Lemma 2.2.11 (e), to the functions $f - m$ and $M - f$.

(b) Let $\mu \in \mathcal{P}(\mathrm{RA}(X))$; since the function ℓ is lower semicontinuous in X, it is Lusin μ-measurable and there exists a sequence of compact sets $K_n \subset X$ with $\lim_{n\to\infty} \mu(X \setminus K_n) = 0$ such that the restriction of ℓ to K_n is continuous. By the previous claim, the restriction of ν to K_n is also continuous.

(c) Let $K := \gamma([0, 1])$; if $y \notin K$ then we can find a function $f \in \mathrm{Lip}_b(X, \tau, \mathsf{d})$ with values in $[0, 1]$ such that $f_{|K} \equiv 0$ and $f(y) = 1$. If $U := \{x \in X : f(x) > 1/2\}$ then there exists $i_0 \in I$ such that $\gamma_i([0, 1]) \cap \overline{U} = \emptyset$ for $i \succeq i_0$. It follows that $\nu_{\gamma_i}(U) = 0$ and therefore

$$\mu(U) \le \liminf_{i \in I} \nu_{\gamma_i}(U) = 0.$$

This shows that $\mathrm{supp}(\mu) \subset K$. If K consists of an isolated point, the thesis then follows. On the other hand, if K contains at least two points and $y \in K$ then for every open neighborhood U of y $\nu_\gamma(U) > 0$ and therefore $\mu(U) > 0$.

(d) The proof of universal measurability follows as in Claim b), by using the continuity property (2.2.53).

Let us now suppose that $\mathbb{X}$ admits an auxiliary topology τ' (thus metrizable and separable) and let us prove that R is Borel from $\mathrm{RA}(X)$ endowed with τ'_{A} to $\mathrm{C}([0, 1]; X)$ endowed with τ_{C} (this implies the same property for the stronger topology τ_{A} on $\mathrm{RA}(X)$). We observe that the map $J : \gamma \to (\gamma, \ell(\gamma))$ is Borel from $(\mathrm{RA}(X), \tau'_{\mathrm{A}})$ to $(\mathrm{RA}(X) \times \mathbb{R}, \tau'_{\mathrm{A}} \times \tau_{\mathbb{R}})$ since the latter topology has a countable base of open sets (thus the Borel σ-algebra coincides with the product of the Borel σ-algebra of the factors) and each component of J is Borel. On the other hand, $G := \{(\gamma, r) \in \mathrm{RA}(X) \times \mathbb{R} : r = \ell(\gamma)\}$ is Borel in $\mathrm{RA}(X) \times \mathbb{R}$ (with the product topology $\tau_{\mathrm{A}} \times \tau_{\mathbb{R}}$) [56, Chapter II, Lemma 12] and Claim (a) shows that the map $\tilde{R} : G \to \mathrm{RA}(X)$, $\tilde{R}(\gamma, r) := R_\gamma$ is continuous in G, so that $R = \tilde{R} \circ J$ is a Borel map. Finally, since $\hat{\mathsf{e}}_t = \mathsf{e}_t \circ R$, the maps $\hat{\mathsf{e}}_t$ are Borel as well.

(e) Let us consider the set $H \subset \mathrm{B}_b(X)$ of functions f such that $\gamma \mapsto \int_\gamma f$ is Borel. H is clearly a vector space and contains the set $C := \{\chi_U,\ U \text{ open in } X\}$, since the map $\gamma \mapsto \int_\gamma \chi_U$ is lower semicontinuous. Since C is closed under multiplication, we can apply the criterium [24, Chap. I, Theorem 21], which shows that $H = \mathrm{B}_b(X)$. A simple truncation argument extends this property to arbitrary nonnegative Borel functions.

(f) The image of $\Gamma_* := R(\Gamma)$ through the arc-length reparametrization R is relatively compact in $\mathrm{C}([0, 1]; (X, \tau))$ by Proposition 2.2.3. Since Γ is the image of Γ_* through the quotient map $\mathsf{q} : \mathrm{C}([0, 1]; X)$ to $\mathrm{A}(X)$, Γ is relatively compact as well.

(g) Let us consider the compactification $(\hat{X}, \hat{\tau}, \hat{\mathsf{d}})$ as in Theorem 2.1.34 and claim (a), and let $[\gamma_i]$ be a net in Γ with $\mu_i := \nu_{\gamma_i}$. We also set $\hat{\mu}_i = \iota_\sharp \mu_i = \nu_{\hat{\gamma}_i}$, $\hat{\gamma}_i = \iota \circ \gamma_i$. It is not restrictive to assume $\gamma_i = R_{\gamma_i}$ so that γ_i is uniformly Lipschitz. We can then apply Proposition 2.2.3 to the net $\hat{\gamma}_i$ in $\mathrm{C}([0, 1]; \hat{X})$ and find a subnet $j \mapsto h(j)$ and a limit curve $\gamma_* \in \mathrm{Lip}([0, 1]; (\hat{X}, \hat{\mathsf{d}}))$ such that $j \mapsto \hat{\gamma}_{h(j)}$ converges to γ_* with respect to $\hat{\tau}_{\mathrm{C}}$. Since the total mass of $\mu_i = \ell(\gamma_i)$ remains bounded, we can also find a further subnet (still denoted by h) and a limit probability measure μ such that $\mu_{h(j)} \rightharpoonup \mu$. Since ι is continuous, we have $\hat{\mu}_{h(j)} \rightharpoonup \hat{\mu} = \iota_\sharp \mu$ with $m := \mu(X) = \hat{\mu}(\hat{X})$.

If $m = 0$ then $\ell(\gamma_*) = 0$ so that γ_* is constant and coincides with a point $\hat{x} \in \hat{X}$. Since the image of every curve γ_i intersects the compact set K we deduce that $\hat{x} = \iota(x)$ for some $x \in K$, so that $\gamma_{h(j)}$ converges to the constant curve γ, $\gamma(t) \equiv x$ w.r.t. τ_{C} and $[\gamma_{h(j)}]$ converges to $[\gamma]$ in $\mathrm{A}(X)$.

If $m > 0$, the uniform tightness condition shows that μ is concentrated on $\cup_{n\in\mathbb{N}} K_{1/n}$ so that $\hat{\mu}(\hat{X} \setminus \iota(X)) = 0$. It follows that $\iota(X)$ is dense in $\mathrm{supp}(\hat{\mu}) = \gamma_*([0, 1])$. Since γ_* is Lipschitz and $\iota(X)$ is complete, and thus d-closed, we conclude that $\gamma_*([0, 1]) \subset \iota(X)$ and therefore $\gamma_* = \iota \circ \gamma$ for a curve $\gamma \in \mathrm{Lip}([0, 1]; X)$. We deduce that $j \mapsto \gamma_{h(j)}$ converges to γ w.r.t. the compact-open topology τ_{C} and therefore $\lim_{j\in J}[\gamma_{h(j)}] = [\gamma]$ w.r.t. τ_{A}.

□

2.2.4 Notes

Section 2.2.1 contains standard material on the compact-open topology (which is well adapted to deal with general topologies τ on X) and its natural role in lifting the metric-topological structure of (X, τ, d) to the space $(\mathrm{C}([a, b]; X), \tau_{\mathrm{C}}, \mathsf{d}_{\mathrm{C}})$. The compactness result of Proposition 2.2.3

combines compactness w.r.t. τ and equicontinuity w.r.t. d, see also [6, Prop. 3.3.1].

Section 2.2.2 devotes some effort to construct a natural notion of invariance by parametrizations for arbitrary continuous curves. Since we did not assume rectifiability, the existence of a canonical arc-length parametrization is not guaranteed and one has to deal with a more general notion where an arbitrary increasing, continuous and surjective change of variable is allowed (see [54] for a similar approach). Here the main properties are provided by Theorem 2.2.4. The construction of an extended metric-topological setting is presented in Proposition 2.2.6: although very natural, it requires a detailed proof. Everything becomes much simpler in the case of Example 2.1.9.

Section 2.2.3 combines the two previous sections to deal with continuous rectifiable arcs. The presentation here slightly differs from [2].

2.3 *Length and Conformal Distances*

2.3.1 The Length Property

To every extended metric space (X, d) it is possible to associate the length distance

$$\mathsf{d}_\ell(x, y) := \inf\Big\{\ell(\gamma) : \gamma \in \mathrm{RA}(X),\ \gamma_0 = x,\ \gamma_1 = y\Big\}; \tag{2.3.1}$$

(X, d) is a length space if $\mathsf{d} = \mathsf{d}_\ell$. (X, d) is a *geodesic space* if for every $x, y \in X$ with $\mathsf{d}(x, y) < \infty$ there exists an arc $\gamma \in \mathrm{RA}(X)$ connecting x to y with $\ell(\gamma) = \mathsf{d}(x, y)$.

It is not difficult to check that the classes of rectifiable arcs for d and for d_ℓ coincide, as well as the corresponding notion of length and integral.

When (X, d) is complete, it is possible to give an equivalent characterization of the length property in terms of the approximate mid-point property: every pair of points $x, y \in X$ with $\mathsf{d}(x, y) < \infty$ admits approximate midpoints

$$\forall \theta > \frac{1}{2}\ \exists z_\theta \in X : \quad \mathsf{d}(x, z_\theta) \vee \mathsf{d}(z_\theta, y) \le \theta \mathsf{d}(x, y). \tag{2.3.2}$$

By iterating the middle point construction, it is possible to show that for every $x, y \in X$ with $\mathsf{d}(x, y) < \infty$ and for every $D > \mathsf{d}(x, y)$ there exists a map $\gamma : \mathbb{D} \to X$ defined on the set of dyadic points in $[0, 1]$, $\mathbb{D} := \{k/2^n : n, k \in \mathbb{N},\ 0 \le k \le 2^n\}$, satisfying

$$\mathsf{d}(\gamma(s), \gamma(t)) \le D|t - s| \quad \text{for every } s, t \in \mathbb{D}. \tag{2.3.3}$$

Thus, if (X, d) is complete the curve γ admits a unique extension to a curve $\tilde{\gamma} \in \mathrm{BVC}([0,1]; X)$ with $\ell(\tilde{\gamma}) \le D$. Since $D > \mathsf{d}(x, y)$ is arbitrary, we conclude that $\mathsf{d}_\ell = \mathsf{d}$.

Notice that if (X, d) satisfies the approximate mid-point property then for every $x, y \in X$, $\varepsilon > 0$, and $L > 1$ there exists a sequence $(x_n)_{n=0}^N \subset X$ such that

$$x_0 = x,\ x_N = y, \qquad \sup_{1\le n\le N} \mathsf{d}(x_{n-1}, x_n) \le \varepsilon, \quad \sum_{n=1}^N \mathsf{d}(x_{n-1}, x_n) \le L\mathsf{d}(x, y). \tag{2.3.4}$$

2.3.2 Conformal Distances

More generally, let $g : X \to (0, \infty)$ be a continuous function satisfying

$$m_g := \inf_X g > 0, \qquad M_g := \sup_X g < \infty. \tag{2.3.5}$$

We can consider g as a conformal metric density, inducing the length distance

$$\mathsf{d}_g(x, y) := \inf\Big\{ \int_\gamma g : \gamma \in \mathrm{RA}(X),\ \gamma(0) = x,\ \gamma(1) = y \Big\}. \tag{2.3.6}$$

It is clear that d_g is an extended distance and satisfies

$$m_g\, \mathsf{d}_\ell(x, y) \le \mathsf{d}_g(x, y) \le M_g\, \mathsf{d}_\ell(x, y) \quad \text{for every } x, y \in X. \tag{2.3.7}$$

By construction, d_g is a length distance, i.e. $(\mathsf{d}_g)_\ell = \mathsf{d}_g$; when $g \equiv 1$ we clearly have $\mathsf{d}_g = \mathsf{d}_\ell$.

We can introduce different inner approximations of d_g. The first one, d'_g, arises by the the following procedure: first of all we set

$$\beta(x, y) := (g(x) \vee g(y))\mathsf{d}(x, y), \qquad \beta_i(x, y) := (g(x) \vee g(y))\mathsf{d}_i(x, y), \tag{2.3.8}$$

where $(\mathsf{d}_i)_{i\in I}$ is a directed family of τ-continuous bounded semidistances generating d by $\mathsf{d}(x, y) = \lim_{i\in I} \mathsf{d}_i(x, y)$ as in Lemma 2.1.4; for every $\varepsilon \in (0, +\infty]$ we first set

$$\begin{aligned}\mathsf{d}_{g,i,\varepsilon}(x, y) := \inf\Big\{ &\sum_{n=1}^N \beta_i(x_{n-1}, x_n) : N \in \mathbb{N},\ (x_n)_{n=0}^N \in X,\\ &x_0 = x,\ x_N = y,\ \mathsf{d}_i(x_{n-1}, x_n) < \varepsilon \Big\} \wedge (M_g \sup \mathsf{d}_i + \varepsilon^{-1}).\end{aligned} \tag{2.3.9}$$

It is not difficult to check that $\mathsf{d}_{g,i,\varepsilon}$ is a bounded τ-continuous semidistance with $\mathsf{d}_{g,i,\varepsilon}(x, y) \le \beta_i(x, y)$ whenever $\mathsf{d}_i(x, y) < \varepsilon$; moreover it is easy to check that if $0 < \varepsilon < \varepsilon'$ and $i \prec j$ we have $m_g\mathsf{d}_i \le \mathsf{d}_{g,i,\varepsilon'} \le \mathsf{d}_{g,j,\varepsilon} \le M_g\mathsf{d}_{j,\ell} \le M_g\mathsf{d}_\ell$. We need a more localized estimate involving the sets

$$D_i(x, y) := \Big\{z \in X : \mathsf{d}_i(z, x) \vee \mathsf{d}_i(z, y) \le \mathsf{d}_i(x, y)\Big\},$$

$$D(x, y) := \Big\{z \in X : \mathsf{d}(z, x) \vee \mathsf{d}(z, y) \le \mathsf{d}(x, y)\Big\},$$

where $x, y \in X$. Notice that $D_i(x, y)$ and $D(x, y)$ are closed sets containing x and y.

Lemma 2.3.1

(a) *For every $x, y \in X$ we have*

$$\mathsf{d}_{g,i,\varepsilon}(x, y) \ge \mathsf{d}_i(x, y) \inf_{D_i(x,y)} g. \tag{2.3.10}$$

(b) *For every $z \in X$, $i \in I$ and $\varepsilon > 0$ the map $h : x \mapsto \mathsf{d}_{g,i,\varepsilon}(x, z)$ belongs to* $\mathrm{Lip}_b(X, \tau, \mathsf{d}_i)$ *with*

$$\mathrm{lip}_{\mathsf{d}_i} h \le g \quad \text{in } X. \tag{2.3.11}$$

(c) *If moreover (X, τ) is compact, then the infimum of g on $D_i(x, y)$ and $D(x, y)$ is attained and*

$$\liminf_{i\in I} \min_{D_i(x,y)} g \ge \min_{D(x,y)} g \quad \text{for every } x, y \in X. \tag{2.3.12}$$

Proof

(a) Let $(x_n)_{n=0}^N$ be any sequence of points connecting x to y as in (2.3.9). If all the points x_n belong to $D_i(x, y)$ then (2.3.10) immediately follows by the inequality

$$\beta_i(x_{n-1}, x_n) \ge \mathsf{d}_i(x_{n-1}, x_n)(g(x_{n-1}) \vee g(x_n)) \ge \mathsf{d}_i(x_{n-1}, x_n) \inf_{D_i(x,y)} g. \tag{2.3.13}$$

If not, there are indexes n such that $\mathsf{d}_i(x_n, x) \vee \mathsf{d}_i(x_n, y) > \mathsf{d}_i(x, y)$. Just to fix ideas, let us suppose that the set of indexes $n \in \{1, \cdots, N-1\}$ such that $\mathsf{d}_i(x_n, x) > \mathsf{d}_i(x, y)$ is not empty and let us call $\bar n$ its minimum, so that $x_n \in D_i(x, y)$ if $0 \le n < \bar n$. It follows that

$$\sum_{n=1}^N \beta_i(x_{n-1}, x_n) \ge \sum_{n=1}^{\bar n} \beta_i(x_{n-1}, x_n) \overset{(2.3.13)}{\ge} \sum_{n=1}^{\bar n} \big(\inf_{D_i(x,y)} g\big)\mathsf{d}_i(x_{n-1}, x_n)$$
$$\ge \big(\inf_{D_i(x,y)} g\big)\mathsf{d}_i(x, x_{\bar n}) \ge \big(\inf_{D_i(x,y)} g\big)\mathsf{d}_i(x, y).$$

A similar argument holds if the set of indexes $n \in \{1, \cdots, N-1\}$ such that $\mathsf{d}_i(x_n, y) > \mathsf{d}_i(x, y)$ is not empty: in this case one can select the largest index.

(b) We first observe that for every $z \in X$ and $\varepsilon > 0$ the map $h : x \mapsto \mathsf{d}_{g,i,\varepsilon}(z, x)$ belongs to $\mathrm{Lip}_b(X, \tau, \mathsf{d}_i)$. In fact the triangle inequality yields

$$|\mathsf{d}_{g,i,\varepsilon}(z, x) - \mathsf{d}_{g,i,\varepsilon}(z, y)| \le \mathsf{d}_{g,i,\varepsilon}(x, y)$$

and

$$\mathsf{d}_{g,i,\varepsilon}(x, y) \le \begin{cases} M_g \mathsf{d}_i(x, y) & \text{if } \mathsf{d}_i(x, y) < \varepsilon; \\ \frac{M_g \sup \mathsf{d}_i + \varepsilon^{-1}}{\varepsilon} \mathsf{d}_i(x, y) & \text{if } \mathsf{d}_i(x, y) \ge \varepsilon. \end{cases}$$

On the other hand, for every $\bar{x} \in X$ the continuity of d_i and of g ensures that there exists a neighborhood $U \in \mathscr{U}_{\bar{x}}$ such that $\mathsf{d}_i(\bar{x}, y) < \varepsilon/2$ and $g(y) \le g(\bar{x}) + \varepsilon$ for every $y \in U$, so that

$$\mathsf{d}_{g,i,\varepsilon}(x, y) \le \beta_i(x, y) \le \mathsf{d}_i(x, y) \sup_U g \le \mathsf{d}_i(x, y)(g(\bar{x}) + \varepsilon) \quad \text{for every } x, y \in U$$

and therefore $\mathrm{Lip}(h, U, \mathsf{d}_i) \le (g(\bar{x}) + \varepsilon)$, $\mathrm{lip}_{\mathsf{d}_i} h(\bar{x}) \le g(\bar{x}) + \varepsilon$. Since $\varepsilon > 0$ is arbitrary we conclude.

(c) Concerning (2.3.12), let $z_i \in D_i(x, y)$, $i \in I$, be a minimizer for g in $D_i(x, y)$, whose existence follows by the compactness of (X, τ) (and therefore of $D_i(x, y)$) and the continuity of d_i. We can find a converging subnet $\alpha \mapsto i(\alpha)$, $\alpha \in A$, such that $z_{i(\alpha)} \to z$, $g(z_{i(\alpha)}) \to g(z) = \liminf_{i \in I} g(z_i)$ and (recalling (2.1.25))

$$\mathsf{d}(z, x) \vee \mathsf{d}(z, y) \le \liminf_{\alpha \in A} \mathsf{d}_{i(\alpha)}(z_{i(\alpha)}, x) \vee \mathsf{d}_{i(\alpha)}(z_{i(\alpha)}, y) \le \liminf_{\alpha \in A} \mathsf{d}_{i(\alpha)}(x, y) = \mathsf{d}(x, y),$$

so that $z \in D(x, y)$. It follows that

$$\liminf_{i \in I} \min_{D_i(x,y)} g = \liminf_{i \in I} g(z_i) = g(z) \ge \min_{D(x,y)} g.$$

□

We then define

$$\mathsf{d}'_g(x, y) := \lim_{\varepsilon \downarrow 0, i \in I} \mathsf{d}_{g,i,\varepsilon}(x, y) = \sup_{\varepsilon > 0, i \in I} \mathsf{d}_{g,i,\varepsilon}(x, y). \tag{2.3.14}$$

Different approximations of d_g are provided by the formula

$$\mathsf{d}''_g(x, y) := \sup \Big\{ |f(x) - f(y)| : \exists i \in I \text{ such that } f \in \mathrm{Lip}_b(X, \tau, \mathsf{d}_i) \text{ and } \mathrm{lip}_{\mathsf{d}_i} f \le g \text{ in } X \Big\} \tag{2.3.15}$$

$$\mathsf{d}_g'''(x,y) := \sup\Big\{|f(x)-f(y)| : \\ f \in \mathrm{Lip}_b(X,\tau,\mathsf{d}) \text{ and } \mathrm{lip}_{\mathsf{d}} f \le g \text{ in } X\Big\}. \tag{2.3.16}$$

When $g \equiv 1$ we will also write $\mathsf{d}_\ell' := \mathsf{d}_1'$, $\mathsf{d}_\ell'' := \mathsf{d}_1''$, $\mathsf{d}_\ell''' := \mathsf{d}_1'''$. In the next Lemma we collect a few results concerning these distances.

Theorem 2.3.2

(a) *If* (X,τ,d) *is an extended metric-topological space, then also* (X,τ,d_g'), (X,τ,d_g'') *and* (X,τ,d_g''') *are extended metric-topological space and we have for every* $x,y \in X$

$$\mathsf{d}_g'(x,y) \le \mathsf{d}_g''(x,y) \le \mathsf{d}_g'''(x,y) \le \mathsf{d}_g(x,y). \tag{2.3.17}$$

(b)

$$(\mathsf{d}_g')_\ell = (\mathsf{d}_g'')_\ell = (\mathsf{d}_g''')_\ell = \mathsf{d}_g. \tag{2.3.18}$$

(c) *If* (X,τ) *is compact then* $\mathsf{d}_g' = \mathsf{d}_g'' = \mathsf{d}_g''' = \mathsf{d}_g$. *In particular,* (X,τ,d_g) *is an extended metric-topological space and* (X,d_g) *is a geodesic space.*

Proof

(a) The fact that we are dealing with extended metric-topological spaces is clear from the construction.

The first inequality $\mathsf{d}_g' \le \mathsf{d}_g''$ in (2.3.17) follows immediately by Lemma 2.3.1(b).

The inequality $\mathsf{d}_g'' \le \mathsf{d}_g'''$ is obvious since the latter is obtained by taking the supremum on a bigger set.

The last inequality $\mathsf{d}_g''' \le \mathsf{d}_g$ easily follows since for every $\gamma \in \mathrm{BVC}([0,1];X)$ and every map $f \in \mathrm{Lip}(X,\tau,\mathsf{d})$ with $\mathrm{lip}_{\mathsf{d}} f \le g$, the composition $\mathrm{f} := f \circ R_\gamma$ is Lipschitz with

$$\big|\mathrm{f}'(t)\big| \le \ell(\gamma)\,\mathrm{lip}_{\mathsf{d}} f(R_\gamma(t)) \le \ell(\gamma) g(R_\gamma(t)) \quad \mathscr{L}^1\text{-a.e. in}[0,1]. \tag{2.3.19}$$

An integration in the interval [0, 1] yields

$$\big|f(\gamma(1)) - f(\gamma(0))\big| \le \int_\gamma g \tag{2.3.20}$$

and a further minimization w.r.t. all the curves γ connecting $x = \gamma(0)$ and $y = \gamma(1)$ yields for every f satisfying (2.3.16)

$$|f(y) - f(x)| \le \mathsf{d}_g(x,y). \tag{2.3.21}$$

Taking the supremum w.r.t. f we conclude.

(b) Since d_g is a length distance, $(\mathsf{d}'_g)_\ell \le \mathsf{d}_g$, so that it is sufficient to prove the converse inequality. Let $x, y \in X$ with $(\mathsf{d}'_g)_\ell(x, y) < D$; we can find $\gamma \in \mathrm{Lip}([0, 1]; (X, \mathsf{d}))$ with $\gamma(0) = x$, $\gamma(1) = y$, and $\mathsf{d}'_g(\gamma(s), \gamma(t)) \le D|s-t|$ for every $0 \le s < t \le 1$. We want to show

$$I := \int_\gamma g \le D. \tag{2.3.22}$$

By (2.3.7) $\mathsf{d}(\gamma(s), \gamma(t)) \le m_g^{-1} D|t-s|$ so that γ is also d-Lipschitz. The map $g \circ \gamma$ is uniformly continuous as well. A standard compactness argument shows that for every $\varepsilon > 0$ there exists $\delta > 0$ such that

$$\inf\Big\{g(z) : z \in X,\ \mathsf{d}(z, x) \le \delta\Big\} \ge g(x) - \varepsilon \quad \text{for every } x \in \gamma([0, 1]). \tag{2.3.23}$$

By (2.2.50) for every $I_1 < I$ and $\varepsilon > 0$ we can find a subdivision $(t_n)_{n=0}^N$ of $[0, 1]$ such that

$$\sum_{n=1}^N \Big(\inf_{[t_{n-1},t_n]} g \circ R_\gamma\Big)\, \mathsf{d}(R_\gamma(t_{n-1}), R_\gamma(t_n)) > I_1, \quad D|t_n - t_{n-1}| \le (\delta \wedge \varepsilon) m_g, \tag{2.3.24}$$

so that, in particular, $\mathsf{d}(R_\gamma(t_{n-1}), R_\gamma(t_n)) \le \varepsilon$. We set

$$m_n := \min_{[t_{n-1},t_n]} g, \quad m_{i,n} := \inf_{D_i(R_\gamma(t_{n-1}), R_\gamma(t_n))} g, \quad i \in I,\ 1 \le n \le N. \tag{2.3.25}$$

We can then find $i_0 \in I$ such that for every $i \succeq i_0$

$$\sum_{n=1}^N m_n\, \mathsf{d}_i(R_\gamma(t_{n-1}), R_\gamma(t_n)) \ge I_1. \tag{2.3.26}$$

Applying (2.3.10) and assuming that $\varepsilon^{-1} > D$, we obtain

$$\begin{aligned} I_1 &\le \sum_{n=1}^N m_{i,n}\, \mathsf{d}_i(R_\gamma(t_{n-1}), R_\gamma(t_n)) + \sum_{n=1}^N (m_n - m_{i,n})\, \mathsf{d}_i(R_\gamma(t_{n-1}), R_\gamma(t_n)) \\ &\le \sum_{n=1}^N m_{i,n}\, \mathsf{d}_i(R_\gamma(t_{n-1}), R_\gamma(t_n)) + m_g^{-1} D \sup_{1\le n\le N} \Big(m_n - m_{i,n}\Big) \end{aligned}$$

$$\leq \sum_{n=1}^{N} \mathsf{d}_{g,i,\varepsilon}(R_\gamma(t_{n-1}), R_\gamma(t_n)) + m_g^{-1} D \sup_{1\le n\le N} \Big(m_n - m_{i,n}\Big)$$

$$\leq D + m_g^{-1} D \sup_{1\le n\le N} \Big(m_n - m_{i,n}\Big)$$

We can now pass to the limit w.r.t. $i \in I$, observing that by (2.3.12) and (2.3.23)

$$\liminf_{i\in I} m_{i,n} \geq \min_{D(R_\gamma(t_{n-1}), R_\gamma(t_n))} g \geq g(R_\gamma(t_n)) - \varepsilon \geq m_n - \varepsilon \tag{2.3.27}$$

since $\mathsf{d}(R_\gamma(t_{n-1}), R_\gamma(t_n)) \leq \delta$. It follows that

$$I_1 \leq D + m_g^{-1} D\varepsilon;$$

since $I_1 < I$ and $\varepsilon > 0$ are arbitrary, we conclude.

(c) We will show that (X, d'_g) *is a geodesic space.* Since (X, τ) is compact, it is sufficient to prove that (X, d'_g) satisfies the approximate mid-point property. In particular $(\mathsf{d}'_g)_\ell = \mathsf{d}'_g$ and the claim will follow by the previous point (b).

Let us fix $x, y \in X$ with $2D := \mathsf{d}'_g(x, y) < \infty$. By definition, for every $\varepsilon > 0$ with $\varepsilon^{-1} > 2D$ we can find $\eta_0 > 0$ with $(M_g \vee 1)\eta_0 < \varepsilon$ and $i_0 \in I$ such that $2D - \varepsilon < \mathsf{d}_{g,i,\eta}(x, y) \leq 2D$ for every $0 < \eta \leq \eta_0$ and $i \succeq i_0$. Therefore, we find points $(x_n)_{n=0}^N \in X$ (depending on i, η) such that $\mathsf{d}_i(x_{n-1}, x_n) < \eta$ and $2D - \varepsilon < \sum_{n=1}^N \beta_i(x_{n-1}, x_n) \leq \mathsf{d}_{g,i,\eta}(x, y) + \varepsilon \leq 2D + \varepsilon$. If $k_0 = \max\{k \leq N : \sum_{n=1}^k \beta_i(x_{n-1}, x_n) \leq D\}$ and $z_{i,\eta} := x_{k_0+1}$, we clearly have

$$\mathsf{d}_{g,i,\eta}(x, z_{i,\eta}) \leq D + \beta_i(x_{k_0}, x_{k_0+1}) \leq D + M_g\eta \leq D + \varepsilon,$$

$$\mathsf{d}_{g,i,\eta}(y, z_{i,\eta}) \leq \sum_{n=k_0+1}^{N} \beta_i(x_{n-1}, x_n) = \sum_{n=1}^{N} \beta_i(x_{n-1}, x_n) - \sum_{n=1}^{k_0+1} \beta_i(x_{n-1}, x_n)$$

$$\leq 2D + \varepsilon - D \leq D + \varepsilon.$$

Let now $(h, k) : J \to \{i \in I : i \succeq i_0\} \times (0, \eta_0)$ be a monotone subnet such that $z_{(h(j),k(j))}$ converges to $z \in X$. Since $(i, \eta) \mapsto \mathsf{d}_{g,i,\eta}$ is monotone, for every $i \in I$ and $\eta > 0$ we have

$$\mathsf{d}_{g,i,\eta}(x, z) = \lim_{j\in J} \mathsf{d}_{g,i,\eta}(x, z_{h(j),k(j)}) \leq \limsup_{j\in J} \mathsf{d}_{g,h(j),k(j)}(x, z_{h(j),k(j)}) \leq D + \varepsilon,$$

$$\mathsf{d}_{g,i,\eta}(y, z) = \lim_{j\in J} \mathsf{d}_{g,i,\eta}(y, z_{h(j),k(j)}) \leq \limsup_{j\in J} \mathsf{d}_{g,h(j),k(j)}(y, z_{h(j),k(j)}) \leq D + \varepsilon.$$

Taking the supremum w.r.t. $i \in I$ and $\eta > 0$ we eventually get

$$\mathsf{d}'_g(x, z) \le D + \varepsilon, \quad \mathsf{d}'_g(y, z) \le D + \varepsilon$$

so that z is an ε-approximate midpoint between x and y.

□

Remark 2.3.3 Notice that when d_g is τ-continuous, then also d is τ-continuous and $\mathsf{d}_g = \mathsf{d}''_g = \mathsf{d}'''_g$. In this case (X, τ, d_g) is an extended metric-topological space.

2.3.3 Duality for Kantorovich-Rubinstein Cost Functionals Induced by Conformal Distances

We apply Theorem 2.3.2 to obtain a useful dual representation for Kantorovich-Rubinstein distances.

Proposition 2.3.4 *Let us suppose that the extended distances d_g and d'_g defined by* (2.3.6) *and* (2.3.14) *coincide (in particular when (X, τ) is compact) and let $\mathsf{K}_{\mathsf{d}_g}$ be the Kantorovich functional induced by d_g. Then for every $\mu_0, \mu_1 \in \mathcal{M}_+(X)$ with the same mass*

$$\mathsf{K}_{\mathsf{d}_g}(\mu_0, \mu_1) = \sup\Big\{ \int \phi_0 \,\mathrm{d}\mu_0 - \int \phi_1 \,\mathrm{d}\mu_1 : \phi_i \in \mathrm{C}_b(X, \tau), \\ \phi_0(x_0) - \phi_1(x_1) \le \mathsf{d}_g(x_0, x_1) \quad \textit{for every } x_0, x_1 \in X \Big\} \tag{2.3.28}$$

$$= \sup\Big\{ \int \phi \,\mathrm{d}(\mu_0 - \mu_1) : \phi \in \mathrm{Lip}_b(X, \tau, \mathsf{d}),\ \mathrm{lip}_{\mathsf{d}}\, \phi \le g \Big\}. \tag{2.3.29}$$

Proof Equation (2.3.28) is a particular case of (2.1.45) for the extended metric-topological space (X, τ, d_g), thanks to Theorem 2.3.2(c).

Concerning (2.3.29), we can first observe that the right hand side is dominated by $\mathsf{K}_{\mathsf{d}_g}$ since every function $\phi \in \mathrm{Lip}_b(X, \tau, \mathsf{d})$ with $\mathrm{lip}_{\mathsf{d}}\, \phi \le g$ belongs to $\mathrm{Lip}_{b,1}(X, \tau, \mathsf{d}_g)$ thanks to (2.3.17) and the very definition of d'''_g given by (2.3.16).

On the other hand, since $\mathsf{d}_g = \mathsf{d}'_g$, the collection $(\mathsf{d}_{g,i,\varepsilon})_{i \in I, \varepsilon > 0}$ is a direct set of continuous and bounded semidistances giving (2.3.14). We can then apply (2.1.44) obtaining

$$\mathsf{K}_{\mathsf{d}_g}(\mu_0, \mu_1) = \lim_{i \in I, \varepsilon \downarrow 0} \mathsf{K}_{\mathsf{d}_{g,i,\varepsilon}}(\mu_0, \mu_1), \tag{2.3.30}$$

so that (2.1.43) yields

$$\mathsf{K}_{\mathsf{d}_g}(\mu_0,\mu_1)=\sup\Big\{\int\phi\,\mathrm{d}(\mu_0-\mu_1):\phi\in\mathrm{Lip}_{b,1}(X,\tau,\mathsf{d}_{g,i,\varepsilon}),\ i\in I,\ \varepsilon>0\Big\}. \tag{2.3.31}$$

On the other hand, using (2.3.11) one immediately sees that

$$\phi\in\mathrm{Lip}_{b,1}(X,\tau,\mathsf{d}_{g,i,\varepsilon}) \quad\Rightarrow\quad \mathrm{lip}_{\mathsf{d}}\,\phi\le\mathrm{lip}_{\mathsf{d}_i}\,\phi\le g.$$

□

2.3.4 Notes

Section 2.3.1 is standard, see e.g. [21]

Section 2.3.2 will play a crucial role in the proof of the identification Theorem for metric Sobolev spaces of Sect. 5.2. One of the main point here is that even in standard metric spaces the length-conformal construction may easily lead to extended distances. Theorem 2.3.2 shows that at least in the compact case we can recover the length-conformal distances by inner approximation with τ-continuous Lipschitz functions. Such kind of constructions and dual representations by local Lipschitz bounds are typical in the study of local properties of Dirichlet forms, see e.g. [15, 59, 60].

Section 2.3.3 contains the natural extension to the Kantorovich distance of the dual characterization $\mathsf{d}_g=\mathsf{d}_g'''$; it will play a crucial role in Sect. 5.2.2. Notice that if d_g is continuous Proposition 2.3.4 could be proven by a more direct argument based on the identity $\mathsf{d}_g=\mathsf{d}_g'''$ and on the classic representation (2.1.46) for d_g.

3 The Cheeger Energy

In all this part we will always refer to this basic setting:

Assumption Let $\mathbb{X}=(X,\tau,\mathsf{d},\mathfrak{m})$ be an extended metric-topological measure space as in Sect. 2.1.2 and let $\mathscr{A}\subset\mathrm{Lip}_b(X,\tau,\mathsf{d})$ be a compatible algebra of functions, according to Definition 2.1.17. For $f\in\mathrm{Lip}_b(X,\tau,\mathsf{d})$ $\mathrm{lip}\,f$ will always refer to the asymptotic Lipschitz constant $\mathrm{lip}_{\mathsf{d}}\,f$ defined in Sect. 2.1.5. We fix an exponent $p\in(1,\infty)$.

3.1 The Strongest Form of the Cheeger Energy

Let us first define the notion of Cheeger energy $\mathsf{CE}_{p,\mathscr{A}}$ associated to $(\mathbb{X}, \mathscr{A})$.

Definition 3.1.1 (Cheeger Energy) For every $\kappa \geq 0$ and $p \in (1, \infty)$ we define the "pre-Cheeger" energy functionals

$$\mathsf{pCE}_p(f) := \int_X \big(\operatorname{lip} f(x)\big)^p \,\mathrm{d}\mathfrak{m} \quad \text{for every } f \in \mathrm{Lip}_b(X, \tau, \mathsf{d}). \tag{3.1.1}$$

The L^p-lower semicontinuous envelope of the restriction to $\mathscr{A}$ of $\mathsf{pCE}_{p,\kappa}$ is the "strong" Cheeger energy

$$\mathsf{CE}_{p,\mathscr{A}}(f) := \inf\Big\{ \liminf_{n\to\infty} \int_X \big(\operatorname{lip} f_n\big)^p \,\mathrm{d}\mathfrak{m} : f_n \in \mathscr{A},\ f_n \to f \text{ in } L^p(X, \mathfrak{m})\Big\}. \tag{3.1.2}$$

When $\mathscr{A} = \mathrm{Lip}_b(X, \tau, \mathsf{d})$ we will simply write $\mathsf{CE}_p(f)$.

Remark 3.1.2 (The Notation CE*)* We used the symbol CE instead of Ch (introduced by [8]) in the previous definition to stress three differences:

- the dependence on the strongest $\operatorname{lip}_d f$ instead of $|\mathrm{D} f|$,
- the restriction to functions in the algebra $\mathscr{A} \subset \mathrm{Lip}_b(X, \tau, \mathsf{d})$,
- the factor 1 instead of $1/p$ in front of the energy integral.

It is not difficult to check that $\mathsf{CE}_p : L^p(X, \mathfrak{m}) \to [0, +\infty]$ is a convex, lower semicontinuous and p-homogeneous functional; it is the greatest L^p-lower semicontinuous functional "dominated" by pCE_p (extended to $+\infty$ whenever a function does not belong to $\mathscr{A}$).

Definition 3.1.3 We denote by $H^{1,p}(\mathbb{X}, \mathscr{A})$ the subset of $L^p(X, \mathfrak{m})$ whose elements f have finite Cheeger energy $\mathsf{CE}_{p,\mathscr{A}}(f) < \infty$: it is a Banach space with norm

$$\|f\|_{H^{1,p}(\mathbb{X},\mathscr{A})} := \Big(\mathsf{CE}_{p,\mathscr{A}}(f) + \|f\|^p_{L^p(X,\mathfrak{m})}\Big)^{1/p}. \tag{3.1.3}$$

When $\mathscr{A} = \mathrm{Lip}_b(X, \tau, \mathsf{d})$ we will simply write $H^{1,p}(\mathbb{X})$.

Remark 3.1.4 ($H^{1,p}(\mathbb{X}, \mathscr{A})$ as Gagliardo Completion [33]) Recall that if $(A, \|\cdot\|_A)$ is a normed vector space continuously imbedded in a Banach space $(B, \|\cdot\|_B)$, the *Gagliardo completion* $A^{B,c}$ is the Banach space defined by

$$A^{B,c} := \Big\{ b \in B : \exists (a_n)_n \subset A,\ \lim_{n\to\infty} \|a_n - b\|_B = 0,\ \sup_n \|a_n\|_A < \infty \Big\} \tag{3.1.4}$$

with norm

$$\|b\|_{A^{B,c}} := \inf\Big\{ \liminf_{n\to\infty} \|a_n\|_A : a_n \in A,\ \lim_{n\to\infty} \|a_n - b\|_B = 0\Big\}. \tag{3.1.5}$$

When $\mathrm{supp}(\mathfrak{m}) = X$, we can identify $\mathscr{A}$ with a vector space A with the norm induced by pCE_p imbedded in $B := L^p(X, \mathfrak{m})$; it is immediate to check that $H^{1,p}(\mathbb{X}, \mathscr{A})$ coincides with the Gagliardo completion of A in B.

Notice that when $\mathfrak{m}$ has not full support, two different elements $f_1, f_2 \in \mathscr{A}$ may give rise to the same equivalence class in $L^p(X, \mathfrak{m})$. In this case, CE_p can be equivalently defined starting from the functional

$$\widetilde{\mathsf{pCE}}_p(f) := \inf\Big\{\mathsf{pCE}_p(\tilde f) : \tilde f \in \mathscr{A},\ \tilde f = f\mathfrak{m}\text{-a.e.}\Big\}, \tag{3.1.6}$$

defined on the quotient space

$$\tilde{\mathscr{A}} := \mathscr{A}/\sim_{\mathfrak{m}}, \quad f_1 \sim_{\mathfrak{m}} f_2 \quad \text{if } f_1 = f_2\mathfrak{m}\text{-a.e.} \tag{3.1.7}$$

3.1.1 Relaxed Gradients and Local Representation of the Cheeger Energy

The Cheeger energy $\mathsf{CE}_{p,\mathscr{A}}$ admits an integral representation in terms of the minimal relaxed gradient $|\mathrm{D}f|_{\star,\mathscr{A}}$: we collect here a series of useful results, which mainly follow by properties (2.1.51a–e) of Lemma 2.1.16 arguing as in [8, Lemma 4.3, 4.4, Prop. 4.8]. Here we have also to take into account the role of the algebra $\mathscr{A}$.

Definition 3.1.5 (Relaxed Gradients) We say that $G \in L^p(X, \mathfrak{m})$ is a $(p, \mathscr{A})$-relaxed gradient of $f \in L^p(X, \mathfrak{m})$ if there exist functions $f_n \in \mathscr{A}$ such that:

(a) $f_n \to f$ in $L^p(X, \mathfrak{m})$ and $\mathrm{lip}\, f_n$ weakly converge to $\tilde G$ in $L^p(X, \mathfrak{m})$;
(b) $\tilde G \le G$ $\mathfrak{m}$-a.e. in X.

We say that G is the minimal $(p, \mathscr{A})$-relaxed gradient of f if its $L^p(X, \mathfrak{m})$ norm is minimal among relaxed gradients. We shall denote by $|\mathrm{D}f|_{\star,\mathscr{A}}$ the minimal relaxed gradient. As usual, we omit the explicit dependence on $\mathscr{A}$ when $\mathscr{A} = \mathrm{Lip}_b(X, \tau, \mathsf{d})$.

Thanks to (2.1.51a) and the reflexivity of $L^p(X, \mathfrak{m})$ one can easily check that

$$S := \Big\{(f, G) \in L^p(X, \mathfrak{m}) \times L^p(X, \mathfrak{m}) : G \text{ is a } (p, \mathscr{A})\text{-relaxed gradient of } f\Big\} \tag{3.1.8}$$

is convex. Its closure follows by the following lemma, which also shows that it is possible to obtain the minimal relaxed gradient as *strong* limit in L^p.

Lemma 3.1.6 (Closure and Strong Approximation of the Minimal Relaxed Gradient)

(a) *If* $(f, G) \in S$ *then there exist functions* $f_n \in \mathscr{A}$, $G_n \in \mathrm{Lip}_b(X, \tau, \mathsf{d})$ $(G_n \in \mathscr{A}$ *if* $\mathscr{A}$ *is adapted) strongly converging to* $f, \tilde{G}$ *in* $L^p(X, \mathfrak{m})$ *with* $\mathrm{lip}\, f_n \le G_n$ *and* $\tilde{G} \le G$.

(b) S *is weakly closed in* $L^p(X, \mathfrak{m}) \times L^p(X, \mathfrak{m})$.

(c) *The collection of all the relaxed gradients of* f *is closed in* $L^p(X, \mathfrak{m})$; *if it is not empty, it contains a unique element of minimal norm and there exist functions* $f_n \in \mathscr{A}$, $G_n \in \mathrm{Lip}_b(X, \tau, \mathsf{d})$ $(G_n \in \mathscr{A}$ *if* $\mathscr{A}$ *is adapted) such that* $G_n \ge \mathrm{lip}\, f_n$ *and*

$$f_n \to f, \quad G_n \to |\mathrm{D}f|_{\star,\mathscr{A}}, \quad \mathrm{lip}\, f_n \to |\mathrm{D}f|_{\star,\mathscr{A}} \quad \textit{strongly in } L^p(X, \mathfrak{m}). \tag{3.1.9}$$

Proof

(a) Since G is a relaxed gradient, we can find functions $h_i \in \mathscr{A}$ such that $h_i \to f$ in $L^p(X, \mathfrak{m})$ and $\mathrm{lip}\, h_i$ weakly converges to $\tilde{G} \le G$ in $L^p(X, \mathfrak{m})$. Since $\mathrm{lip}\, h_i$ are bounded, nonnegative and upper semicontinuous, by Corollary 2.1.38 we can find functions $g_i \in \mathscr{A}(\mathbb{X})$ $(g_i \in \mathscr{A}$ if $\mathscr{A}$ is adapted) such that $g_i \ge \mathrm{lip}\, h_i$ and $\|g_i - \mathrm{lip}\, h_i\|_{L^p(X,\mathfrak{m})} \le 2^{-i}$ so that $\tilde{G}$ is also the weak limit of g_i in $L^p(X, \mathfrak{m})$. By Mazur's lemma we can find a sequence of convex combinations G_n of g_i (thus belonging to $\mathscr{A}$), starting from an index $i(n) \to \infty$, strongly convergent to $\tilde{G}$ in $L^p(X, \mathfrak{m})$; the corresponding convex combinations of h_i, that we shall denote by f_n, still belong to $\mathscr{A}$, converge in $L^p(X, \mathfrak{m})$ to f and $\mathrm{lip}\, f_n$ is bounded from above by G_n, thanks to (2.1.51a).

(b) Let us prove now the weak closure in $L^p(X, \mathfrak{m}) \times L^p(X, \mathfrak{m})$ of S. Since S is convex, it is sufficient to prove that S is strongly closed. If $S \ni (f^i, G^i) \to (f, G)$ strongly in $L^p(X, \mathfrak{m}) \times L^p(X, \mathfrak{m})$, we can find sequences of functions $(f^i_n)_n \in \mathscr{A}$ and of nonnegative functions $(G^i_n)_n \in L^p(X, \mathfrak{m})$ such that

$$f^i_n \overset{n\to\infty}{\longrightarrow} f^i, \quad G^i_n \overset{n\to\infty}{\longrightarrow} \tilde{G}^i \text{ strongly in } L^p(X, \mathfrak{m}), \quad \mathrm{lip}\, f^i_n \le G^i_n, \quad \tilde{G}^i \le G^i.$$

Possibly extracting a suitable subsequence, we can assume that $\tilde{G}^i \rightharpoonup \tilde{G}$ weakly in $L^p(X, \mathfrak{m})$ with $\tilde{G} \le G$; by a standard diagonal argument we can find an increasing sequence $i \mapsto n(i)$ such that $f^i_{n(i)} \to f$, $G^i_{n(i)} \rightharpoonup \tilde{G}$ in $L^p(X, \mathfrak{m})$ and $\mathrm{lip}\, f^i_{n(i)}$ is bounded in $L^p(X, \mathfrak{m})$. By the reflexivity of $L^p(X, \mathfrak{m})$ we can also assume, possibly extracting a further subsequence, that $\mathrm{lip}\, f^i_{n(i)} \rightharpoonup H$. It follows that $H \le \tilde{G} \le G$ so that G is a relaxed gradient for f.

(c) The closure of the collection of the relaxed gradients of f follows by the previous claim. Since the L^p-norm is strictly convex, if it is not empty it contains a unique element of minimal norm.

Let us consider now the minimal relaxed gradient $G := |\mathrm{D}f|_{\star,\mathscr{A}}$ and let f_n, G_n be sequences in $L^p(X, \mathfrak{m})$ as in the first part of the present Lemma. Since $\operatorname{lip} f_n$ is uniformly bounded in $L^p(X, \mathfrak{m})$ it is not restrictive to assume that it is weakly convergent to some limit $H \in L^p(X, \mathfrak{m})$ with $0 \le H \le \tilde{G} \le G$. This implies at once that $H = \tilde{G} = G$ and $\operatorname{lip} f_n$ weakly converges to $|\mathrm{D}f|_{\star,\mathscr{A}}$ (because any limit point in the weak topology of $\operatorname{lip} f_n$ is a relaxed gradient with minimal norm) and that the convergence is strong, since

$$\limsup_{n\to\infty} \int_X (\operatorname{lip} f_n)^p \,\mathrm{d}\mathfrak{m} \le \limsup_{n\to\infty} \int_X G_n^p \,\mathrm{d}\mathfrak{m} = \int_X G^p \,\mathrm{d}\mathfrak{m} = \int_X H^p \,\mathrm{d}\mathfrak{m}.$$

□

Corollary 3.1.7 (Representation of the Cheeger Energy) *A function $f \in L^p(X, \mathfrak{m})$ belongs to $H^{1,p}(\mathbb{X}, \mathscr{A})$ if and only if it admits a p-relaxed gradients. In this case*

$$\mathsf{CE}_{p,\mathscr{A}}(f) = \int_X |\mathrm{D}f|_{\star,\mathscr{A}}^p \,\mathrm{d}\mathfrak{m}. \tag{3.1.10}$$

Remark 3.1.8 (Dependence of $|\mathrm{D}f|_\star$ with respect to p) Notice that $|\mathrm{D}f|_\star$ may depend on p, even for Lipschitz functions, see e.g. [7]. Since in these notes we will keep the exponent p fixed, we will omit to denote this dependence in the notation for $|\mathrm{D}f|_\star$.

We want to show now that if $f \in H^{1,p}(\mathbb{X}, \mathscr{A})$ satisfies the uniform bound $a \le f \le b$ $\mathfrak{m}$-a.e. in X, then there exists a sequence $f_n \in \mathscr{A}$ satisfying (3.1.9) and the same uniform bounds of f. This result is trivial if $\mathscr{A}$ is the algebra of bounded Lipschitz functions, since truncations operates on $\mathscr{A}$. In the general case we use the approximated truncation polynomials of Corollary 2.1.24.

Corollary 3.1.9 *Let $f \in H^{1,p}(\mathbb{X}, \mathscr{A})$ be satisfying the uniform bounds $\alpha \le f \le \beta$ $\mathfrak{m}$-a.e. in X. Then there exists a sequence $(f_n) \subset \mathscr{A}$ satisfying* (3.1.9) *such that $\alpha \le f_n \le \beta$ in X for every $n \in \mathbb{N}$.*

Proof Let $(f_n)_{n\in\mathbb{N}}$ be a sequence in $\mathscr{A}$ as in (3.1.9). Since functions in $\mathscr{A}$ are bounded, we can find a sequence $c_n > 0$ such that $f_n(X) \subset [-c_n, c_n]$. Let us choose a vanishing sequence $\varepsilon_n \downarrow 0$ and consider the truncation polynomials $P_n = P_{\varepsilon_n}^{c_n,\alpha,\beta}$ of Corollary 2.1.24 corresponding to $c := c_n$ and satisfying (2.1.62). We can then define the functions $\tilde{f}_n := P_n \circ f_n$ taking values in $[\alpha, \beta]$ and $h_n := -c_n \vee f \wedge c_n$ taking values in $[\alpha, \beta] \cap [-c_n, c_n]$; since $|P_n(r) - P_n(s)| \le |r - s|$ for every $r, s \in [-c_n, c_n]$ we have as $n \to \infty$

$$\begin{aligned}
\|\tilde{f}_n - f\|_{L^p} &\le \|P_n \circ f_n - P_n \circ h_n\|_{L^p} + \|P_n \circ h_n - h_n\|_{L^p} + \|h_n - f\|_{L^p} \\
&\le \|f_n - h_n\|_{L^p} + \mathfrak{m}(X)^{1/p}\varepsilon_n + \|h_n - f\|_{L^p} \\
&\le \|f_n - f\|_{L^p} + \mathfrak{m}(X)^{1/p}\varepsilon_n + 2\|h_n - f\|_{L^p} \to 0 \quad \text{as } n \uparrow \infty.
\end{aligned}$$

On the other hand (2.1.51d) yields $\operatorname{lip} \tilde{f}_n = |P_n' \circ f| \operatorname{lip} f_n \le \operatorname{lip} f_n$ so that

$$\limsup_{n\to\infty} \int_X |\operatorname{lip} \tilde{f}_n|^p \,\mathrm{dm} \le \limsup_{n\to\infty} \int_X |\operatorname{lip} f_n|^p \,\mathrm{dm} = \int_X |\mathrm{D}f|^p_{\star,\mathscr{A}} \,\mathrm{dm}.$$

Since $|\mathrm{D}f|_{\star,\mathscr{A}}$ is the minimal $(p, \mathscr{A})$-relaxed gradient, we also have

$$\liminf_{n\to\infty} \int_X |\operatorname{lip} \tilde{f}_n|^p \,\mathrm{dm} \ge \int_X |\mathrm{D}f|^p_{\star,\mathscr{A}} \,\mathrm{dm},$$

so that the sequence $\tilde{f}_n$ satisfies the properties stated by the Lemma. □

Corollary 3.1.10 (Lebnitz Rule) *For every $f, g \in H^{1,p}(\mathbb{X}, \mathscr{A}) \cap L^\infty(X, \mathfrak{m})$ we have $fg \in H^{1,p}(\mathbb{X}, \mathscr{A})$ and*

$$|\mathrm{D}(fg)|_{\star,\mathscr{A}} \le |f|\, |\mathrm{D}g|_{\star,\mathscr{A}} + |g|\, |\mathrm{D}f|_{\star,\mathscr{A}}. \tag{3.1.11}$$

Proof It is sufficient to approximate f, g by two uniformly bounded sequences $f_n, g_n \in \mathscr{A}$ thanks to Corollary 3.1.9 and then pass to the limit in (2.1.51b). □

Let us now consider the locality property of the minimal p-relaxed gradient, by adapting the proof of [8] to the case of an arbitrary algebra $\mathscr{A}$.

Lemma 3.1.11 (Locality) *Let G_1, G_2 be $(p, \mathscr{A})$-relaxed gradients of f. Then $\min\{G_1, G_2\}$ and $\chi_B G_1 + \chi_{X\setminus B} G_2$, $B \in \mathscr{B}(X)$, are relaxed gradients of f as well. In particular, for any $(p, \mathscr{A})$-relaxed gradient G of f it holds*

$$|\mathrm{D}f|_{\star,\mathscr{A}} \le G \qquad \mathfrak{m}\text{-a.e. in } X. \tag{3.1.12}$$

Proof It is sufficient to prove that if $B_i \in \mathscr{B}(X)$ with $B_1 \cap B_2 = \emptyset$ and $B_1 \cup B_2 = X$ then $\chi_{B_1} G_1 + \chi_{B_2} G_2$ is a relaxed gradient of f. By approximation, taking into account the closure of the class of relaxed gradients and the inner regularity of $\mathfrak{m}$, we can assume with no loss of generality that B_2 is a compact set (and, in particular, B_1 is open). We can also approximate B_1 by an increasing sequence of compact sets $B_{n,1} \subset B_1$ such that $\mathfrak{m}(B_1 \setminus B_{n,1}) \to 0$.

Let us fix an integer n and consider the compact set $K_n := B_{n,1} \cup B_2$; since $\mathscr{A}$ contains the constants and separates the points of K_n, the restriction of $\mathscr{A}$ to K_n is uniformly dense in $\mathrm{C}(K_n)$ by Stone-Weierstrass Theorem. Being $B_{n,1}$ and B_2 compact and disjoint, the function

$$\chi_n(x) := \begin{cases} 1 & \text{if } x \in B_{n,1} \\ 0 & \text{if } x \in B_2 \end{cases}$$

belongs to $\mathrm{C}(K_n)$ so that for every $\varepsilon > 0$ we can find $\tilde{\chi}_{n,\varepsilon} \in \mathscr{A}$ such that $\sup_{K_{n,1}} |\tilde{\chi}_{n,\varepsilon} - \chi_n| \le \varepsilon/2$. If we compose $\tilde{\chi}_{n,\varepsilon}$ with the truncation polynomial

$P = P^{c,0,1}_{\varepsilon/2}$ of Corollary 2.1.24 corresponding $c := 1 + \sup |\chi_{n,\varepsilon}|$, we obtain the function $\chi_{n,\varepsilon} := P \circ \tilde{\chi}_{n,\varepsilon}$ taking values in [0, 1] and satisfying

$$\sup_{K_n} |\chi_{n,\varepsilon} - P \circ \chi_n| \le \varepsilon/2, \quad \sup_{K_n} |P \circ \chi_n - \chi_n| \le \varepsilon/2$$

since $0 \vee \chi_n \wedge 1 = \chi_n$ on K_n. We deduce that

$$0 \le \chi_{n,\varepsilon} \le 1, \quad 0 \le \chi_{n,\varepsilon} \le \varepsilon \text{ on } B_2, \quad 1 - \varepsilon \le \chi_{n,\varepsilon} \le 1 \text{ on } B_{n,1}. \tag{3.1.13}$$

Let now $h_{k,i} \in \mathscr{A}, i = 1, 2$, functions converging to f in L^p as $k \to \infty$ with $\operatorname{lip} h_{k,i}$ weakly converging to $\tilde{G}_i \le G_i$, and set $f_{k,n,\varepsilon} := \chi_{n,\varepsilon} h_{k,1} + (1 - \chi_{n,\varepsilon}) h_{k,2} \in \mathscr{A}$. Passing first to the limit as $k \uparrow +\infty$, since $f_{k,n,\varepsilon} \to f$, (2.1.51c) immediately gives that $G_{n,\varepsilon} := \chi_{n,\varepsilon} G_1 + (1 - \chi_{n,\varepsilon}) G_2 \ge \chi_{n,\varepsilon} \tilde{G}_1 + (1 - \chi_{n,\varepsilon}) \tilde{G}_2$ is a relaxed gradient of f.

We can now select a vanishing sequence $(\varepsilon_j)_{j \in \mathbb{N}}$ and we pass to the limit as $j \uparrow +\infty$, obtaining (possibly extracting a further subsequence) a limit function χ_n taking values in [0, 1] such that $G_n := \chi_n G_1 + (1 - \chi_n) G_2 \ge \chi_n \tilde{G}_1 + (1 - \chi_n) \tilde{G}_2$ is a relaxed gradient and $\chi_n|_{B_2} = 0$, $\chi_n|_{B_{n,1}} = 1$. We can finally pass to the limit as $n \to \infty$, observing that χ_n converges pointwise $\mathfrak{m}$-a.e. to the characteristic function of B.

For the second part of the statement we argue by contradiction: let G be a relaxed gradient of f and assume that there exists a Borel set B with $\mathfrak{m}(B) > 0$ on which $G < |\mathrm{D} f|_{\star,\mathscr{A}}$. Consider the relaxed gradient $G\chi_B + |\mathrm{D} f|_{\star,\mathscr{A}} \chi_{X \setminus B}$: its L^p norm is strictly less than the L^p norm of $|\mathrm{D} f|_{\star,\mathscr{A}}$, which is a contradiction. □

Theorem 3.1.12 *For every $f, g \in H^{1,p}(\mathbb{X}, \mathscr{A})$ we have*

(a) *(Pointwise sublinearity) For* $|\mathrm{D}(\alpha f + \beta g)|_{\star,\mathscr{A}} \le \alpha |\mathrm{D} f|_{\star,\mathscr{A}} + \beta |\mathrm{D} g|_{\star,\mathscr{A}}$.
(b) *(Locality) For any Borel set $N \subset \mathbb{R}$ with $\mathscr{L}^1(N) = 0$ we have*

$$|\mathrm{D} f|_{\star,\mathscr{A}} = 0 \quad \mathfrak{m}\text{-a.e. on } f^{-1}(N). \tag{3.1.14}$$

In particular for every constant $c \in \mathbb{R}$

$$|\mathrm{D} f|_{\star,\mathscr{A}} = |\mathrm{D} g|_{\star,\mathscr{A}} \ \mathfrak{m}\text{-a.e. on } \{f - g = c\}. \tag{3.1.15}$$

(c) *(Chain rule) If $\phi \in \mathrm{Lip}(\mathbb{R})$ then $\phi \circ f \in H^{1,p}(\mathbb{X}, \mathscr{A})$ with*

$$|\mathrm{D}(\phi \circ f)|_{\star,\mathscr{A}} \le |\phi'(f)| \, |\mathrm{D} f|_{\star,\mathscr{A}}. \tag{3.1.16}$$

Equality holds in (3.1.16) *if ϕ is monotone or* C^1.

(d) *(Normal contractions) If $\phi : \mathbb{R} \to \mathbb{R}$ is a nondecreasing contraction and $\tilde{f} = f + \phi(g - f)$, $\tilde{g} = g + \phi(f - g)$ then*

$$|\mathrm{D}\tilde{f}|^p_{\star,\mathscr{A}} + |\mathrm{D}\tilde{g}|^p_{\star,\mathscr{A}} \le |\mathrm{D}f|^p_{\star,\mathscr{A}} + |\mathrm{D}g|^p_{\star,\mathscr{A}}. \tag{3.1.17}$$

Proof

(a) follows immediately by the convexity of the set S defined by (3.1.8) and (3.1.12).

(b) We first claim that for $\phi : \mathbb{R} \to \mathbb{R}$ continuously differentiable whose derivative ϕ' is Lipschitz on the image of f it holds

$$|\mathrm{D}\phi(f)|_{\star,\mathscr{A}} \le |\phi' \circ f||\mathrm{D}f|_{\star,\mathscr{A}}, \qquad \mathfrak{m}\text{-a.e. in } X, \tag{3.1.18}$$

for any $f \in H^{1,p}(\mathbb{X}, \mathscr{A})$. Equation (3.1.18) easily follows by approximation from (2.1.51d) whenever f is bounded and ϕ is a polynomial: it is sufficient to apply Corollary 3.1.9.

Still assuming the boundedness of f, arbitrary C^1 functions ϕ can be approximated by a sequence of polynomials P_n with respect to the C^1-norm induced by a compact interval containing $f(X)$. Thanks to the weak closure of (3.1.8) we can pass to the limit in (3.1.18) written for P_n and obtain the same bound for ϕ. In particular, for every $f \in \mathscr{A}$ we get

$$|\mathrm{D}(\phi \circ f)|_{\star,\mathscr{A}} \le |\phi' \circ f||\mathrm{D}f|_{\star,\mathscr{A}} \le |\phi' \circ f| \operatorname{lip} f \quad \mathfrak{m}\text{-a.e. in } X. \tag{3.1.19}$$

If now $\phi \in \mathrm{C}^1(\mathbb{R})\cap\mathrm{Lip}(\mathbb{R})$ and $f \in H^{1,p}(\mathbb{X}, \mathscr{A})$ we can use the approximation (3.1.9) and (3.1.19) to obtain a sequence $f_n \in \mathscr{A}$ such that

$$\begin{aligned} \phi \circ f_n &\to \phi \circ f && \text{in } L^p(X, \mathfrak{m}),\\ |\mathrm{D}(\phi \circ f_n)|_{\star,\mathscr{A}} &\rightharpoonup G && \text{in } L^p(X, \mathfrak{m}),\\ |\phi' \circ f_n||\mathrm{D}f_n|_{\star,\mathscr{A}} &\to |\phi' \circ f||\mathrm{D}f|_{\star,\mathscr{A}} && \text{in } L^p(X, \mathfrak{m}), \end{aligned}$$

so that $|\mathrm{D}(\phi \circ f)|_{\star,\mathscr{A}} \le G \le |\phi' \circ f||\mathrm{D}f|_{\star,\mathscr{A}}$.

Now, assume that N is compact. In this case, let $A_n \subset \mathbb{R}$ be open sets such that $A_n \downarrow N$ and $\mathscr{L}^1(A_1) < \infty$. Also, let $\psi_n : \mathbb{R} \to [0, 1]$ be a continuous function satisfying $\chi_N \le \psi_n \le \chi_{A_n}$, and define $\phi_n : \mathbb{R} \to \mathbb{R}$ by

$$\begin{cases} \phi_n(0) = 0,\\ \phi_n'(z) = 1 - \psi_n(z). \end{cases}$$

The sequence (ϕ_n) uniformly converges to the identity map, and each ϕ_n is 1-Lipschitz and C^1. Therefore $\phi_n \circ f$ converge to f in L^2. Taking into account that $\phi_n' = 0$ on N and (3.1.18) we deduce

$$\int_X |\mathrm{D}f|^p_{\star,\mathscr{A}}\,\mathrm{dm} \le \liminf_{n\to\infty}\int_X |\mathrm{D}\phi_n(f)|^p_{\star,\mathscr{A}}\,\mathrm{dm} \le \liminf_{n\to\infty}\int_X |\phi_n'\circ f|^p|\mathrm{D}f|^p_{\star,\mathscr{A}}\,\mathrm{dm}$$
$$= \liminf_{n\to\infty}\int_{X\setminus f^{-1}(N)} |\phi_n'\circ f|^p|\mathrm{D}f|^p_{\star,\mathscr{A}}\,\mathrm{dm} \le \int_{X\setminus f^{-1}(N)} |\mathrm{D}f|^p_{\star,\mathscr{A}}\,\mathrm{dm}.$$

It remains to deal with the case when N is not compact. In this case we consider the finite measure $\mu := f_\sharp\mathfrak{m}$. Then there exists an increasing sequence (K_n) of compact subsets of N such that $\mu(K_n)\uparrow\mu(N)$. By the result for the compact case we know that $|\mathrm{D}f|_{\star,\mathscr{A}} = 0$ $\mathfrak{m}$-a.e. on $\cup_n f^{-1}(K_n)$, and by definition of push forward we know that $\mathfrak{m}(f^{-1}(N\setminus\cup_n K_n)) = 0$.

Equation (3.1.15) then follows if g is identically 0. In the general case we notice that $|\mathrm{D}(f-g)|_{\star,\mathscr{A}}+|\mathrm{D}g|_{\star,\mathscr{A}}$ is a relaxed gradient of f, hence on $\{f-g=c\}$ we conclude that $\mathfrak{m}$-a.e. it holds $|\mathrm{D}f|_{\star,\mathscr{A}} \le |\mathrm{D}g|_{\star,\mathscr{A}}$. Reversing the roles of f and g we conclude.

(c) By 2. and Rademacher Theorem we know that the right hand side is well defined, so that the statement makes sense (with the convention to define $|\phi'\circ f|$ arbitrarily at points x such that ϕ' does not exist at $f(x)$). Also, by (3.1.18) we know that the thesis is true if ϕ is C^1. For the general case, just approximate ϕ with a sequence (ϕ_n) of equi-Lipschitz and C^1 functions, such that $\phi_n'\to\phi'$ a.e. on the image of f.

Let us now consider the monotone case; with no loss of generality we can assume that $0\le\phi'\le 1$. We know that $(1-\phi'(f))|\mathrm{D}f|_{\star,\mathscr{A}}$ and $\phi'(f)|\mathrm{D}f|_{\star,\mathscr{A}}$ are relaxed gradients of $f-\phi(f)$ and f respectively. Since

$$|\mathrm{D}f|_{\star,\mathscr{A}} \le |\mathrm{D}(f-\phi(f))|_{\star,\mathscr{A}} + |\mathrm{D}\phi(f)|_{\star,\mathscr{A}}$$
$$\le \Big((1-\phi'(f)) + \phi'(f)\Big)|\mathrm{D}f|_{\star,\mathscr{A}} = |\mathrm{D}f|_{\star,\mathscr{A}}$$

it follows that all inequalities are equalities $\mathfrak{m}$-a.e. in X.

When ϕ is C^1 we can use the locality property.

(d) Applying Lemma 3.1.6 we find two optimal sequences (f_n), (g_n) of bounded Lipschitz functions satisfying (3.1.9) (w.r.t. f and g respectively). When ϕ is of class C^1, passing to the limit in the inequality (2.1.51e) written for f_n and g_n we easily get (3.1.17). In the general case, we first approximate ϕ by a sequence ϕ_n of nondecreasing contraction of class C^1 converging to ϕ pointwise and then pass to the limit in (3.1.17) written for ϕ_n.

□

Corollary 3.1.13 *If $f_1, \cdots, f_M \in H^{1,p}(\mathbb{X}, \mathscr{A})$ then also the functions $f_+ := f_1 \vee f_2 \vee \cdots \vee f_M$ and $f_- := f_1 \wedge f_2 \wedge \cdots \wedge f_M$ belong to $H^{1,p}(\mathbb{X}, \mathscr{A})$ and*

$$
\begin{aligned}
|\mathrm{D}f_+|_{\star,\mathscr{A}} &= |\mathrm{D}f_j|_{\star,\mathscr{A}} \quad \text{on } A_j := \{x \in X : f_+ = f_j\},\\
|\mathrm{D}f_-|_{\star,\mathscr{A}} &= |\mathrm{D}f_j|_{\star,\mathscr{A}} \quad \text{on } B_j := \{x \in X : f_- = f_j\}.
\end{aligned}
\tag{3.1.20}
$$

3.1.2 Invariance w.r.t. Restriction and Completion

It is obvious that the Cheeger energy and the minimal relaxed gradient are invariant with respect to isomorphisms of e.m.t.m. structures $(\mathbb{X}, \mathscr{A})$, according to Definition 2.1.28. Here we state two simple (and very preliminary) results concerning the behaviour of the Cheeger energy w.r.t. a general measure-preserving embedding ι of $(\mathbb{X}, \mathscr{A})$ into $(\mathbb{X}', \mathscr{A}')$: we keep the same notation of Sect. 2.1.7. We will state a much deeper result in the last Section of these notes, see Theorem 5.3.3.

Lemma 3.1.14 *For every $f' \in H^{1,p}(\mathbb{X}', \mathscr{A}')$ the function $f := \iota^* f'$ belongs to $H^{1,p}(\mathbb{X}, \mathscr{A})$ and*

$$
|\mathrm{D}f|_{\star,\mathscr{A}} \le \iota^*(|\mathrm{D}f'|_{\star,\mathscr{A}'}) \quad \mathfrak{m}\text{-a.e. in } X. \tag{3.1.21}
$$

Proof Let us first observe that if $f' \in \mathrm{Lip}_b(X', \tau', \mathsf{d}')$ and $G' \ge \mathrm{lip}_{\mathsf{d}'} f'$, then

$$
\iota^* G' \ge \mathrm{lip}_{\mathsf{d}}(\iota^* f'). \tag{3.1.22}
$$

In fact, setting $f := \iota^*(f')$ and choosing arbitrary sets $U \in X$ and $U' \in X'$ containing $\iota(U)$, if $L = \mathrm{Lip}(f', U', \mathsf{d}')$ we have

$$
|f(x) - f(y)| = |f'(\iota(x)) - f'(\iota(y))| \le L\mathsf{d}'(\iota(x), \iota(y)) = L\mathsf{d}(x, y) \quad \text{for every } x, y \in U
$$

so that

$$
\mathrm{Lip}(f, U, \mathsf{d}) \le \mathrm{Lip}(f', U', \mathsf{d}'). \tag{3.1.23}
$$

Recalling the definition (2.1.48) and considering the collection of all the open neighborhood of $\iota(x)$ in X', we get (3.1.22).

In order to obtain (3.1.21) it is now sufficient to take an optimal sequence f_n', G_n' as in (3.1.9) for the $(p, \mathscr{A}')$-minimal relaxed gradient $|\mathrm{D}f'|_{\star,\mathscr{A}'}$ of $f' \in H^{1,p}(\mathbb{X}', \mathscr{A}')$ observing that

$$
\iota^* f_n' \to f, \quad \iota^* G_n' \to \iota^* |\mathrm{D}f'|_\star \quad \text{strongly in } L^p(X, \mathfrak{m}),
$$

so that $\iota^* |\mathrm{D}f'|_{\star,\mathscr{A}'}$ is a relaxed gradient for f. □

When $\iota(X)$ is d'-dense in X' (in particular, when X' is a completion according to the definition given in Corollary 2.1.36), we have a better behaviour.

Proposition 3.1.15 *Suppose that $\iota : X \to X'$ is a measure-preserving embedding of $(\mathbb{X}, \mathscr{A})$ into $(X', \mathscr{A}')$ such that*

$$\iota(X) \text{ is } \mathsf{d}'\text{-dense in } X', \quad \iota^*(\mathscr{A}') = \mathscr{A}. \tag{3.1.24}$$

Then ι^* is an isomorphism of $H^{1,p}(\mathbb{X}', \mathscr{A}')$ onto $H^{1,p}(\mathbb{X}, \mathscr{A})$ and for every $f = \iota^* f'$

$$|\mathrm{D}f|_{\star,\mathscr{A}} = \iota^*(|\mathrm{D}f'|_{\star,\mathscr{A}'}) \quad \mathfrak{m}\text{-a.e. in } X. \tag{3.1.25}$$

Proof Let $f \in H^{1,p}(\mathbb{X}, \mathscr{A})$ and let $f_n \in \mathscr{A}$ be an optimal approximating sequence as in (3.1.9).

We want to show that $f' = \iota_* f \in H^{1,p}(\mathbb{X}', \mathscr{A}')$; by Lemma 2.1.31 and (3.1.24), we can find $f'_n \in \mathscr{A}'$ such that $f_n = \iota^* f'_n$. Since ι_* is an L^p-isometry, we know that $f'_n \to f'$ strongly in $L^p(X', \mathfrak{m}')$. Since ι is a homeomorphism between X and $\iota(X)$, if $x' = \iota(x)$ and $g > \mathrm{lip}_{\mathsf{d}} f_n(x)$, we can find an open neighborhood U' of x' such that setting $U := \iota^{-1}(U')$ we have

$$|f_n(z) - f_n(y)| = |f'_n(\iota(z)) - f'_n(\iota(y))| \le g\mathsf{d}'(\iota(z), \iota(y)) = g\mathsf{d}(z, y) \quad \text{for every } z, y \in U. \tag{3.1.26}$$

On the other hand, the d'-density of $\iota(X)$ in X' guarantees that for every $z', y' \in U' \setminus (\iota(U))$ there exist d' balls $B_\delta(z')$, $B_\delta(y')$ of radius δ such that

$$B_\delta(z') \subset U', \ B_\delta(y') \subset U', \quad B_\delta(z') \cap \iota(U) \neq \emptyset, \ B_\delta(y') \cap \iota(U) \neq \emptyset,$$

so that (3.1.26) extends to U' as

$$|f'_n(z') - f'_n(y')| \le g\mathsf{d}(y', z') \quad \text{for every } z', y' \in U'. \tag{3.1.27}$$

We deduce that

$$\mathrm{lip}_{\mathsf{d}'} f'_n(\iota(x)) \le \mathrm{lip}\, f_n(x) \quad \text{for every } x \in X \tag{3.1.28}$$

and therefore

$$\begin{aligned}\limsup_{n\to\infty} \int_{X'} \big(\mathrm{lip}_{\mathsf{d}'} f'_n\big)^p \,\mathrm{d}\mathfrak{m}' &= \limsup_{n\to\infty} \int_{X'} \big(\mathrm{lip}_{\mathsf{d}'} f'_n(\iota(x)\big)^p \,\mathrm{d}\mathfrak{m}(x) \\ &\le \lim_{n\to\infty} \int_X \big(\mathrm{lip}\, f_n(x)\big)^p \,\mathrm{d}\mathfrak{m}(x) = \mathsf{CE}_{p,\mathscr{A}}(f)\end{aligned}$$

We obtain that

$$f' \in H^{1,p}(\mathbb{X}', \mathscr{A}'), \quad \mathsf{CE}_{p,\mathscr{A}''}(f') \le \mathsf{CE}_{p,\mathscr{A}}(f).$$

Thanks to (3.1.21) we also get (3.1.25). □

As an immediate application we obtain that the class of complete e.m.t.m. spaces is the natural setting for the Cheeger energy.

Corollary 3.1.16 (Invariance of the Cheeger Energy by Completion) *If* $\bar{\mathbb{X}} = (\bar{X}, \bar{\tau}, \bar{\mathsf{d}}, \bar{\mathfrak{m}})$ *is the completion of* $\mathbb{X}$ *induced by* $\iota : X \to \bar{X}$ *of Corollary 2.1.36 and* $\bar{\mathscr{A}} = \{\bar{f} : f \in \mathscr{A}\}$*, then* ι^* *is an isomorphism of* $H^{1,p}(\bar{\mathbb{X}}, \bar{\mathscr{A}})$ *onto* $H^{1,p}(\mathbb{X}, \mathscr{A})$ *and*

$$|\mathrm{D}f|_{\star,\mathscr{A}} = \iota^*(|\mathrm{D}\bar{f}|_{\star,\bar{\mathscr{A}}}) \quad \mathfrak{m}\text{-a.e. in } X \quad \text{for every } f = \iota^*\bar{f}. \tag{3.1.29}$$

We conclude with another easy application of the previous results to restrictions, as in Example 2.1.30(c). Recall that if $Y \subset X$ is a τ-dense $\mathfrak{m}$-measurable subset satisfying $\mathfrak{m}(X \setminus Y) = 0$, the restriction to Y is an isomorphism of $L^p(X, \mathfrak{m})$ with $L^p(Y, \mathfrak{m}_Y)$, so that one can compare the Sobolev spaces $H^{1,p}(\mathbb{X}, \mathscr{A})$ and $H^{1,p}(\mathbb{Y}, \mathscr{A}_Y)$. We will denote by $|\mathrm{D}f|_{\star,\mathbb{Y},\mathscr{A}_Y}$ the $(p, \mathscr{A}_Y)$ minimal relaxed gradient in $H^{1,p}(\mathbb{Y}, \mathscr{A}_Y)$.

Corollary 3.1.17 (Restriction) *Let* $\mathbb{X} = (X, \tau, \mathsf{d}, \mathfrak{m})$ *be an e.m.t.m. space and let* $Y \subset X$ *be a* τ*-dense* $\mathfrak{m}$*-measurable subset satisfying* $\mathfrak{m}(X \setminus Y) = 0$*. With the above notation, (the restriction to* Y *of) every function* $f \in H^{1,p}(\mathbb{X}, \mathscr{A})$ *belongs to* $H^{1,p}(\mathbb{Y}, \mathscr{A}_Y)$ *and*

$$|\mathrm{D}f|_{\star,\mathbb{Y},\mathscr{A}_Y} \le |\mathrm{D}f|_{\star,\mathscr{A}} \quad \mathfrak{m}\text{-a.e.} \tag{3.1.30}$$

If moreover Y *is* d*-dense in* X*, then the converse property is also true: for every* $f \in L^p(X, \mathfrak{m})$

$$f \in H^{1,p}(\mathbb{X}, \mathscr{A}) \quad \Leftrightarrow \quad H^{1,p}(\mathbb{Y}, \mathscr{A}_Y), \qquad |\mathrm{D}f|_{\star,\mathbb{Y},\mathscr{A}_Y} = |\mathrm{D}f|_{\star,\mathscr{A}} \quad \mathfrak{m}\text{-a.e.} \tag{3.1.31}$$

3.1.3 Notes

Section 3.1.1 is strongly inspired by Cheeger's work [22] (where the energy is obtained starting from upper gradients instead of the local Lipschitz constants) and follows quite closely the presentation of [7, 8], with the required adjustments due to the presence of a compatible algebra $\mathscr{A}$ instead of $\mathrm{Lip}_b(X, \tau, \mathsf{d})$. Corollary 3.1.9 and the crucial locality Lemma 3.1.11 take advantage of the approximation tools presented in Sect. 2.1.6. Even if a posteriori the Cheeger energy will be independent of $\mathscr{A}$, the role of the algebra should be considered as a technique to get new density results. Moreover, it allows for simpler constructions in many cases, where a distinguished algebra provides better structural properties of the energy, see the final Sect. 5.3.

Section 3.1.2 contains some preliminary facts about the behaviour of the Cheeger energy with respect to measure-preserving embeddings of e.m.t.m. spaces (in particular w.r.t. completion). The possibility to modify the topological and the algebraic properties of the e.m.t.m. setting is one of its strength point.

3.2 Invariance of the Cheeger Energy with Respect to the Core Algebra: The Compact Case

The aim of this section is to study the property of the Cheeger energy with respect to the choice of the core algebra $\mathscr{A}$ in the case of a compact ambient space (X, τ). An important tool is provided by the (generalized) Hopf-Lax flow, which we collect in the next section.

3.2.1 The Metric Hopf-Lax Flow in Compact Spaces

Let (X, τ, d) be a compact extended metric-topological space and let $\delta : X \times X \to [0, +\infty]$ be a τ-l.s.c. continuous extended semidistance (our main examples will be the extended distance d and the continuous semidistances d_i as in Lemma 2.1.4). For every $f \in \mathrm{C}_b(X)$, $x, y \in X$ and $t > 0$ we set

$$F^\delta(t, x, y) := f(y) + \frac{\delta^q(x, y)}{q\, t^{q-1}}, \qquad F(t, x, y) := F^{\mathsf{d}}(t, x, y). \tag{3.2.1}$$

F^δ is a l.s.c. (continuous, if δ is continuous) function bounded from below.

Let us also fix a compact set $K \subset X$ such that there exists a constant $S = S(K, \delta) \in [0, +\infty[$ satisfying

$$\min_{y \in K} \delta(x, y) \le S \quad \text{for every } x \in X. \tag{3.2.2}$$

Equation (3.2.2) is always satisfied if δ is continuous or if $K = X$ (and in this case $S = 0$).

The modified Hopf-Lax evolution is defined by the formula

$$\mathsf{Q}_t^{K,\delta} f(x) := \min_{y \in K} F^\delta(t, x, y) \quad t > 0, \tag{3.2.3}$$

where we will omit to indicate the explicit dependence on K (resp. on δ) when $K = X$ (resp. when $\delta = \mathsf{d}$), thus setting

$$\mathsf{Q}_t^\delta f := \mathsf{Q}_t^{X,\delta} f, \qquad \mathsf{Q}_t f := \mathsf{Q}_t^{X,\mathsf{d}} f. \tag{3.2.4}$$

Since K is compact and $F^\delta(t, x, \cdot)$ takes at least one finite value in K by (3.2.2), the minimum in (3.2.3) is attained: for every $x \in X$ we also set

$$\mathsf{J}_t^{K,\delta} f(x) := \Big\{ y \in K : f(y) + \frac{\delta^q(x, y)}{p t^{q-1}} = \mathsf{Q}_t^{K,\delta} f(x) \Big\}, \quad \mathsf{J}_t^\delta := \mathsf{J}_t^{X,\delta}, \quad \mathsf{J}_t := \mathsf{J}_t^{X,\mathsf{d}} \tag{3.2.5}$$

and

$$\mathsf{D}_t^{K,\delta,+} f(x) := \max_{y \in \mathsf{J}_t^{K,\delta} f(x)} \delta(x, y), \quad \mathsf{D}_t^{K,\delta,-} f(x) := \min_{y \in \mathsf{J}_t^{K,\delta} f(x)} \delta(x, y). \tag{3.2.6}$$

As usual, we set $\mathrm{Osc}(f, X) := \sup_X f - \inf_X f$.

Lemma 3.2.1 (Basic Estimates) *Let $f \in \mathrm{C}_b(X)$ and let $f_t(x) := \mathsf{Q}_t^{K,\delta} f(x)$, $J_t(x) := \mathsf{J}_t^{K,\delta} f$, $D_t^\pm(x) := \mathsf{D}_t^{K,\delta,\pm} f(x)$ be defined as* (3.2.3), (3.2.5), (3.2.6) *for $t > 0$. For every $x, y \in X$, $0 < s < t$, $x' \in J_t(x)$, $y' \in J_s(y)$ we have*

$$\min_X f \le f_t(x) \le \max_X f + \frac{1}{q\, t^{q-1}} S^q \quad \textit{for every } t > 0,\ x \in X, \tag{3.2.7}$$

$$\Big(\frac{D_t^+(x)}{t}\Big)^q \le \min\Big(q\, t^{-1} \mathrm{Osc}(f, X), \big(q \,\mathrm{Lip}(f, X, \delta)\big)^p\Big), \tag{3.2.8}$$

$$\Big(\frac{1}{q s^{q-1}} - \frac{1}{q t^{q-1}}\Big) D_s^+(x) \le f_s(x) - f_t(x) \le \Big(\frac{1}{q s^{q-1}} - \frac{1}{q t^{q-1}}\Big) D_t^-(x), \tag{3.2.9}$$

$$-\Big(\frac{\delta(x, y')}{t}\Big)^{q-1} \delta(x, y) + \frac{1}{p}\Big(\frac{\delta(y, y')}{t}\Big)^q (t - s) \le f_s(y) - f_t(x) \tag{3.2.10}$$

$$\le \Big(\frac{\delta(y, x')}{s}\Big)^{q-1} \delta(x, y) + \frac{1}{p}\Big(\frac{\delta(x, x')}{s}\Big)^q (t - s). \tag{3.2.11}$$

Proof Equation (3.2.7) is immediate. In order to prove (3.2.8) we simply observe that for every $x' \in J_t(x)$

$$\frac{\delta^q(x, x')}{t^{q-1}} \le q\big(f(x) - f(x')\big) \le q \min\big(\mathrm{Osc}(f, X), \mathrm{Lip}(f, X, \delta)\delta(x, x')\big)$$

thus obtaining

$$\Big(\frac{\delta(x,x')}{t}\Big)^q \le \frac{q}{t}\operatorname{Osc}(f,X),\qquad \Big(\frac{\delta(x,x')}{t}\Big)^{q-1} \le q\operatorname{Lip}(f,X,\delta).$$

Let us now check (3.2.11): selecting $x' \in J_t(x)$

$$\begin{aligned} f_s(y) - f_t(x) &\le F^\delta(s,y,x') - F^\delta(t,x,x') = \frac{\delta^q(y,x')}{qs^{q-1}} - \frac{\delta^q(x,x')}{qt^{q-1}} \\ &= \frac{\delta^q(y,x')}{qs^{q-1}} - \frac{\delta^q(x,x')}{qs^{q-1}} + \Big(\frac{1}{qs^{q-1}} - \frac{1}{qt^{q-1}}\Big)\delta^q(x,x') \\ &\le \Big(\frac{\delta(y,x')}{s}\Big)^{q-1}|\delta(y,x') - \delta(x,x')| + \Big(\frac{1}{qs^{q-1}} - \frac{1}{qt^{q-1}}\Big)\delta^q(x,x'). \end{aligned} \tag{3.2.12}$$

Applying the triangle inequality for δ and the elementary inequality (arising from the convexity of $r \mapsto 1/r^{q-1}$ in $(0,\infty)$)

$$\frac{1}{pt^q}(t-s) \le \frac{1}{qs^{q-1}} - \frac{1}{qt^{q-1}} \le \frac{1}{ps^q}(t-s) \quad \text{for every } s,t \in (0,+\infty), \tag{3.2.13}$$

we obtain (3.2.11). (3.2.10) will follow by switching the role of (x,t) and (y,s).

Concerning (3.2.9), the right inequality can be easily obtained by choosing $y = x$ in (3.2.12) and minimizing with respect to $x' \in J_t(x)$. Inverting the role of s and t (notice that (3.2.12) does not require $s < t$) and maximizing with respect to $x' \in J_s(x)$ we get the left inequality of (3.2.9). □

We collect further properties in the next Lemma.

Lemma 3.2.2 *Let us assume that δ is continuous. For every $f \in \mathrm{C}(X)$ we have*

(a) *The map $(x,t) \mapsto \mathsf{Q}_t^{K,\delta} f(x)$ is continuous in $X \times (0,+\infty)$ and for every $0 < s < t$ it satisfies the estimate*

$$\frac{1}{p}\Big(\frac{\mathsf{D}_s^{K,\delta,+} f(x)}{t}\Big)^q (t-s) \le \mathsf{Q}_s^{K,\delta} f(x) - \mathsf{Q}_t^{K,\delta} f(x) \le \frac{1}{p}\Big(\frac{\mathsf{D}_t^{K,\delta,-} f(x)}{s}\Big)^q (t-s). \tag{3.2.14}$$

(b) *The map $(x,t) \mapsto \mathsf{D}_t^{K,\delta,+} f(x)$ (resp. $(x,t) \mapsto \mathsf{D}_t^{K,\delta,-} f(x)$) is upper (resp. lower) semicontinuous in $X \times (0,\infty)$ and there holds*

$$\mathsf{D}_s^{K,\delta,-} f(x) \le \mathsf{D}_s^{K,\delta,+} f(x) \le \mathsf{D}_t^{K,\delta,-} f(x) \quad \text{if} \quad 0 < s < t. \tag{3.2.15}$$

(c) *If $(K_\lambda)_{\lambda\in\Lambda}$ is an increasing net with $\cup_{\lambda\in\Lambda} K_\lambda$ dense in K, then for every $x \in X$ the net $\lambda \mapsto \mathsf{Q}_t^{K_\lambda,\delta} f(x)$ is decreasing and converging to $\mathsf{Q}_t^{K,\delta} f(x)$.*

(d) *If* $(K_\lambda)_{\lambda\in\Lambda}$ *is an increasing net with* $K_\lambda \subset K$*, then*

$$\lim_{\lambda\in\Lambda} \mathsf{Q}_t^{K_\lambda,\delta} f(x) = \mathsf{Q}_t^{K,\delta} f(x) \quad \Rightarrow \quad \limsup_{\lambda\in\Lambda} \mathsf{D}_t^{K_\lambda,\delta,+} f(x) \le \mathsf{D}_t^{K,\delta,+} f(x). \tag{3.2.16}$$

Proof

(a) Continuity follows by (3.2.10) and (3.2.11). Equation (3.2.14) is a consequence of (3.2.9) and (3.2.13).

(b) Equation (3.2.15) follows immediately from (3.2.9). It is not difficult to check that $(t,x) \mapsto \mathsf{D}_t^{K,\delta,+} f(x)$ is upper semicontinuous: if (x_λ, t_λ), $\lambda \in \Lambda$, is a net converging to $(x,t) \in X \times (0,+\infty)$ with $\mathsf{D}_{t_\lambda}^{K,\delta,+}(x_\lambda) \ge c$ and $y_\lambda \in \mathsf{J}_{t_\lambda}^{K,\delta}(x_\lambda)$ such that

$$F^\delta(t_\lambda, x_\lambda, y_\lambda) = \mathsf{Q}_{t_\lambda}^{K,\delta} f(x_\lambda), \quad \delta(x_\lambda, y_\lambda) = \mathsf{D}_{t_\lambda}^{K,\delta,+}(x_\lambda) \ge c, \tag{3.2.17}$$

we can find a subnet $j \mapsto \lambda(j)$, $j \in J$, such that $j \mapsto y_{\lambda(j)}$ converges to a point $y \in K$ with

$$F^\delta(t,x,y) = \lim_{j\in J} F^\delta(t_{\lambda(j)}, x_{\lambda(j)}, y_{\lambda(j)}) = \lim_{j\in J} \mathsf{Q}_{t_{\lambda(j)}}^{K,\delta} f(x_{\lambda(j)}) = \mathsf{Q}_t^{K,\delta} f(x),$$

showing that $y \in \mathsf{J}_t^{K,\delta} f(x)$. Since

$$\delta(x,y) = \lim_{j\in J} \delta(x_{\lambda(j)}, y_{\lambda(j)}) \ge c,$$

we obtain that $\mathsf{D}_t^{K,\delta,+}(x) \ge c$. A similar argument holds for the lower semicontinuity of $\mathsf{D}_t^{K,\delta,-} f$.

(c) The decreasing property of Q w.r.t. λ is obvious; in particular it yields

$$\mathsf{Q}_t^{K_\lambda,\delta} f(x) \ge \mathsf{Q}_t^{K,\delta} f(x). \tag{3.2.18}$$

On the other hand, by the density of $\cup_\lambda K_\lambda$ in K and the continuity of F^δ, for every $y \in \mathsf{J}_t^{K,\delta} f(x)$ and $\varepsilon > 0$ we can find $\lambda_\varepsilon \in \Lambda$ and $y_\varepsilon \in K_{\lambda_\varepsilon}$ such that $F^\delta(t,x,y_\varepsilon) \le F^\delta(t,x,y) + \varepsilon$ so that for every $\lambda \succ \lambda_\varepsilon$

$$\mathsf{Q}_t^{K_\lambda,\delta} f(x) \le F^\delta(t,x,y_\varepsilon) \le F^\delta(t,x,y) + \varepsilon = \mathsf{Q}_t^{K,\delta} f(x) + \varepsilon.$$

Since $\varepsilon > 0$ is arbitrary we obtain the proof of the claim.

(d) We argue as in the proof of the second claim: we select $y_\lambda \in \mathsf{J}_t^{K_\lambda,\delta} f(x) \subset K_\lambda$ so that

$$F(t,x,y_\lambda) = \mathsf{Q}_t^{K_\lambda,\delta} f(x), \quad \mathsf{D}_t^{K_\lambda,\delta,+} f(x) = \delta(x, y_\lambda). \tag{3.2.19}$$

We can find a subnet $j \mapsto y_{\lambda(j)}$, $j \in J$, converging to some $y \in K$ with $S := \limsup_{\lambda\in\Lambda} \mathsf{D}_t^{K_\lambda,\delta,+} f(x) = \lim_{j\in J} \delta(x, y_{\lambda(j)}) = \delta(x, y)$. It follows that

$$\mathsf{Q}_t^{K,\delta} f(x) \overset{(3.2.16)}{=} \lim_{j\in J} \mathsf{Q}_t^{K_{\lambda(j)},\delta} f(x) = \lim_{j\in J} F^\delta(t, x, y_{\lambda(j)}) = F^\delta(t, x, y)$$

so that $y \in \mathsf{J}_t^{K,\delta} f(x)$. This yields

$$\mathsf{D}_t^{K,\delta,+} f(x) \geq \delta(x, y) = \limsup_{\lambda\in\Lambda} \mathsf{D}_t^{K_\lambda,\delta,+} f(x).$$

□

We consider now the behaviour of $\mathsf{Q}^\delta = \mathsf{Q}^{X,\delta}$ with respect to δ.

Proposition 3.2.3 *Let* $(\mathsf{d}_i)_{i\in I}$ *be a directed family of continuous semidistances as in* (2.1.24a,b,c,d) *and let* $f \in C_b(X)$. *For every* $x \in X$ *the net* $i \mapsto \mathsf{Q}_t^{\mathsf{d}_i} f(x)$ *is monotonically converging to* $\mathsf{Q}_t f(x)$ *and*

$$\limsup_{i\in I} \mathsf{D}_t^{\mathsf{d}_i,+} f(x) \leq \mathsf{D}_t^+ f(x). \tag{3.2.20}$$

More generally, if $j \mapsto i(j)$, $j \in J$, *is an increasing net (but not necessarily a subnet)*

$$\lim_{j\in J} \mathsf{Q}_t^{\mathsf{d}_{i(j)}} f(x) = \mathsf{Q}_t f(x) \quad \Rightarrow \quad \limsup_{j\in J} \mathsf{D}_t^{\mathsf{d}_{i(j)},+} f(x) \leq \mathsf{D}_t^+ f(x). \tag{3.2.21}$$

Proof Since $i \prec j$ yields $\mathsf{d}_i \leq \mathsf{d}_j$, it is clear that $i \mapsto \mathsf{Q}_t^{\mathsf{d}_i} f(x)$ is increasing. For every fixed t, x, y we have $\lim_{i\in I} F^{\mathsf{d}_i}(t, x, y) = F^{\mathsf{d}}(t, x, y)$ monotonically. The first statement then follows by a standard application of Γ-convergence of a family of increasing real functions in a compact set. Equation (3.2.20) is a particular case of (3.2.21) for the identity map in the directed set I.

Let us now assume that $\lim_{j\in J} \mathsf{Q}_t^{\mathsf{d}_{i(j)}} f(x) = \mathsf{Q}_t f(x)$ along an increasing net $j \mapsto i(j)$. We can select $y_j \in X$ such that

$$\mathsf{Q}_t^{\mathsf{d}_{i(j)}} f(x) = f(y_j) + \frac{1}{qt^{q-1}} \mathsf{d}_{i(j)}^q(x, y_j), \quad \mathsf{D}_t^{\mathsf{d}_{i(j)},+} f(x) = \mathsf{d}_{i(j)}(x, y_j). \tag{3.2.22}$$

We can find a further subnet $h \mapsto j(h)$, $h \in H$, such that $(y_{j(h)})_{h\in H}$ is convergent to $y \in X$ and

$$\limsup_{j\in J} \mathsf{D}_t^{\mathsf{d}_{i(j)},+} f(x) = \lim_{h\in H} \mathsf{d}_{i(j(h))}(x, y_{j(h)}). \tag{3.2.23}$$

Passing to the limit in the first equation of (3.2.22) and using the assumption of (3.2.21) we get

$$\mathsf{Q}_t f(x) = f(y) + \frac{1}{qt^{q-1}} \lim_{h\in H} \mathsf{d}^q_{i(j(h))}(x, y_{j(h)}) \overset{(2.1.25)}{\geq} f(y) + \frac{1}{qt^{q-1}} \mathsf{d}^q(x, y) \geq \mathsf{Q}_t f(x) \tag{3.2.24}$$

where the last inequality follows by the very definition of $\mathsf{Q}_t f(x)$. We deduce that

$$y \in \mathsf{J}_t f(x), \quad \lim_{h\in H} \mathsf{d}^q_{i(j(h))}(x, y_{j(h)}) = \mathsf{d}^q(x, y). \tag{3.2.25}$$

Since $\mathsf{d}(x, y) \leq \mathsf{D}^+_t f(x)$, by (3.2.23) we get (3.2.21). □

Notice that the upper semicontinuity property of (3.2.20) and of (3.2.21) are not immediately obvious as in the case of, e.g., (3.2.16), since d is typically just lower semicontinuous along τ-converging sequences. In the proof we used in an essential way the minimality of y_j and the continuity of f.

We conclude this section with the main structural properties for the Hopf-Lax evolution generated by d.

Theorem 3.2.4 *Let $f \in \mathrm{Lip}_b(X, \tau, \mathsf{d})$ and let $\mathsf{Q}_t f$, $\mathsf{J}_t f$, $\mathsf{D}^\pm_t f$ be defined as* (3.2.4), (3.2.5), (3.2.6) *for $t > 0$.*

(a) *The functions $(x, t) \mapsto \mathsf{Q}_t f(x)$, $\mathsf{D}^-_t f(x)$ are lower semicontinuous in $X \times (0, +\infty)$ and*

$$\min_X f \leq \mathsf{Q}_t f(x) \leq \max_X f \quad \text{for every } t > 0,\ x \in X. \tag{3.2.26}$$

(b) *For every $x \in X$*

$$\lim_{t\downarrow 0} \mathsf{Q}_t f(x) = \mathsf{Q}_0 f(x) := f(x), \tag{3.2.27}$$

the map $t \mapsto \mathsf{Q}_t f(x)$ is Lipschitz in $[0, \infty)$ and satisfies

$$\frac{\mathrm{d}}{\mathrm{d}t}\mathsf{Q}_t f(x) = -\frac{1}{p}\Big(\frac{\mathsf{D}^\pm_t f(x)}{t}\Big)^q \quad \text{for } t > 0 \text{with at most countable exceptions.} \tag{3.2.28}$$

(c) *For every $x \in X$ and $t > 0$*

$$f(x) - \mathsf{Q}_t f(x) = \frac{t}{p}\int_0^1 \Big(\frac{\mathsf{D}^+_{tr} f(x)}{tr}\Big)^p \,\mathrm{d}r, \tag{3.2.29}$$

$$\limsup_{t\downarrow 0} \frac{f(x) - \mathsf{Q}_t f(x)}{t} \leq \frac{1}{p}|\mathrm{D} f|^p(x) \leq \frac{1}{p}\big(\mathrm{lip}\, f(x)\big)^p. \tag{3.2.30}$$

Proof

(a) Lower semicontinuity of $\mathsf{Q}f$ is a consequence of the joint lower semicontinuity of F and of the compactness of (X, τ); it can also be obtained by Proposition 3.2.3, which characterizes $Q_t f$ as a supremum of continuous functions. The bound (3.2.26) is immediate.

(b) As for the proof of (3.2.14), we get from (3.2.9) and (3.2.13)

$$\frac{1}{p}\Big(\frac{\mathsf{D}_s^+ f(x)}{t}\Big)^q (t-s) \le \mathsf{Q}_s f(x) - \mathsf{Q}_t f(x) \le \frac{1}{p}\Big(\frac{\mathsf{D}_t^- f(x)}{s}\Big)^q (t-s), \tag{3.2.31}$$

and (3.2.8) yields the uniform bound

$$\Big(\frac{\mathsf{D}_t^+ f(x)}{t}\Big)^q \le \big(q\,\mathrm{Lip}(f, X)\big)^p. \tag{3.2.32}$$

Since $t \mapsto \mathsf{Q}_t f(x)$ is decreasing, we obtain

$$\frac{|\mathsf{Q}_s f(x) - \mathsf{Q}_t f(x)|}{|t-s|} \le \frac{1}{p}\Big(\frac{t}{s}\Big)^q \big(q\,\mathrm{Lip}(f, X)\big)^p \quad \text{for every } 0 < s < t. \tag{3.2.33}$$

Equation (3.2.33) shows that $t \mapsto \mathsf{Q}_t f(x)$ is Lipschitz in every compact interval of $(0, \infty)$ with

$$\Big|\frac{\mathrm{d}}{\mathrm{d}t}\mathsf{Q}_t f(x)\Big| \le \frac{1}{p}\big(q\,\mathrm{Lip}(f, X)\big)^p \quad \text{for a.e. } t > 0, \tag{3.2.34}$$

so that $t \mapsto \mathsf{Q}_t f(x)$ is Lipschitz in $(0, +\infty)$. In order to prove (3.2.27) we simply observe that for every $x' \in \mathsf{J}_t f(x)$, $x' \ne x$,

$$f(x) - \mathsf{Q}_t f(x) = f(x) - f(x') - \frac{\mathsf{d}^q(x, x')}{q\, t^{q-1}} \le \mathsf{d}(x, x')\Big(\frac{f(x) - f(x')}{\mathsf{d}(x, x')} - \frac{1}{q}\frac{\mathsf{d}^{q-1}(x, x')}{t^{q-1}}\Big) \tag{3.2.35}$$

so that

$$0 \le f(x) - \mathsf{Q}_t f(x) \le \mathsf{D}_t^- f(x)\,\mathrm{Lip}(f, X) \tag{3.2.36}$$

and the right hand side vanishes as $t \downarrow 0$ thanks to (3.2.32).

Equation (3.2.28) follows from (3.2.37) and the monotonicity property (a consequence of (3.2.9))

$$\mathsf{D}_s^- f(x) \le \mathsf{D}_s^+ f(x) \le \mathsf{D}_t^- f(x) \quad \text{for every } x \in X,\ 0 < s < t, \tag{3.2.37}$$

which in particular shows that $\mathsf{D}_t^- f(x) = \mathsf{D}_t^+ f(x)$ for every $t > 0$ with at most countable exceptions.

(c) (3.2.29) follows by integrating (3.2.28).

Dividing (3.2.35) by t we get for every $x' \in \mathsf{J}_t f(x) \setminus \{x\}$

$$\frac{f(x) - \mathsf{Q}_t f(x)}{t} = \frac{\mathsf{d}(x,x')}{t}\frac{f(x) - f(x')}{\mathsf{d}(x,x')} - \frac{1}{q}\frac{\mathsf{d}^q(x,x')}{t^q} \le \frac{1}{p}\Big(\frac{f(x) - f(x')}{\mathsf{d}(x,x')}\Big)^p;$$

passing to the limit as $t \downarrow 0$ and observing that $\lim_{t\downarrow 0} \mathsf{D}_t^+ f(x) = 0$ we obtain (3.2.30).

□

We conclude this section with a discussion of the measurability properties of the maps $\mathsf{D}^\pm f$. In the case when d is continuous, $\mathsf{D}^+ f$ (resp. $\mathsf{D}^- f$) is upper- (resp. lower-) semicontinuous by Proposition 3.2.3. In the general case we can anyway prove that they are $\mathfrak{m} \times \mathscr{L}^1$ measurable.

Lemma 3.2.5 (Conditional Semicontinuity and Measurability of $\mathsf{D}^\pm f$) *Under the same assumptions of Theorem 3.2.4:*

(a) *for every net $(x_\lambda, t_\lambda)_{\lambda\in\Lambda}$ in $X \times (0,\infty)$ such that $\lim_{\lambda\in\Lambda}(x_\lambda, t_\lambda) = (x,t) \in X \times (0,\infty)$ we have*

$$\lim_{\lambda\in\Lambda} \mathsf{Q}_{t_\lambda} f(x_\lambda) = \mathsf{Q}_t f(x) \quad \Rightarrow \quad \begin{cases} \limsup_{\lambda\in\Lambda} \mathsf{D}_{t_\lambda}^+ f(x_\lambda) \le \mathsf{D}_t^+ f(x), \\ \liminf_{\lambda\in\Lambda} \mathsf{D}_{t_\lambda}^- f(x_\lambda) \ge \mathsf{D}_t^- f(x). \end{cases} \tag{3.2.38}$$

(b) *The maps $(x,t) \mapsto \mathsf{D}_t^\pm f(x)$ are Lusin $\mathfrak{m} \otimes \mathscr{L}^1$-measurable in $X \times (0,\infty)$; moreover, for every $t > 0$ the maps $x \mapsto \mathsf{D}_t^\pm f(x)$ are Lusin $\mathfrak{m}$-measurable in X.*

Proof

(a) Let us check the upper semicontinuity of $\mathsf{D}_\cdot^+ f$, by arguing as in the proof of Lemma 3.2.2(b) (the proof of the conditional lower semicontinuity of $\mathsf{D}_\cdot^- f$ is completely analogous). We fix $c \ge 0$ and we suppose that for some $\lambda_0 \in \Lambda$ $\mathsf{D}_{t_\lambda}^+(x_\lambda) \ge c$ for every $\lambda \succ \lambda_0$. We pick $y_\lambda \in \mathsf{J}_{t_\lambda}(x_\lambda)$ such that

$$F(t_\lambda, x_\lambda, y_\lambda) = \mathsf{Q}_{t_\lambda} f(x_\lambda), \quad \mathsf{d}(x_\lambda, y_\lambda) = \mathsf{D}_{t_\lambda}^+(x_\lambda) \ge c. \tag{3.2.39}$$

We can find a subnet $j \mapsto \lambda(j)$, $j \in J$, such that $j \mapsto y_{\lambda(j)}$ converges to a point $y \in X$ with

$$F(t,x,y) \le \liminf_{j\in J} F(t_{\lambda(j)}, x_{\lambda(j)}, y_{\lambda(j)}) = \liminf_{j\in J} \mathsf{Q}_{t_{\lambda(j)}} f(x_{\lambda(j)}) \overset{(3.2.38)}{=} \mathsf{Q}_t f(x),$$

showing that $y \in \mathsf{J}_t f(x)$ and $\lim_{j\in J} \mathsf{d}(x_{\lambda(j)}, y_{\lambda(j)}) = \mathsf{d}(x, y) \ge c$. It follows that $\mathsf{D}_t^+ f(x) \ge \mathsf{d}(x, y) \ge c$.

(b) Since the map $\mathsf{Q}f$ is lower semicontinuous, it is Lusin $\mathfrak{m} \otimes \mathscr{L}^1$ measurable in $X \times (0,\infty)$ [56, I.1.5, Theorem 5]. For every compact set $K \subset (0,\infty)$ and every $\varepsilon > 0$ we can find a compact subset $H_\varepsilon \subset X \times K$ such that the restriction of $\mathsf{Q}f$ to H_ε is continuous and $\mathfrak{m}\otimes\mathscr{L}^1\big((X\times K)\setminus H_\varepsilon\big) \le \varepsilon/2$. By the previous claim, we deduce that the restriction of $\mathsf{D}^\pm f$ to H_ε are semicontinuous, and thus Lusin $\mathfrak{m} \otimes \mathscr{L}^1$-measurable: therefore we can find a further compact subset $H'_\varepsilon \subset H_\varepsilon$ such that the restriction of $\mathsf{D}^\pm f$ to H'_ε are continuous and $\mathfrak{m} \otimes \mathscr{L}^1\big(H_\varepsilon \setminus H'_\varepsilon\big) \le \varepsilon/2$, so that $\mathfrak{m} \otimes \mathscr{L}^1\big((X \times K) \setminus H'_\varepsilon\big) \le \varepsilon$. We conclude that $\mathsf{D}^\pm f$ are Lusin $\mathfrak{m} \otimes \mathscr{L}^1$-measurable. The second statement can be proved by the same argument.

□

3.2.2 Invariance of the Cheeger Energy with Respect to $\mathscr{A}$ When (X, τ) Is Compact

As a preliminary obvious remark, we observe that if $\mathscr{A}' \subset \mathscr{A}'' \subset \mathrm{Lip}_b(X, \tau, \mathsf{d})$ are two algebras of Lipschitz functions compatible with the metric-topological measure structure $\mathbb{X} = (X, \tau, \mathsf{d}, \mathfrak{m})$ we have

$$H^{1,p}(\mathbb{X}, \mathscr{A}') \subset H^{1,p}(\mathbb{X}, \mathscr{A}'') \subset H^{1,p}(\mathbb{X}), \tag{3.2.40}$$

and for every $f \in H^{1,p}(\mathbb{X}, \mathscr{A}')$

$$\mathsf{CE}_{p,\mathscr{A}'}(f) \ge \mathsf{CE}_{p,\mathscr{A}''}(f) \ge \mathsf{CE}_p(f), \quad |\mathrm{D}f|_{\star,\mathscr{A}'} \ge |\mathrm{D}f|_{\star,\mathscr{A}''} \ge |\mathrm{D}f|_\star \quad \mathfrak{m}\text{-a.e. in } X. \tag{3.2.41}$$

We will see that (3.2.40) and (3.2.41) can be considerably refined, obtaining the complete independence of the choice of $\mathscr{A}$. In this section we will focus on the case when (X, τ) is a compact topological space; we suppose that $\mathscr{A}$ is an algebra compatible with $\mathbb{X}$, we denote by I the directed set of all the finite collections $i \subset \mathscr{A}_1$ satisfying $f \in i \Rightarrow -f \in i$ and we set

$$\mathsf{d}_i(x, y) := \sup_{f\in i} f(x) - f(y), \quad i \text{ is a finite subset of } \mathscr{A}_1 \text{satisfying } f \in i \Rightarrow -f \in i. \tag{3.2.42}$$

Lemma 3.2.6 *Let us suppose that (X, τ) is compact and $\mathscr{A}$ is an algebra compatible with $\mathbb{X}$ generating the bounded continuous semidistances $(\mathsf{d}_i)_{i\in I}$ as in (3.2.42).*

(a) *For every $y \in X$ and $i \in I$ the map $h_i^y : x \to \mathsf{d}_i(x, y)$ belongs to $H^{1,p}(\mathbb{X}, \mathscr{A})$ and $|\mathrm{D}h_i^y|_{\star,\mathscr{A}} \le 1$ $\mathfrak{m}$-a.e.*

(b) *For every* $f \in \mathrm{C}(X)$, $t > 0$, $i \in I$, *the function* $\mathsf{Q}_t^{\mathsf{d}_i} f$ *belongs to* $H^{1,p}(\mathbb{X}, \mathscr{A})$ *and*

$$|\mathrm{D}\mathsf{Q}_t^{\mathsf{d}_i} f|_{\star,\mathscr{A}} \le t^{-1}\mathsf{D}_t^{\mathsf{d}_i,+} f \quad \mathfrak{m}\text{-}a.e.. \tag{3.2.43}$$

(c) *For every* $f \in \mathrm{C}(X)$ *and* $t > 0$ *the function* $\mathsf{Q}_t f$ *belongs to* $H^{1,p}(\mathbb{X}, \mathscr{A})$ *and*

$$|\mathrm{D}\mathsf{Q}_t f|_{\star,\mathscr{A}} \le t^{-1}\mathsf{D}_t^{+} f \quad \mathfrak{m}\text{-}a.e.. \tag{3.2.44}$$

Proof Claim **(a)** immediately follows from Corollary 3.1.13.
(b) Let us denote by Π the directed family of finite subsets of X. For every $\pi \in \Pi$, the definition of $\mathsf{Q}_t^{\pi,\mathsf{d}_i}$ given in (3.2.3), the chain rule (3.1.16), the previous claim, and Corollary 3.1.13 yield

$$\mathsf{Q}_t^{\pi,\mathsf{d}_i} f \in H^{1,p}(\mathbb{X}, \mathscr{A}), \quad |\mathrm{D}\mathsf{Q}_t^{\pi,\mathsf{d}_i} f|_{\star,\mathscr{A}} \le g_t^{\pi,\mathsf{d}_i} := \Big(t^{-1}\mathsf{D}_t^{\pi,\mathsf{d}_i+} f\Big)^{q-1} \tag{3.2.45}$$

By Lemma 3.2.2(c), for every $x \in X$ and $t > 0$ $\pi \mapsto \mathsf{Q}_t^{\pi,\mathsf{d}_i} f(x)$ is a decreasing net, converging to $\mathsf{Q}_t^{\mathsf{d}_i} f(x)$. Thanks to (2.1.8) and to the uniform bound (3.2.7) (where $S := \max_{x,y\in X} \mathsf{d}_i(x, y)$) we have

$$\lim_{\pi\in\Pi} \int_X \big|\mathsf{Q}_t^{\pi,\mathsf{d}_i} f - \mathsf{Q}_t^{\mathsf{d}_i} f\big| \,\mathrm{d}\mathfrak{m} = \lim_{\pi\in\Pi} \int_X \mathsf{Q}_t^{\pi,\mathsf{d}_i} f \,\mathrm{d}\mathfrak{m} - \int_X \mathsf{Q}_t^{\mathsf{d}_i} f \,\mathrm{d}\mathfrak{m} = 0. \tag{3.2.46}$$

We can thus find an increasing sequence $n \mapsto \pi_n \in \Pi$ such that

$$\lim_{n\to\infty} \int_X \big|\mathsf{Q}_t^{\pi_n,\mathsf{d}_i} f - \mathsf{Q}_t^{\mathsf{d}_i} f\big| \,\mathrm{d}\mathfrak{m} = \lim_{n\to\infty} \int_X \big|\mathsf{Q}_t^{\pi_n,\mathsf{d}_i} f - \mathsf{Q}_t^{\mathsf{d}_i} f\big|^p \,\mathrm{d}\mathfrak{m} = 0 \tag{3.2.47}$$

and a $\mathfrak{m}$-negligible subset $N \subset X$ such that

$$\lim_{n\to\infty} \mathsf{Q}_t^{\pi_n,\mathsf{d}_i} f(x) = \mathsf{Q}_t^{\mathsf{d}_i} f(x) \quad \text{for every } x \in X \setminus N. \tag{3.2.48}$$

Applying (3.2.16) we deduce

$$\limsup_{n\to\infty} g_t^{\pi_n,\mathsf{d}_i}(x) \le g_t^{\mathsf{d}_i}(x) := \Big(t^{-1}\mathsf{D}_t^{\mathsf{d}_i+} f\Big)^{q-1} \quad \text{for every } x \in X \setminus N. \tag{3.2.49}$$

By (3.2.45), it follows that any weak limit point in $L^p(X, \mathfrak{m})$ of the (uniformly bounded) sequence $(|\mathrm{D}\mathsf{Q}_t^{\pi_n,\mathsf{d}_i} f|_{\star,\mathscr{A}})_{n\in\mathbb{N}}$ will be bounded by $t^{-1}\mathsf{D}_t^{\mathsf{d}_i+} f$. By (3.2.46) we obtain (3.2.43).

(c) The argument is very similar to the previous one, but now using Proposition 3.2.3 and the net $i \mapsto \mathsf{d}_i$. □

Theorem 3.2.7 *If (X, τ) is a compact space and $\mathscr{A} \subset \mathrm{Lip}(X, \tau, \mathsf{d})$ is a compatible algebra according to Definition 2.1.17, then $H^{1,p}(\mathbb{X}) = H^{1,p}(\mathbb{X}, \mathscr{A})$ with equal minimal relaxed gradient (and therefore equal Cheeger energy). Equivalently, for every $f \in H^{1,p}(\mathbb{X})$ there exists a sequence $f_n \in \mathscr{A}$ such that*

$$f_n \to f, \quad \mathrm{lip}\, f_n \to |\mathrm{D} f|_\star \quad \text{strongly in } L^p(X, \mathfrak{m}). \tag{3.2.50}$$

Proof Let us denote by $|\mathrm{D}f|_{\star,\mathscr{A}}$ (resp. $|\mathrm{D}f|_\star$) the minimal relaxed gradient induced by $\mathscr{A}$ (resp. by $\mathrm{Lip}(X, \tau, \mathsf{d})$). It is clear that $H^{1,p}(X, \mathscr{A}) \subset H^{1,p}(X)$ and $|\mathrm{D}f|_\star \le |\mathrm{D}f|_{\star,\mathscr{A}}$ for every $f \in H^{1,p}(X, \mathscr{A})$; in order to prove the Theorem it is sufficient to show that every $f \in \mathrm{Lip}(X, \tau, \mathsf{d})$ belongs to $H^{1,p}(\mathbb{X}, \mathscr{A})$ with $|\mathrm{D}f|_{\star,\mathscr{A}} \le \mathrm{lip}\, f$ $\mathfrak{m}$-a.e.

We select an arbitrary Borel set $B \subset X$; by using the uniform bound (a consequence of (3.2.34))

$$\frac{f(x) - \mathsf{Q}_t f(x)}{t} \le \frac{1}{p}\big(q\,\mathrm{Lip}(f, X)\big)^p \quad \text{for every } x \in X,$$

the superior limit (3.2.30) and Fatou's Lemma, we obtain

$$\frac{1}{p}\int_B \big(\mathrm{lip}\, f(x)\big)^p \,\mathrm{d}\mathfrak{m}(x) \ge \limsup_{t\downarrow 0} \int_B \frac{f(x) - \mathsf{Q}_t f(x)}{t} \,\mathrm{d}\mathfrak{m}(x). \tag{3.2.51}$$

On the other hand, (3.2.29), the measurability of $\mathsf{D}^+ f$ given by Lemma 3.2.5, and Fubini's Theorem yield

$$\begin{aligned}\int_B \frac{f(x) - \mathsf{Q}_t f(x)}{t} \,\mathrm{d}\mathfrak{m}(x) &= \frac{1}{p}\int_{B\times(0,1)} \Big(\frac{\mathsf{D}^+_{tr} f(x)}{tr}\Big)^p \,\mathrm{d}(\mathfrak{m}\otimes\mathscr{L}^1)(x, r)\\ &= \frac{1}{p}\int_0^1 \Big(\int_B \Big(\frac{\mathsf{D}^+_{tr} f(x)}{tr}\Big)^p \,\mathrm{d}\mathfrak{m}(x)\Big)\,\mathrm{d}r.\end{aligned}$$

A further application of Fatou's Lemma yields

$$\liminf_{t\downarrow 0} \int_B \frac{f(x) - \mathsf{Q}_t f(x)}{t} \,\mathrm{d}\mathfrak{m}(x) \ge \frac{1}{p}\liminf_{s\downarrow 0} \int_B \Big(\frac{\mathsf{D}^+_s f(x)}{s}\Big)^p \,\mathrm{d}\mathfrak{m}(x), \tag{3.2.52}$$

where we used the fact that for every $r \in (0, 1)$

$$\liminf_{t\downarrow 0} \int_B \Big(\frac{\mathsf{D}^+_{tr} f(x)}{tr}\Big)^p \,\mathrm{d}\mathfrak{m}(x) = \liminf_{s\downarrow 0} \int_B \Big(\frac{\mathsf{D}^+_s f(x)}{s}\Big)^p \,\mathrm{d}\mathfrak{m}(x).$$

Recalling (3.2.43) and the fact that $\mathsf{Q}_s f \to f$ in $L^p(X, \mathfrak{m})$ as $s \downarrow 0$ we get

$$\liminf_{s\downarrow 0} \int_B \Big(\frac{\mathsf{D}_s^+ f}{s}\Big)^p \,\mathrm{d}\mathfrak{m} \ge \liminf_{s\downarrow 0} \int_B |\mathrm{D}\mathsf{Q}_s f|^p_{\star,\mathscr{A}} \,\mathrm{d}\mathfrak{m} \ge \int_B |\mathrm{D}f|^p_{\star,\mathscr{A}} \,\mathrm{d}\mathfrak{m}. \tag{3.2.53}$$

Combining (3.2.51), (3.2.52) and (3.2.53) we deduce that

$$\int_B \big(\mathrm{lip}\, f(x)\big)^p \,\mathrm{d}\mathfrak{m}(x) \ge \int_B |\mathrm{D}f|^p_{\star,\mathscr{A}}(x) \,\mathrm{d}\mathfrak{m}(x) \tag{3.2.54}$$

for every Borel subset B, so that $\mathrm{lip}\, f(x) \ge |\mathrm{D}f|_{\star,\mathscr{A}}(x)$ for $\mathfrak{m}$-a.e. $x \in X$. □

3.2.3 Notes

Section 3.2.1 collects all the basic estimates concerning the Hopf-Lax flow in a general extended metric-topological setting. We followed the approach of [8, § 3] with some differences: we assumed compactness of (X, τ) (as in [7, 39]), we considered general exponents $q = p' \in (1, \infty)$ as in [7], and we devoted some effort to study the dependence of the Hopf-Lax formula on the distance and on the minimizing set, a point of view that has also been used in [3, 11]. In this respect, compactness plays an essential role. Differently from [8], the Hopf-Lax flow is not used as a crucial ingredient for the so-called Kuwada Lemma [8, § 6], [47], but as a powerful approximation tool of general functions by elements in the algebra $\mathscr{A}$.

Section 3.2.2 contains the crucial results which justify the study of the Hopf-Lax formula in our setting: Lemma 3.2.6 provides a crucial estimate of $|\mathrm{D}\mathsf{Q}_t f|_{\star,\mathscr{A}}$ for the regularized functions and Theorem 3.2.7 shows that compatible algebra are dense (in energy) in $H^{1,p}(\mathbb{X})$. Both the results are inspired by the techniques of [11, Theorem 3.12] and of [3, § 12].

4 p-Modulus and Nonparametric Dynamic Plans

Assumption As in the previous section, we consider an extended metric-topological measure space $\mathbb{X} = (X, \tau, \mathsf{d}, \mathfrak{m})$ and we fix an exponent $p \in (1, +\infty)$. $\mathrm{RA}(X) = \mathrm{RA}(X, \mathsf{d})$ is the space of rectifiable arcs with the quotient topology τ_{A} studied in Sect. 2.2.2, see Proposition 2.2.6.

4.1 p-Modulus of a Family of Measures and of a Family of Rectifiable Arcs

4.1.1 p-Modulus of a Family of Radon Measures

Given $\Sigma \subset \mathcal{M}_+(X)$ we define (with the usual convention $\inf \emptyset = \infty$)

$$\mathsf{Mod}_p(\Sigma) := \inf\left\{\int_X f^p \,\mathrm{d}\mathfrak{m} \;:\; f \in \mathcal{L}^p_+(X, \mathfrak{m}),\ \int_X f \,\mathrm{d}\mu \geq 1 \quad \text{for all } \mu \in \Sigma\right\}, \tag{4.1.1}$$

$$\mathsf{Mod}_{p,c}(\Sigma) := \inf\left\{\int_X f^p \,\mathrm{d}\mathfrak{m} \;:\; f \in \mathrm{C}_b(X),\ \int_X f \,\mathrm{d}\mu \geq 1 \quad \text{for all } \mu \in \Sigma\right\}. \tag{4.1.2}$$

Since the infimum in (4.1.2) is unchanged if we restrict the minimization to nonnegative functions $f \in \mathrm{C}_b(X)$ we get $\mathsf{Mod}_{p,c}(\Sigma) \geq \mathsf{Mod}_p(\Sigma)$. Also, whenever Σ contains the null measure, we have $\mathsf{Mod}_{p,c}(\Sigma) = \mathsf{Mod}_p(\Sigma) = \infty$, whereas $\mathsf{Mod}_{p,c}(\emptyset) = 0$.

Definition 4.1.1 (**Mod_p-Negligible Sets and Properties Mod_p-a.e.**) A set $\Sigma \subset \mathcal{M}_+(X)$ is said to be Mod_p-negligible if $\mathrm{Mod}_p(\Sigma) = 0$.

We say that a property P on $\mathcal{M}_+(X)$ holds Mod_p-a.e. if the set of measures where P fails is Mod_p-negligible.

The next result collects various well known properties of the Modulus, see e.g. [16], [45, § 5.2].

Proposition 4.1.2 *The set functions $\Sigma \subset \mathcal{M}_+(X) \mapsto \mathsf{Mod}_p(\Sigma)$, $\Sigma \subset \mathcal{M}_+(X) \mapsto \mathsf{Mod}_{p,c}(\Sigma)$ satisfy the following properties:*

(a) *both are monotone and subadditive.*
(b) *If $g \in \mathcal{L}^p_+(X, \mathfrak{m})$ then $\int_X g \,\mathrm{d}\mu < \infty$ for Mod_p-almost every μ; conversely, if $\mathsf{Mod}_p(\Sigma) = 0$ then there exists $g \in \mathcal{L}^p_+(X, \mathfrak{m})$ such that $\int_X g \,\mathrm{d}\mu = \infty$ for every $\mu \in \Sigma$.*
(c) [Fuglede's Lemma] *If $(f_n) \subset \mathcal{L}^p_+(X, \mathfrak{m})$ converges in $L^p(X, \mathfrak{m})$ seminorm to $f \in \mathcal{L}^p_+(X, \mathfrak{m})$, there exists a subsequence $(f_{n(k)})$ such that*

$$\lim_{k\to\infty} \int_X |f_{n(k)} - f| \,\mathrm{d}\mu = 0 \qquad \mathsf{Mod}_p\text{-a.e. in } \mathcal{M}_+(X). \tag{4.1.3}$$

(d) *For every $\Sigma \subset \mathcal{M}_+(X)$ with $\mathsf{Mod}_p(\Sigma) < \infty$ there exists $f \in \mathcal{L}^p_+(X, \mathfrak{m})$, unique up to $\mathfrak{m}$-negligible sets, such that $\int_X f \,\mathrm{d}\mu \geq 1$ Mod_p-a.e. on Σ and $\|f\|_p^p = \mathsf{Mod}_p(\Sigma)$.*
(e) *If Σ_n are nondecreasing subsets of $\mathcal{M}_+(X)$ then $\mathsf{Mod}_p(\Sigma_n) \uparrow \mathsf{Mod}_p(\cup_n \Sigma_n)$.*

(f) *If K_n are nonincreasing compact subsets of $\mathcal{M}_+(X)$ then* $\mathsf{Mod}_{p,c}(K_n) \downarrow \mathsf{Mod}_{p,c}(\bigcap_n K_n)$.

Proof We repeat almost word by word the arguments of [2, Proposition 2.2].

(a) Monotonicity is an obvious consequence of the definition. For the subadditivity, if we take two sets $A, B \subset \mathcal{M}_+(X)$ and two functions $f, g \in \mathcal{L}^p_+(X, \mathfrak{m})$ with $\int_X f \,\mathrm{d}\mu \geq 1$ for every $\mu \in A$ and $\int_X g \,\mathrm{d}\mu \geq 1$ for every $\mu \in B$, then the function $h := (f^p + g^p)^{1/p} \geq \max(f, g)$ still satisfies $\int_X h \,\mathrm{d}\mu \geq 1$ for every $\mu \in A \cup B$, hence

$$\mathsf{Mod}_p(A \cup B) \leq \int_X h^p \,\mathrm{d}\mathfrak{m} = \int_X f^p \,\mathrm{d}\mathfrak{m} + \int_X g^p \,\mathrm{d}\mathfrak{m}.$$

Minimizing over f and g we get the subadditivity.

(b) If we consider the set where the property fails

$$\Sigma = \left\{\mu \in \mathcal{M}_+(X) : \int_X g \,\mathrm{d}\mu = \infty\right\},$$

then it is clear that for every $k > 0$ we have $\Sigma = \left\{\mu \in \mathcal{M}_+(X) : \int_X kg \,\mathrm{d}\mu = \infty\right\}$ so that $\mathsf{Mod}_p(\Sigma) \leq k^p \|g\|_p^p$ for every $\kappa > 0$ and we deduce that $\mathsf{Mod}_p(\Sigma) = 0$.

Conversely, if $\mathsf{Mod}_p(A) = 0$ for every $n \in \mathbb{N}$ we can find $g_n \in \mathcal{L}^p_+(X, \mathfrak{m})$ with $\int_X g_n \,\mathrm{d}\mu \geq 1$ for every $\mu \in A$ and $\int_X g_n^p \,\mathrm{d}\mathfrak{m} \leq 2^{-np}$. Thus $g := \sum_n g_n$ satisfies the required properties.

(c) Let $f_{n(k)}$ be a subsequence such that $\|f - f_{n(k)}\|_p \leq 2^{-k}$. If we set

$$g(x) := \sum_{k=1}^{\infty} |f(x) - f_{n(k)}(x)|$$

we have that $g \in \mathcal{L}^p_+(X, \mathfrak{m})$ and $\|g\|_{L^p(X,\mathfrak{m})} \leq 1$; in particular we have, for Claim (b) above, that $\int_X g \,\mathrm{d}\mu$ is finite for Mod_p-almost every μ. For those μ we get

$$\sum_{k=1}^{\infty} \int_X |f - f_{n(k)}| \,\mathrm{d}\mu < \infty$$

which yields (4.1.3).

(d) Claim (b) shows in particular that

$$\mathsf{Mod}_p(\Sigma) = \inf\left\{\int_X f^p \,\mathrm{d}\mathfrak{m} : \int_X f \,\mathrm{d}\mu \geq 1 \quad \text{for } \mathsf{Mod}_p\text{-a.e. } \mu \in \Sigma\right\}, \tag{4.1.4}$$

so that by Claim (c) the class of admissible functions f involved in the variational definition od Mod_p is a convex and closed subset of the Lebesgue space $L^p(X, \mathfrak{m})$. Hence, uniqueness follows by the strict convexity of the L^p-norm.

(e) By the monotonicity, it is clear that $\mathsf{Mod}_p(A_n)$ is an increasing sequence and that setting $M := \lim_{n\to\infty} \mathsf{Mod}_p(A_n)$ we have $M \geq \mathsf{Mod}_p(\cup_n A_n)$.

If $M = \infty$ there is nothing to prove, otherwise, we need to show that $\mathsf{Mod}_p(\cup_n A_n) \leq M$; let $(f_n) \subset \mathcal{L}^p_+(X, \mathfrak{m})$ be a sequence of functions such that $\int_X f \,\mathrm{d}\mu_n \geq 1$ on A_n and $\|f_n\|^p_{L^p(X,\mathfrak{m})} \leq \mathsf{Mod}_p(A_n) + \frac{1}{n}$. In particular we get that $\limsup_n \|f_n\|^p_p = M < \infty$ and so, possibly extracting a subsequence, we can assume that f_n weakly converge to some $f \in \mathcal{L}^p_+(X, \mathfrak{m})$. By Mazur lemma we can find convex combinations

$$\hat{f}_n = \sum_{k=n}^{\infty} \kappa_{k,n} f_k$$

such that $\hat{f}_n$ converge strongly to f in $L^p(X, \mathfrak{m})$; furthermore we have that $\int_X f_k \,\mathrm{d}\mu \geq 1$ on A_n if $k \geq n$ and so

$$\int_X \hat{f}_n \,\mathrm{d}\mu = \sum_{k=n}^{\infty} \kappa_{k,n} \int_X f \,\mathrm{d}\mu_k \geq 1 \qquad \text{on } A_n.$$

By Claim (c) above we obtain a subsequence $n(k)$ and a Mod_p-negligible set $\Sigma \subset \mathcal{M}_+(X)$ such that $\int_X \hat{f}_{n(k)} \,\mathrm{d}\mu \to \int_X f \,\mathrm{d}\mu$ outside Σ; in particular $\int_X f \,\mathrm{d}\mu \geq 1$ on $\cup_n A_n \setminus \Sigma$.

By Claim (b) we can find $g \in \mathcal{L}^p_+(X, \mathfrak{m})$ such that $\int_X g \,\mathrm{d}\mu = \infty$ on Σ, so that we have $\int_X (f + \varepsilon g) \,\mathrm{d}\mu \geq 1$ on $\cup_n A_n$ and

$$\begin{aligned}\mathsf{Mod}_p(\cup_n A_n)^{1/p} &\leq \|\varepsilon g + f\|_p \leq \varepsilon \|g\|_p + \|f\|_p \\ &\leq \varepsilon \|g\| + \liminf \|f_n\|_p \leq \varepsilon \|g\|_p + M^{1/p}.\end{aligned}$$

Letting $\varepsilon \to 0$ and taking the p-th. power the inequality $\mathsf{Mod}_p(A) \leq \sup_n \mathsf{Mod}_p(A_n)$ follows.

(f) By the monotonicity we get $\mathsf{Mod}_{p,c}(K) \leq \mathsf{Mod}_{p,c}(K_n)$; if $C := \lim_{n\to\infty} \mathsf{Mod}_{p,c}(K_n)$, we only have to prove $\mathsf{Mod}_{p,c}(K) \geq C$ and it is not restrictive to assume $C > 0$. We argue by contradiction: if $\mathsf{Mod}_{p,c}(K) < C$ we can find a nonnegative $\psi \in C_b(X)$ such that $\int_X \psi \,\mathrm{d}\mu \geq 1$ for every $\mu \in K$ and $\alpha^p = \int_X \psi^p \,\mathrm{dm} < C$. Setting $\phi := \alpha^{-1}\psi$ we obtain a function $\phi \in C_b(X)$ with $\|\phi\|_p = 1$ and

$$\inf_{\mu \in K} \int_X \phi \,\mathrm{d}\mu > \alpha^{-1}.$$

By the compactness of K the infimum above is a minimum; since K_n is decreasing,

$$\min_{\mu\in K_n}\int_X \phi\,\mathrm{d}\mu \to \min_{\mu\in K}\int_X \phi\,\mathrm{d}\mu.$$

It follows that there exists $\bar n \in \mathbb{N}$ such that $\min\limits_{\mu\in K_n}\int_X \phi\,\mathrm{d}\mu \ge \alpha^{-1}$ for every $n \ge \bar n$ so that using $\psi = \alpha\phi$ we deduce that $\mathsf{Mod}_{p,c}(K_n) \le \alpha^p < C$ for $n \ge \bar n$, a contradiction.

□

Another important property is the tightness of Mod_p in $\mathcal{M}_+(X)$.

Lemma 4.1.3 (Tightness of Mod_p) *For every $\varepsilon > 0$ there exists $K_\varepsilon \subset \mathcal{M}_+(X)$ compact such that $\mathsf{Mod}_p(\mathcal{M}_+(X) \setminus K_\varepsilon) \le \varepsilon$.*

Proof Since $\mathfrak{m}$ is a Radon measure we can find a nondecreasing family of τ-compact sets $K_n \subset X$ such that $m_n = \mathfrak{m}(X \setminus K_n) > 0$, $\lim_{n\to\infty}\mathfrak{m}(X \setminus K_n) = 0$. We set

$$\delta_n = (\sqrt{m_n} + \sqrt{m_{n+1}})^{1/p}, \quad a_n := \delta_n^{-1}, \tag{4.1.5}$$

observing that $\delta_n \downarrow 0$ and $a_n \uparrow +\infty$ as $n \to \infty$. For $k \in \mathbb{N}$ let us now define the sets

$$E_k := \Big\{\mu \in \mathcal{M}_+(X) : \mu(X) \le k,\ \ \mu(X \setminus K_n) \le \delta_n \text{ for every } n \ge k\Big\}, \tag{4.1.6}$$

which are compact in $\mathcal{M}_+(X)$ by Theorem 2.1.2.

To evaluate $\mathsf{Mod}_p(\mathcal{M}_+(X) \setminus E_k)$ we introduce the functions

$$f_k(x) := \begin{cases} 0 & \text{if } x \in K_k,\\ a_n & \text{if } x \in K_{n+1}\setminus K_n \text{ and } n \ge k,\\ +\infty & \text{otherwise.}\end{cases} \tag{4.1.7}$$

We observe that if $\mu \in \mathcal{M}_+(X) \setminus E_k$ then we have either $\mu(X) > k$ or $\mu(X \setminus K_n) > \delta_n$ for some $n \ge k$. In either case the integral of the function $f_k + \frac{1}{k}$ along μ is larger than or equal to 1:

- if $\mu(X) > k$ then

$$\int_X \Big(f_k + \frac{1}{k}\Big)\,\mathrm{d}\mu \ge \frac{1}{k}\ell(\mu) \ge 1;$$

- if $\mu(X \setminus K_n) > \delta_n$ for some $n \geq k$ we have that

$$\int_X \Big(f_k + \frac{1}{k}\Big)\,\mathrm{d}\mu \geq \int_{X\setminus K_n} f_k\,\mathrm{d}\mu \geq \int_{X\setminus K_n} a_n\,\mathrm{d}\mu > \delta_n a_n = 1.$$

So we have that $\mathsf{Mod}_p(\mathcal{M}_+(X) \setminus E_k) \leq \|f_k + \frac{1}{k}\|_p^p \leq (\|f_k\|_p + \|1/k\|_p)^p$. But

$$\int_X f_k^p\,\mathrm{dm} = \sum_{n=k}^{\infty} \int_{K_{n+1}\setminus K_n} a_n^p\,\mathrm{dm} = \sum_{n=k}^{\infty} \frac{m_n - m_{n+1}}{\sqrt{m_n} + \sqrt{m_{n+1}}} = \sum_{n=k}^{\infty}(\sqrt{m_n} - \sqrt{m_{n+1}}) = \sqrt{m_k}, \tag{4.1.8}$$

$$\|f_k + \frac{1}{k}\| \leq (m_k)^{1/(2p)} + (\mathrm{m}(X))^{1/p}/k \tag{4.1.9}$$

and therefore we obtain $\mathsf{Mod}_p(\mathcal{M}_+(X)\setminus E_k) \leq \Big((m_k)^{1/(2p)} + (\mathrm{m}(X))^{1/p}/k\Big)^p \to 0.$ □

4.1.2 *p*-Modulus of a Family of Rectifiable Arcs

There is a natural way to lift the notion of Modulus for a family of Radon measures in $\mathcal{M}_+(X)$ to a corresponding Modulus for a collection of rectifiable arcs: it is sufficient to assign a map $M : \mathrm{RA}(X) \to \mathcal{M}_+(X)$ and for every $\Gamma \subset \mathrm{RA}(X)$ set $\mathrm{Mod}_{p,M}(\Gamma) := \mathsf{Mod}_p(M(\Gamma))$. Clearly such a notion depends on the choice of M; in these notes we will consider two (slightly) different situations: the first one correspond to the most classic and widely used choice of the $\mathscr{H}^1$-measure carried by γ of (2.2.44)

$$\mathrm{M}\gamma := \nu_\gamma, \quad \nu_\gamma := \ell(\gamma)(R_\gamma)_\sharp(\mathscr{L}^1 \llcorner [0,1]). \tag{4.1.10}$$

In this case we will keep the standard notation of $\mathrm{Mod}_p(\Gamma)$, $\mathrm{Mod}_{p,c}(\Gamma)$; e.g. in the case of Mod_p (4.1.1) reads

$$\mathrm{Mod}_p(\Gamma) := \mathsf{Mod}_p(\mathrm{M}(\Gamma)) = \inf\Big\{\int_X f^p\,\mathrm{dm} : f \in \mathcal{L}_+^p(X,\mathrm{m}),\ \int_\gamma f \geq 1 \quad \text{for all } \gamma \in \Gamma\Big\}, \tag{4.1.11}$$

with obvious modification for $\mathrm{Mod}_{p,c}(\Gamma)$. The second choice corresponds to

$$\tilde{\mathrm{M}}\gamma = \tilde{\nu}_\gamma := \nu_\gamma + \delta_{\gamma_0} + \delta_{\gamma_1}, \quad \int_X f\,\mathrm{d}(\tilde{\nu}_\gamma) = f(\gamma_0) + f(\gamma_1) + \int_\gamma f, \tag{4.1.12}$$

where as usual we write the initial and final points of γ as $\gamma_i = \mathsf{e}_i(\gamma) = R_\gamma(i)$, $i = 0, 1$. We will denote by $\widetilde{\mathrm{Mod}}_p(\Gamma)$ the corresponding modulus,

$$\begin{aligned}\widetilde{\mathrm{Mod}}_p(\Gamma) &:= \mathsf{Mod}_p(\tilde{\mathrm{M}}(\Gamma))\\ &= \inf\left\{\int_X f^p \,\mathrm{dm} \,:\, f \in \mathcal{L}^p_+(X, \mathfrak{m}),\ f(\gamma_0) + f(\gamma_1) + \int_\gamma f \geq 1 \quad \text{for all } \gamma \in \Gamma\right\}.\end{aligned} \tag{4.1.13}$$

It is clear that

$$\widetilde{\mathrm{Mod}}_p(\Gamma) \leq \mathrm{Mod}_p(\Gamma), \quad \widetilde{\mathrm{Mod}}_p(\Gamma) \leq 2^{-p}\mathfrak{m}(X) \quad \text{for every } \Gamma \subset \mathrm{RA}(X). \tag{4.1.14}$$

One main difference between Mod_p and $\widetilde{\mathrm{Mod}}_p$ is the behaviour on constant arcs: if Γ contains a constant arc than it is clear that $\mathrm{Mod}_p(\Gamma) = +\infty$, whereas for a collection $\Gamma_A = \{\gamma_x : x \in A\}$ of constant arcs parametrized by a Borel set $A \subset X$ we have $\widetilde{\mathrm{Mod}}_p(\Gamma_A) = 2^{-p}\mathfrak{m}(A)$.

The notions of Mod_p- or $\widetilde{\mathrm{Mod}}_p$-negligible set of arcs (and of properties which hold Mod_p- or $\widetilde{\mathrm{Mod}}_p$-a.e.) follow accordingly from Definition 4.1.1. Properties (a–e) of Proposition 4.1.2 have an obvious version for arcs. The only statements that require some care are Proposition 4.1.2(f) and Lemma 4.1.3, since compactness in $\mathrm{RA}(X)$ for subsets Γ of arcs is not equivalent to compactness for the corresponding subsets $\mathrm{M}(\Gamma)$, $\tilde{\mathrm{M}}(\Gamma)$ in $\mathcal{M}_+(X)$.

Concerning the validity of Proposition 4.1.2(f), it is sufficient to note that for every nonnegative $\phi \in \mathrm{C}_b(X)$ the maps

$$\gamma \mapsto \int \phi \,\mathrm{d}(\mathrm{M}\gamma) = \int_\gamma \phi$$

$$\gamma \mapsto \int \phi \,\mathrm{d}(\tilde{\mathrm{M}}\gamma) = \phi(\gamma_0) + \phi(\gamma_1) + \int_\gamma \phi,$$

are lower semicontinuous, thanks to Theorem 2.2.13, so that the argument of the proof works as well.

Corollary 4.1.4 *If K_n is a nonincreasing sequence of compact sets in* $\mathrm{RA}(X)$ *we have* $\mathrm{Mod}_{p,c}(K_n) \downarrow \mathrm{Mod}_{p,c}(\cap_n K_n)$, $\widetilde{\mathrm{Mod}}_{p,c}(K_n) \downarrow \widetilde{\mathrm{Mod}}_{p,c}(\cap_n K_n)$.

Concerning the tightness Lemma 4.1.3 we have:

Lemma 4.1.5 (Tightness of Mod_p and $\widetilde{\mathrm{Mod}}_p$) *Let us suppose that (X, d) is complete.*

(a) *For every $\varepsilon > 0$ there exists $K_\varepsilon \subset \mathrm{RA}(X)$ compact such that* $\widetilde{\mathrm{Mod}}_p(\mathrm{RA}(X) \setminus K_\varepsilon) \leq \varepsilon$.

(b) *For every* $\eta, \varepsilon > 0$ *there exists* $K_\varepsilon \subset \mathrm{RA}(X)$ *compact such that* $\mathrm{Mod}_p(\mathrm{RA}_\eta(X) \setminus K_\varepsilon) \le \varepsilon$ *(where* $\mathrm{RA}_\eta(X)$ *has been defined in* (2.2.51)*).*

Proof

(a) We can repeat verbatim the proof of Lemma 4.1.3: keeping the same notation, the main point is that the sets

$$\tilde{\mathrm{M}}^{-1}(E_k) := \Big\{\gamma \in \mathrm{RA}(X) : 2 + \ell(\gamma) \le k,\ \ \tilde{\mathrm{M}}\gamma(X \setminus K_n) \le \delta_n \text{ for every } n \ge k\Big\}, \tag{4.1.15}$$

are compact in $\mathrm{RA}(X)$ by Theorem 2.2.13(g): in fact, since $\tilde{\mathrm{M}}\gamma = \delta_{\gamma_0} + \delta_{\gamma_1} + \nu_\gamma$, whenever $\delta_n < 1$ condition (4.1.15) yields $\gamma_0, \gamma_1 \in K$, so that also the assumption 2. of Theorem 2.2.13(g) holds.

(b) In this case the set $\mathrm{E}_k := \mathrm{M}^{-1}(E_k)$ cannot be compact in general, since it contains all the constant curves, so that there is no hope to construct inner compact approximations K_ε such that $\mathrm{Mod}_p(\mathrm{RA}(X)\setminus K_\varepsilon) \downarrow 0$. We thus replace $\mathrm{RA}(X)$ with $\mathrm{RA}_\eta(X)$, $\eta > 0$, and modify the definition of the sets E_k by requiring that the support of μ intersect K_k; in terms of $\mathrm{E}_k = \mathrm{M}^{-1}(E_k)$ this corresponds to

$$\mathrm{E}_k := \Big\{\gamma \in \mathrm{RA}(X) : \ell(\gamma) \le k,\ \ \mathsf{e}(\gamma) \cap K_k \neq \emptyset,\ \ \nu_\gamma(X \setminus K_n)) \le \delta_n \text{ for every } n \ge k\Big\}, \tag{4.1.16}$$

which are compact in $\mathrm{RA}(X)$ by Theorem 2.2.13(g).

To evaluate $\mathrm{Mod}_p(\mathrm{RA}_\eta(X) \setminus E_k)$ we introduce the functions f_k as in (4.1.7) and we observe that if $\gamma \in \mathrm{RA}_\eta(X) \setminus E_k$ then we have either $\ell(\gamma) > k$ or $\mathsf{e}(\gamma) \subset X \setminus K_k$ or $\nu_\gamma(X \setminus K_n) > \delta_n$ for some $n \ge k$. In either case the integral of the function $f_k + \frac{1}{k}$ along γ is greater or equal to 1: we have just to check the case $\mathsf{e}(\gamma) \cap K_k = \emptyset$, for which

$$\int_\gamma \Big(f_k + \frac{1}{k}\Big) \ge a_k \ell(\gamma) \ge \eta a_k \ge 1 \quad \text{if} \quad a_k \ge \eta^{-1};$$

For sufficiently big k we thus obtain the same estimates (4.1.8) and (4.1.9).

□

4.1.3 Notes

The notion of p-Modulus has been introduced by Fuglede [30] in the natural framework of collection of positive measures, as in [2]. Its application to the metric theory of Sobolev spaces has been proposed in [46] and further studied in [57], where the definition of Newtonian spaces has been introduced. We refer to [16, 45] for a comprehensive presentation of this topic.

The tightness estimate for Mod_p in $\mathcal{M}_+(X)$ has been introduced by [2], where it plays a crucial role. Here we used the same approach to derive tightness estimates directly in $\mathrm{RA}(X)$, for the two relevant embeddings of $\mathrm{RA}(X)$ in $\mathcal{M}_+(X)$ giving raise to Mod_p and $\widetilde{\mathrm{Mod}}_p$.

4.2 (Nonparametric) Dynamic Plans with Barycenter in $L^q(X, \mathfrak{m})$

Let us keep the main Assumption of page 205. We denote by $q = p' = p/(p-1) \in (1, \infty)$ the conjugate exponent of p.

Definition 4.2.1 ((Nonparametric) Dynamic Plans) A (nonparametric) dynamic plan is a Radon measure $\boldsymbol{\pi} \in \mathcal{M}_+(\mathrm{RA}(X))$ on $\mathrm{RA}(X)$ such that

$$\boldsymbol{\pi}(\ell) := \int_{\mathrm{RA}(X)} \ell(\gamma)\,\mathrm{d}\boldsymbol{\pi}(\gamma) < \infty. \tag{4.2.1}$$

Since $\mathrm{RA}(X)$ is a F_σ-subset of $\mathrm{A}(X)$, a dynamic plan can also be considered as the restriction of a Radon measure $\boldsymbol{\pi}'$ on $\mathrm{A}(X)$ satisfying $\int \ell\,\mathrm{d}\boldsymbol{\pi}' < \infty$; in particular $\boldsymbol{\pi}'$ is concentrated on $\mathrm{RA}(X)$, i.e. $\ell(\gamma) < \infty$ for $\boldsymbol{\pi}'$-a.e. γ. Using the universally Lusin-measurable map $G : \mathrm{RA}(X) \to \mathrm{BVC}_c([0, 1]; X)$ (2.2.41) we can also lift $\boldsymbol{\pi}$ to a Radon measure $\tilde{\boldsymbol{\pi}} = R_\sharp\boldsymbol{\pi}$ on $\mathrm{C}([0, 1]; X)$ concentrated on the set $\mathrm{BVC}_c([0, 1]; X)$ (2.2.37). Conversely, any Radon measure $\tilde{\boldsymbol{\pi}}$ on $\mathrm{C}([0, 1]; X)$ concentrated on $\mathrm{BVC}([0, 1]; X)$ yields the Radon measure $\boldsymbol{\pi} := \mathsf{q}_\sharp\tilde{\boldsymbol{\pi}}$ on $\mathrm{RA}(X)$. Notice that $\mathsf{q}_\sharp(R_\sharp\boldsymbol{\pi}) = \boldsymbol{\pi}$.

If $\boldsymbol{\pi}$ is a dynamic plan in $\mathcal{M}_+(\mathrm{RA}(X))$, thanks to Theorem 2.2.13(e) and Fubini's Theorem [24, Chap. II-14], we can define the Borel measure $\mu_{\boldsymbol{\pi}} := \mathrm{Proj}(\boldsymbol{\pi}) \in \mathcal{M}_+(X)$ by the formula

$$\int f\,\mathrm{d}\mu_{\boldsymbol{\pi}} := \iint_\gamma f\,\mathrm{d}\boldsymbol{\pi}(\gamma) \quad \text{for every bounded Borel function } f : X \to \mathbb{R}. \tag{4.2.2}$$

It is not difficult to show that $\mu_{\boldsymbol{\pi}}$ is a Radon measure with total mass $\boldsymbol{\pi}(\ell)$ given by (4.2.1): in fact, setting $\mathrm{RA}^L(X) := \{\gamma \in \mathrm{RA}(X) : \ell(\gamma) \le L\}$, for every $\varepsilon > 0$ we can find a length $L > 0$ such that $\boldsymbol{\pi}(\mathrm{RA}(X) \setminus \mathrm{RA}^L(X)) \le \varepsilon/2$. Since $\boldsymbol{\pi}$ is Radon and R is Lusin $\boldsymbol{\pi}$-measurable, we can also find a compact set $\mathscr{K}_\varepsilon \subset \mathrm{RA}^L(X)$ on which R is continuous and $\boldsymbol{\pi}(\mathrm{RA}(X) \setminus \mathscr{K}_\varepsilon) \le \varepsilon/(2L)$. We deduce that $\mu_{\boldsymbol{\pi}}$ is ε-concentrated on the compact $K_\varepsilon := \{R_\gamma(t) : t \in [0, 1],\ \gamma \in \mathscr{K}_\varepsilon\} = \mathsf{e}([0, 1] \times R(\mathscr{K}_\varepsilon))$, i.e. $\mu_{\boldsymbol{\pi}}(X \setminus K_\varepsilon) \le \varepsilon$, since

$$\begin{aligned}\mu_{\boldsymbol{\pi}}(X \setminus K_\varepsilon) &= \int \Big(\int_\gamma \chi_{X\setminus K_\varepsilon}\Big)\,\mathrm{d}\boldsymbol{\pi}(\gamma) = \int \Big(\ell(\gamma)\int_0^1 \chi_{X\setminus K_\varepsilon}(R_\gamma(t))\,\mathrm{d}t\Big)\,\mathrm{d}\boldsymbol{\pi}(\gamma)\\ &\le L\int \Big(\int_0^1 \chi_{X\setminus K_\varepsilon}(R_\gamma(t))\,\mathrm{d}t\Big)\,\mathrm{d}\boldsymbol{\pi}(\gamma) = L(\mathscr{L}^1 \otimes \boldsymbol{\pi})\{(t,\gamma) : R_\gamma(t) \notin K_\varepsilon\}\\ &\le L\boldsymbol{\pi}(\{\gamma \in \mathrm{RA}(X) : \gamma \notin \mathscr{K}_\varepsilon\}) \le \varepsilon/2.\end{aligned}$$

Notice that $\mu_{\boldsymbol{\pi}}$ can be considered as the integral w.r.t. $\boldsymbol{\pi}$ of the Borel family of measures ν_γ, $\gamma \in \mathrm{RA}(X)$ [24, Chap. II-13], in the sense that

$$\int_X f\,\mathrm{d}\mu_{\boldsymbol{\pi}}(x) = \int_{\mathrm{RA}(X)} \Big(\int_X f\,\mathrm{d}\nu_\gamma\Big)\,\mathrm{d}\boldsymbol{\pi}(\gamma). \tag{4.2.3}$$

Definition 4.2.2 We say that $\boldsymbol{\pi} \in \mathcal{M}_+(\mathrm{RA}(X))$ has barycenter in $L^q(X,\mathfrak{m})$ if there exists $h \in L^q(X,\mathfrak{m})$ such that $\mu_{\boldsymbol{\pi}} = h\mathfrak{m}$, or, equivalently, if

$$\int\int_\gamma f\,\mathrm{d}\boldsymbol{\pi}(\gamma) = \int fh\,\mathrm{d}\mathfrak{m} \quad \text{for every } f \in \mathrm{B}_b(X), \tag{4.2.4}$$

and we call $\mathrm{Bar}_q(\boldsymbol{\pi}) := \|h\|_{L^q(X,\mathfrak{m})}$ the barycentric q-entropy of $\boldsymbol{\pi}$. We will denote by $\mathcal{B}_q$ the set of all plans with barycenter in $L^q(X,\mathfrak{m})$ and we will set $\mathrm{Bar}_q(\boldsymbol{\pi}) := +\infty$ if $\boldsymbol{\pi} \notin \mathcal{B}_q$.

$\mathrm{Bar}_q : \mathcal{M}_+(\mathrm{RA}(X)) \to [0,+\infty]$ is a convex and positively 1-homogeneous functional. When $q = 1$ $\boldsymbol{\pi}$ has barycenter in $L^1(X,\mathfrak{m})$ if and only if $\mu_{\boldsymbol{\pi}} \ll \mathfrak{m}$ and in this case $\mathrm{Bar}_1(\boldsymbol{\pi}) = \boldsymbol{\pi}(\ell) = \int \ell\,\mathrm{d}\boldsymbol{\pi}$.

If $q > 1$ (which corresponds to our setting, when q is the dual of p) then Bar_q is also lower semicontinuous w.r.t. the weak topology of $\mathcal{M}_+(\mathrm{RA}(X))$, a property which can be easily deduced by the equivalent representation formula (4.2.7) below. Notice that $\mathrm{Bar}_q(\boldsymbol{\pi}) = 0$ iff $\boldsymbol{\pi}$ is concentrated on the set of constant arcs in $\mathrm{RA}(X)$.

$\mathrm{Bar}_q(\boldsymbol{\pi})$ has two equivalent representation. The first one is related to the L^q entropy of the projected measure $\mu_{\boldsymbol{\pi}} = \mathrm{Proj}(\boldsymbol{\pi})$ with respect to $\mathfrak{m}$:

$$\frac{1}{q}\mathrm{Bar}_q^q(\boldsymbol{\pi}) = \mathscr{L}_q(\mu_{\boldsymbol{\pi}}|\mathfrak{m}) \tag{4.2.5}$$

where for an arbitrary $\mu \in \mathcal{M}_+(X)$

$$\mathscr{L}_q(\mu|\mathfrak{m}) := \begin{cases} \dfrac{1}{q}\displaystyle\int_X \Big(\frac{\mathrm{d}\mu}{\mathrm{d}\mathfrak{m}}\Big)^q \mathrm{d}\mathfrak{m} & \text{if } \mu \ll \mathfrak{m},\\ +\infty & \text{otherwise.}\end{cases} \tag{4.2.6}$$

A second interpretation arises from the dual characterization of $\mathscr{L}_q$, since $q = p' \in (1, +\infty)$ [48, Thm. 2.7, Rem. 2.8]

$$\mathscr{L}_q(\mu|\mathfrak{m}) = \sup\Big\{\int_X f\,\mathrm{d}\mu - \frac{1}{p}\int_X f^p\,\mathrm{d}\mathfrak{m} : f \in \mathrm{C}_b(X),\quad f \ge 0\Big\}. \tag{4.2.7}$$

We immediately obtain

$$\frac{1}{q}\mathrm{Bar}_q^q(\boldsymbol{\pi}) = \sup\Big\{\int\int_\gamma f\,\mathrm{d}\boldsymbol{\pi}(\gamma) - \frac{1}{p}\int_X f^p\,\mathrm{d}\mathfrak{m} : f \in \mathrm{C}_b(X),\quad f \ge 0\Big\}. \tag{4.2.8}$$

Similarly (see Lemma A.7 in the Appendix) we can easily check that

Lemma 4.2.3 *If $q \in (1, \infty)$, $p = q'$, a Radon measure $\boldsymbol{\pi}$ on* $\mathrm{RA}(X)$ *has barycenter in $L^q(X, \mathfrak{m})$ if there exists $c \in [0, \infty)$ such that*

$$\int_{\mathrm{RA}(X)}\int_\gamma f\,\mathrm{d}\boldsymbol{\pi}(\gamma) \le c\|f\|_{L^p(X,\mathfrak{m})} \qquad \textit{for every } f \in \mathcal{L}_+^p(X, \mathfrak{m}). \tag{4.2.9}$$

In this case $\mathrm{Bar}_q(\boldsymbol{\pi})$ *is the minimal constant c in* (4.2.9)*. Moreover, it is equivalent to check* (4.2.9) *on nonnegative functions $f \in \mathrm{C}_b(X)$.*

Definition 4.2.4 ($\mathcal{B}_q$-Negligible Sets) We say that a set $\Gamma \subset \mathrm{RA}(X)$ is $\mathcal{B}_q$-negligible if $\boldsymbol{\pi}(\Gamma) = 0$ for every $\boldsymbol{\pi} \in \mathcal{B}_q$. Similarly, a property P on the set of arcs $\mathrm{RA}(X)$ holds $\mathcal{B}_q$-a.e. if $\{\gamma \in \mathrm{RA}(X) : P(\gamma)$ does not hold$\}$ is contained in a $\mathcal{B}_q$-negligible set.

It is easy to check that for every Borel set $B \subset X$ with $\mathfrak{m}(B) = 0$ the set

$$\big\{\gamma \in \mathrm{RA}(X) : \nu_\gamma(B) > 0\big\} \quad \text{is } \mathcal{B}_q\text{-negligible.} \tag{4.2.10}$$

There is a simple duality inequality, involving the minimization in the definition (4.1.1) of Mod_p and a maximization among all $\boldsymbol{\pi}$'s with barycenter in $L^q(X, \mathfrak{m})$. To see it, let's take $f \in \mathcal{L}_+^p(X, \mathfrak{m})$ such that $\int_\gamma f \ge 1$ on $\Gamma \subset \mathcal{M}_+(X)$. Then, if Γ is universally Lusin measurable we may take any plan $\boldsymbol{\pi}$ with barycenter in $L^q(X, \mathfrak{m})$ to obtain

$$\boldsymbol{\pi}(\Gamma) \le \int\int_\gamma f\,\mathrm{d}\boldsymbol{\pi}(\gamma) \le \mathrm{Bar}_q(\boldsymbol{\pi})\|f\|_p \quad \text{if } \int_\gamma f \ge 1 \text{ for every } \gamma \in \Gamma. \tag{4.2.11}$$

By the definition of Mod_p we obtain

$$\boldsymbol{\pi}(\Gamma) \le \mathrm{Bar}_q(\boldsymbol{\pi})\,\mathrm{Mod}_p^{1/p}(\Gamma). \tag{4.2.12}$$

In particular we have

$$\mathrm{Mod}_p(\Gamma) = 0 \quad \Longrightarrow \quad \boldsymbol{\pi}(\Gamma) = 0 \qquad \text{for all } \boldsymbol{\pi} \in \mathcal{B}_q. \tag{4.2.13}$$

Lemma 4.2.5 (An Equi-Tightness Criterium) *Let us suppose that (X, d) is complete and let $\mathcal{K}$ be a subset of $\mathcal{B}_q$ satisfying the following conditions:*

(T1) *There exist constants $C_1, C_2 > 0$ such that*

$$\boldsymbol{\pi}(\mathrm{RA}(X)) \le C_1, \quad \mathrm{Bar}_q(\boldsymbol{\pi}) \le C_2 \quad \textit{for every } \boldsymbol{\pi} \in \mathcal{K}. \tag{4.2.14}$$

(T2) *For every $\varepsilon > 0$ there exists a τ-compact set $H_\varepsilon \subset X$ such that*

$$\boldsymbol{\pi}\big(\{\gamma \in \mathrm{RA}(X) : \mathsf{e}(\gamma) \cap H_\varepsilon = \emptyset\}\big) \le \varepsilon \quad \textit{for every } \boldsymbol{\pi} \in \mathcal{K}. \tag{4.2.15}$$

Then $\mathcal{K}$ is relatively compact in $\mathcal{M}_+(\mathrm{RA}(X))$.

Proof We want to apply Prokhorov's Theorem 2.1.2 so that for every $\varepsilon > 0$ we have to exhibit a compact set $K_\varepsilon \subset \mathrm{RA}(X)$ such that $\boldsymbol{\pi}(\mathrm{RA}(X) \setminus K_\varepsilon) \le \varepsilon$.

Let $K_n,\ m_n,\ a_n,\ \delta_n,\ f_k$ be defined as in the proof of Lemma 4.1.3 and let us set $\mathrm{E}_{k,\xi} := \mathrm{F}_k \cap \mathrm{G}_\xi$ where

$$\begin{aligned} \mathrm{F}_k &:= \Big\{\gamma \in \mathrm{RA}(X) : \ell(\gamma) \le k,\ \ \nu_\gamma(X \setminus K_n)) \le \delta_n \text{ for every } n \ge k\Big\}, \\ \mathrm{G}_\xi &:= \Big\{\gamma \in \mathrm{RA}(X) : \mathsf{e}(\gamma) \cap H_\xi \ne \emptyset\Big\}. \end{aligned} \tag{4.2.16}$$

For every $k \in \mathbb{N}$ and $\xi > 0$ $\mathrm{E}_{k,\xi}$ are compact by Theorem 2.2.13(g).

Let us estimate $\boldsymbol{\pi}(\mathrm{RA}(X) \setminus \mathrm{E}_{k,\xi})$ for $\boldsymbol{\pi} \in \mathcal{K}$. By (T2) we know that $\boldsymbol{\pi}(\mathrm{RA}(X) \setminus \mathrm{G}_\xi) \le \xi$. On the other hand, since $\int_\gamma (f_k + 1/k) \ge 1$ for every $\gamma \in \mathrm{RA}(X) \setminus \mathrm{F}_k$ we have

$$\boldsymbol{\pi}(\mathrm{RA}(X) \setminus \mathrm{F}_k) \overset{(4.2.11)}{\le} C_2 \| f_k + 1/k \|_{L^p(X,\mathfrak{m})} \overset{(4.1.8)}{\le} C_2 M_k,$$

where $M_k := (m_k)^{1/(2p)} + (\mathfrak{m}(X))^{1/p}/k \downarrow 0$ as $k \to \infty$. We deduce $\boldsymbol{\pi}(\mathrm{RA}(X) \setminus \mathrm{E}_{k,\xi}) \le \xi + C_2 M_k$ and the thesis follows by choosing $K_\varepsilon := \mathrm{E}_{k,\xi}$ for $\xi + C_2 M_k < \varepsilon$. □

It is easy to check that (T2) is also a necessary condition for the equi-tightness of $\mathcal{K}$. In fact, if $\mathcal{K}$ is equi-tight in $\mathcal{M}_+(\mathrm{RA}(X))$ then the collection $\mathcal{K}' := \{(\mathsf{e}_0)_\sharp \boldsymbol{\pi} : \boldsymbol{\pi} \in \mathcal{K}\}$ is equi-tight in $\mathcal{M}_+(X)$ (since e_0 is a continuous map from $\mathrm{RA}(X)$ to X). Therefore, for every $\varepsilon > 0$ there exists a compact set $K_\varepsilon \subset X$ such that $\boldsymbol{\pi}(\{\gamma \in \mathrm{RA}(X) : \gamma_0 \notin K_\varepsilon\}) \le \varepsilon$, which clearly yields (4.2.15).

It is interesting to notice that if $\mathcal{K} \subset \mathcal{M}_+(\mathrm{RA}(X))$ satisfies the property (T1) above and

$$\ell(\gamma) \geq C_3 > 0 \quad \boldsymbol{\pi}\text{-a.e.} \quad \text{for every } \boldsymbol{\pi} \in \mathcal{K}, \tag{4.2.17}$$

then (T2) is satisfied as well. In fact, if $K_n \subset X$ is a compact set with $\mathfrak{m}(X \setminus K_n) \leq m_n$ and $\mathrm{G}_n := \{\gamma \in \mathrm{RA}(X) : \mathsf{e}(\gamma) \cap K_n = \emptyset\}$ for every $\boldsymbol{\pi} \in \mathcal{K}$ we have

$$\begin{aligned}\boldsymbol{\pi}\big(\mathrm{G}_n\big) &\leq \frac{1}{C_3}\int_{\mathrm{G}_n} \ell(\gamma)\,\mathrm{d}\boldsymbol{\pi}(\gamma) \leq \frac{1}{C_3}\int \Big(\int_\gamma \chi_{X\setminus K_n}\Big)\,\mathrm{d}\boldsymbol{\pi}(\gamma)\\ &= \frac{1}{C_3}\mu_{\boldsymbol{\pi}}(X \setminus K_n) \leq \frac{1}{C_3}\mathrm{Bar}_q(\boldsymbol{\pi})m_n^{1/p} \leq \frac{C_2}{C_3}m_n^{1/p} \downarrow 0 \quad \text{as } n \to \infty.\end{aligned}$$

The inequality (4.2.11) motivates the next definition.

Definition 4.2.6 (p-Content) If $\Gamma \subset \mathrm{RA}(X)$ is a universally measurable set we say that Γ has finite content if there exists a constant $c \geq 0$ such that

$$\boldsymbol{\pi}(\Gamma) \leq c\,\mathrm{Bar}_q(\boldsymbol{\pi}) \quad \text{for every } \boldsymbol{\pi} \in \mathcal{M}_+(\mathrm{RA}(X)). \tag{4.2.18}$$

In this case, the p-content of Γ $\mathrm{Cont}_p(\Gamma)$ is the minimal constant c satisfying (4.2.18). If Γ has not finite content we set $\mathrm{Cont}_p(\Gamma) := +\infty$.

Notice that if Γ contains a constant arc we get $\mathrm{Cont}_p(\Gamma) = +\infty$; conversely, $\mathrm{Cont}_p(\Gamma) = 0$ if and only if Γ is $\mathcal{B}_q$-negligible. We can formulate (4.2.18) in the equivalent form

$$\mathrm{Cont}_p(\Gamma) = \sup_{\boldsymbol{\pi} \in \mathcal{M}_+(\mathrm{RA}(X)),\ \mathrm{Bar}_q(\boldsymbol{\pi})>0} \frac{\boldsymbol{\pi}(\Gamma)}{\mathrm{Bar}_q(\boldsymbol{\pi})}, \tag{4.2.19}$$

and we can also limit the sup in (4.2.19) and the condition (4.2.18) to probability plans (i.e. $\boldsymbol{\pi}(\mathrm{RA}(X)) = 1$) concentrated on Γ.

By Lemma A.7 we easily find the equivalent characterizations of Cont_p:

$$\frac{1}{p}\mathrm{Cont}_p^p(\Gamma) = \sup_{\boldsymbol{\pi} \in \mathcal{M}_+(\mathrm{RA}(X))} \boldsymbol{\pi}(\Gamma) - \frac{1}{q}\mathrm{Bar}_q^q(\boldsymbol{\pi}) \tag{4.2.20}$$

showing that $\frac{1}{p}\mathrm{Cont}_p^p$ is in fact the Legendre transform of $\frac{1}{q}\mathrm{Bar}_q^q$.

Let us now address the question of existence of an optimal dynamic plan attaining the supremum in (4.2.19) (or, equivalently, in (4.2.20)). The next result corresponds to [2, Lemma 4.4], where however the condition concerning the closure of Γ is missing. See also the comments in the Notes 4.2.1 at the end of this section.

Lemma 4.2.7 *Let us suppose that* (X, d) *is complete and let* $\Gamma \subset \mathrm{RA}(X)$ *be a closed set such that* $0 < \mathrm{Cont}_p(\Gamma) < +\infty$. *If there exists a compact set* $K \subset X$

such that $\mathsf{e}(\gamma) \cap K \neq \emptyset$ *for every* $\gamma \in \Gamma$ *(in particular if* Γ *is compact), then there exists an optimal plan* $\boldsymbol{\pi}_\Gamma$ *with barycenter in* $L^q(X, \mathfrak{m})$ *attaining the supremum in* (4.2.20)*.* $\boldsymbol{\pi}_\Gamma$ *is concentrated on* Γ *and satisfies*

$$\boldsymbol{\pi}_\Gamma(\Gamma) = \mathrm{Cont}_p^p(\Gamma) = \mathrm{Bar}_q^q(\boldsymbol{\pi}_\Gamma). \tag{4.2.21}$$

In particular, $\tilde{\boldsymbol{\pi}}_\Gamma := (\boldsymbol{\pi}_\Gamma(\Gamma))^{-1}\boldsymbol{\pi}_\Gamma$ *is a probability plan and*

$$\mathrm{Cont}_p(\Gamma)\,\mathrm{Bar}_q(\tilde{\boldsymbol{\pi}}_\Gamma) = 1.$$

Proof Taking perturbations of the form $\boldsymbol{\pi} \to \kappa\boldsymbol{\pi}$, $\kappa > 0$, we immediately see that we can restrict the maximization to plans satisfying

$$\boldsymbol{\pi}(\Gamma) = \boldsymbol{\pi}(\mathrm{RA}(X)) = \mathrm{Bar}_q^q(\boldsymbol{\pi}) \leq \mathrm{Cont}_p^p(\Gamma). \tag{4.2.22}$$

It is also easy to see that it is possible to restrict the maximization to plans concentrated on Γ, since the restriction $\boldsymbol{\pi} \mapsto \bar{\boldsymbol{\pi}} = \chi_\Gamma\boldsymbol{\pi}$ satisfies $\bar{\boldsymbol{\pi}}(\Gamma) = \boldsymbol{\pi}(\Gamma)$ and $\mathrm{Bar}_q(\bar{\boldsymbol{\pi}}) \leq \mathrm{Bar}_q(\boldsymbol{\pi})$. Since Γ is closed the functional of (4.2.20) is upper semicontinuous and therefore it admits a maximum on the compact set defined by (4.2.22). □

4.2.1 Notes

The notions of barycentric entropy and content have been introduced in [2] for measures in $\mathcal{M}_+(X)$, in order to provide an equivalent measure-theoretic characterization of the modulus Mod_p. Here we decided to focus mainly on nonparametric dynamic plans and to develop the main properties in the more restrictive setting characterized by the embedding $\mathrm{M} : \mathrm{RA}(X) \to \mathcal{M}_+(X)$ of (4.1.10), which is well adapted to the classic modulus Mod_p on arcs. In Sect. 5.1.2 we will also briefly discuss the notions of barycentric entropy and content related to $\widetilde{\mathrm{Mod}}_p$ and to the embedding $\tilde{\mathrm{M}}$ of (4.1.12).

The equi-tightness criterium Lemma 4.2.5 requires slightly more restrictive assumptions than in [11] since here compactness is obtained directly in $\mathcal{M}_+(\mathrm{RA}(X))$ instead of $\mathcal{M}_+(\mathcal{M}_+(X))$. Notice that the class of constant arcs is homeomorphic to X in $\mathrm{RA}(X)$, whereas it is identified with the null measure in $\mathcal{M}_+(X)$.

The existence of an optimal plan attaining (4.2.21) requires at least the closure of Γ: this condition should also be added to Lemma 4.4, Corollary 5.2(b) and Theorem 7.2 of [11]. Notice however that the main consequences [2, Theorem 8.3, Corollary 8.7] of Theorem 7.2 in [2] still hold, since they only require the existence of a nontrivial dynamic plan in $\mathcal{B}_q$ giving positive mass to Γ whenever $\mathrm{Mod}_p(\Gamma) > 0$: thanks to Choquet theorem this property holds for an arbitrary Souslin set Γ

and does not require its closedness. We will also discuss these aspects in the next Sect. 4.3, see Theorem 4.3.2.

4.3 *Equivalence Between* Cont_p *and* Mod_p

In this section we always refer to the main Assumption of page 205. We have seen that Cont_p, Mod_p, $\mathrm{Mod}_{p,c}$ satisfy the property

$$\mathrm{Cont}_p^p \le \mathrm{Mod}_p \le \mathrm{Mod}_{p,c} \qquad \text{on universally measurable subsets of } \mathrm{RA}(X). \tag{4.3.1}$$

We first prove that (4.3.1) is in fact an identity if Γ is compact.

Theorem 4.3.1 *If* Γ *is a compact subset of* $\mathrm{RA}(X)$ *we have*

$$\mathrm{Cont}_p^p(\Gamma) = \mathrm{Mod}_p(\Gamma) = \mathrm{Mod}_{p,c}(\Gamma). \tag{4.3.2}$$

Proof We will set $\mathcal{M}_+(\Gamma) := \{\boldsymbol{\pi} \in \mathcal{M}_+(\mathrm{RA}(X)) : \mathrm{supp}(\boldsymbol{\pi}) \subset \Gamma\}$. Since ℓ is a lower semicontinuous map, the minimum $\ell_0 := \min_{\gamma\in\Gamma} \ell(\gamma)$ is attained. If $\ell_0 = 0$ Γ contains a constant arc and (4.3.2) is trivially satisfied since the common value is $+\infty$. We can thus assume $\ell_0 > 0$.

We will prove (4.3.2) by using a minimax argument by applying Von Neumann Theorem A.8.

First of all we observe that for every $f \in \mathrm{C}_b(X)$

$$\int_\gamma f \ge 1 \text{ for every } \gamma \in \Gamma \quad \Leftrightarrow \quad \sup\Big\{\int \Big(1 - \int_\gamma f\Big)\,\mathrm{d}\boldsymbol{\pi}(\gamma) : \boldsymbol{\pi} \in \mathcal{M}_+(\Gamma)\Big\} = 0, \tag{4.3.3}$$

so that

$$\frac{1}{p}\mathrm{Mod}_{p,c}(\Gamma) = \inf_{f\in\mathrm{C}_b(X)} \sup_{\boldsymbol{\pi}\in\mathcal{M}_+(\Gamma)} \mathcal{L}(\boldsymbol{\pi}, f),$$

$$\mathcal{L}(\boldsymbol{\pi}, f) := \frac{1}{p}\int_X f^p\,\mathrm{dm} + \int \Big(1 - \int_\gamma f\Big)\,\mathrm{d}\boldsymbol{\pi}(\gamma).$$

By choosing $f_\star \equiv k \ge 2/\ell_0$ we clearly have

$$\mathcal{L}(\boldsymbol{\pi}, f_*) = \frac{1}{p}\int_X f^p\,\mathrm{dm} + \int \Big(1 - \int_\gamma f\Big)\,\mathrm{d}\boldsymbol{\pi}(\gamma) \le c_k - \boldsymbol{\pi}(\Gamma), \quad c_k := \frac{1}{p}\mathfrak{m}(X)k^p,$$

so that choosing $D_\star < \frac{1}{p}\operatorname{Mod}_{p,c}(\Gamma)$ and c_k sufficiently big, the set $\{\boldsymbol{\pi} \in \mathcal{M}_+(\Gamma) : \mathcal{L}(\boldsymbol{\pi}, f_*) \geq D_*\}$ is not empty (it contains the null plan) and it is contained in the compact set $\{\boldsymbol{\pi} \in \mathcal{M}_+(\Gamma) : \boldsymbol{\pi}(\Gamma) \leq c_k - D_*\}$. Condition (A.12) is thus satisfied and we deduce

$$\begin{aligned}
\frac{1}{p}\operatorname{Mod}_{p,c}(\Gamma) &= \max_{\boldsymbol{\pi}\in\mathcal{M}_+(\Gamma)} \inf_{f\in \mathrm{C}_b(X)} \mathcal{L}(\boldsymbol{\pi}, f)\\
&= \max_{\boldsymbol{\pi}\in\mathcal{M}_+(\Gamma)} \boldsymbol{\pi}(\Gamma) - \sup \int\int_\gamma f\,\mathrm{d}\boldsymbol{\pi}(\gamma) - \frac{1}{p}\int_X f^p\,\mathrm{dm}\\
&= \max_{\boldsymbol{\pi}\in\mathcal{M}_+(\Gamma)} \boldsymbol{\pi}(\Gamma) - \frac{1}{q}\operatorname{Bar}_q^q(\boldsymbol{\pi}) \overset{(4.2.20)}{=} \operatorname{Cont}_p^p(\Gamma).
\end{aligned}$$

□

Theorem 4.3.1 has an important implication in terms of Choquet capacity; we refer to [24, Chap. III, § 2] and to the brief account given in Sect. A.4 of the Appendix. Recall that $\mathscr{B}(Y)$ (resp. $\mathscr{K}(Y)$) will denote the collection of all the Borel (resp. compact) subsets of a Hausdorff space Y. The definition and the main properties of Souslin and Analytic sets are briefly recalled in Sect. A.3.

Theorem 4.3.2 *Let $\mathbb{X} = (X, \tau, \mathsf{d}, \mathfrak{m})$ be an e.m.t.m. space.*

(a) Mod_p *is a Choquet* $\mathscr{K}(\mathrm{RA}(X), \tau_{\mathrm{A}})$*-capacity in* $\mathrm{RA}(X)$.
(b) *For every universally measurable* $\Gamma \subset \mathrm{RA}(X)$

$$\operatorname{Cont}_p(\Gamma) = \sup\Big\{\operatorname{Cont}_p(\mathrm{K}) : \mathrm{K} \subset \Gamma, \mathrm{K}\ \textit{compact}\Big\}. \tag{4.3.4}$$

(c) *If* (X, d) *is complete and* (X, τ) *is Souslin, then every* $\mathscr{B}(\mathrm{RA}(X), \tau_{\mathrm{A}})$*-analytic set* Γ *(in particular, every Souslin set) is* Mod_p*-capacitable*
(d) *If* (X, d) *is complete and* (X, τ) *is Souslin, then every* $\mathscr{B}(\mathrm{RA}(X), \tau_{\mathrm{A}})$*-analytic set* Γ *satisfies* $\operatorname{Mod}_p(\Gamma) = \operatorname{Cont}_p^p(\Gamma)$. *In particular* Γ *is* Mod_p*-negligible if and only if it is* $\mathcal{B}_q$*-negligible.*

Proof

(a) Proposition 4.1.2(e,f) and the fact that $\operatorname{Mod}_{p,c} = \operatorname{Mod}_p$ if the set is compact by Theorem 4.3.1 give us that Mod_p is a $\mathscr{K}$-capacity in $(\mathrm{RA}(X), \tau_{\mathrm{A}})$.
(b) By (4.2.20) for every $S < \operatorname{Cont}_p(\Gamma)$ we can find $\boldsymbol{\pi} \in \mathcal{B}_q$ such that

$$\frac{1}{p}S^p < \boldsymbol{\pi}(\Gamma) - \frac{1}{q}\operatorname{Bar}_q(\boldsymbol{\pi}).$$

Since Γ is $\boldsymbol{\pi}$-measurable and $\boldsymbol{\pi}$ is Radon, we can find a compact set $\mathrm{K} \subset \Gamma$ such that

$$\frac{1}{p} S^p < \boldsymbol{\pi}(\mathrm{K}) - \frac{1}{q} \mathrm{Bar}_q(\boldsymbol{\pi}) \le \frac{1}{p} \mathrm{Cont}_p(\mathrm{K}),$$

which eventually yields (4.3.4) since S is arbitrary.

(c) Let us now assume that (X, d) is complete and (X, τ) is Souslin and let us prove that every $\mathscr{B}$-analytic set is capacitable. By Choquet's Theorem A.6, it is sufficient to prove that every Borel set is capacitable. By Corollary 2.2.7 we know that $(\mathrm{RA}(X), \tau_{\mathrm{A}}, \mathsf{d}_{\mathrm{A}})$ admits an auxiliary topology τ'_{A}.

Let Γ be a Borel subset of $\mathrm{RA}(X)$. If Γ contains a constant arc there is nothing to prove, so that we can assume that $\Gamma \subset \mathrm{RA}_0(X) = \{\gamma \in \mathrm{RA}(X) : \ell(\gamma) > 0\}$. Recalling the definition of the open sets $\mathrm{RA}_\eta(X)$ given in Lemma 4.1.3 and Proposition 4.1.2(e), we know that $\mathrm{Mod}_p(\Gamma) = \lim_{\eta \downarrow 0} \mathrm{Mod}_p(\Gamma_\eta)$, where $\Gamma_\eta := \Gamma \cap \mathrm{RA}_\eta(X)$. It is therefore sufficient to prove that every Γ_η is capacitable. Let us fix $\eta > 0$; by applying Lemma 4.1.3 for every $\varepsilon > 0$ we can find a compact set $\mathrm{K}_\varepsilon \subset \mathrm{RA}(X)$ such that $\mathrm{Mod}_p(\Gamma_\eta \setminus \mathrm{K}_\varepsilon) \le \mathrm{Mod}_p(\mathrm{RA}_\eta(X) \setminus \mathrm{K}_\varepsilon) \le \varepsilon$. Since Mod_p is subadditive, it remains to prove that $\Gamma_\eta \cap \mathrm{K}_\varepsilon$ is capacitable. Notice that K_ε is also compact with respect to the coarser (metrizable and separable) topology τ'_{A} and the restriction of τ'_{A} to K_ε coincides with τ_{A}, so that $(\mathrm{K}_\varepsilon, \tau_{\mathrm{A}})$ is a Polish space. Since $\Gamma_\eta \cap \mathrm{K}_\varepsilon$ is Borel in K_ε, it is also $\mathscr{F}$-analytic and therefore (being K_ε compact) $\mathscr{K}$-analytic. By claim (a) above we deduce that $\Gamma_\eta \cap \mathrm{K}_\varepsilon$ is capacitable.

(d) It is an immediate consequence of the previous claims, recalling that every $\mathscr{B}$-analytic set is universally measurable.

□

4.3.1 Notes

Theorem 4.3.1 has been proved in [2] by using a different argument based on Hahn-Banach theorem. The proof presented here, based on Von Neumann theorem, shows more clearly that the definitions of Mod_p and of Cont_p rely on dual optimization problems, so that their equality is a nice application of a min-max argument.

Theorem 4.3.2 strongly relies on Choquet's Theorem. It is interesting to note that the possibility to separate distance and topology in e.m.t.m. space expands the range of application and covers the case of general Souslin spaces: the use of an auxiliary topology overcomes the difficulty related to the unknown Souslin character of path spaces (see [24, page 46-III]).

Theorem 4.3.1 and Theorem 4.3.2 could be directly stated at the level of modulus and contents on $\mathcal{M}_+(X)$ instead of $\mathrm{RA}(X)$, see [2]. We will discuss another important case in Sect. 5.1.2.

5 Weak Upper Gradients and Identification of Sobolev Spaces

5.1 (Nonparametric) Weak Upper Gradients and Weak Sobolev Spaces

In this section we introduce a notion of weak upper gradient modeled on $\mathcal{T}_q$-test plans, in the usual setting stated at page 205.

5.1.1 $\mathcal{T}_q$-Test Plans and $\mathcal{T}_q$-Weak Upper Gradients

Recall that the (stretched) evaluation maps $\hat{\mathsf{e}}_t : \mathrm{RA}(X) \to X$ are defined by $\hat{\mathsf{e}}_t(\gamma) := R_\gamma(t)$, see (2.2.52). We also introduce the restriction maps $\mathrm{Restr}_s^t : \mathrm{RA}(X) \to \mathrm{RA}(X)$, $0 \le s < t \le 1$, given by

$$\mathrm{Restr}_s^t(\gamma) := \mathsf{q}(R_\gamma^{s\to t}), \quad R_\gamma^{s\to t}(r) := R_\gamma((1-r)s + rt) \quad r \in [0,1], \tag{5.1.1}$$

where q is the projection map from $\mathrm{C}([0,1]; X)$ to $\mathrm{A}(X)$. Restr_s^t restricts the arc-length parametrization R_γ of the arc γ to the interval $[s,t]$ and then "stretches" it on the whole of $[0,1]$, giving back the equivalent class in $\mathrm{RA}(X)$. Notice that for every $\gamma \in \mathrm{RA}(X)$

$$\int_{\mathrm{Restr}_s^t(\gamma)} f = (t-s)\ell(\gamma)\int_0^1 f(R_\gamma^{s\to t}(r))\,\mathrm{d}r = \ell(\gamma)\int_s^t f(R_\gamma(r))\,\mathrm{d}r. \tag{5.1.2}$$

Definition 5.1.1 (Nonparametric $\mathcal{T}_q$-Test Plans) Let $q = p' \in (1,\infty)$. We call $\mathcal{T}_q = \mathcal{T}_q(\mathbb{X}) \subset \mathcal{M}_+(\mathrm{RA}(X))$ the collection of all (nonparametric) dynamic plans $\boldsymbol{\pi} \in \mathcal{M}_+(\mathrm{RA}(X))$ such that

$$\mathrm{Bar}_q(\boldsymbol{\pi}) < \infty, \quad \mathscr{L}_q((\mathsf{e}_i)_\sharp\boldsymbol{\pi}|\mathfrak{m}) < \infty \quad i = 0, 1; \tag{5.1.3}$$

Dynamic plans in $\mathcal{T}_q$ will be also called $\mathcal{T}_q$-test plans.
We call $\mathcal{T}_q^*$ the subset of $\mathcal{T}_q$ whose plans $\boldsymbol{\pi}$ satisfy the following property: there exists a constant $c > 0$ and a compact set $\mathrm{K} \subset \mathrm{RA}(X)$ (depending on $\boldsymbol{\pi}$) such that

$$(\mathsf{e}_i)_\sharp\boldsymbol{\pi} \le c\mathfrak{m},\ i = 0, 1, \quad \ell_{|\mathrm{K}} \text{ is bounded, continuous and strictly positive,}$$
$$\boldsymbol{\pi}(\mathrm{RA}(X) \setminus \mathrm{K}) = 0. \tag{5.1.4}$$

We say that $\boldsymbol{\pi}$ is a *stretchable* $\mathcal{T}_q$-test plan if

$$\mathrm{Bar}_q(\boldsymbol{\pi}) < \infty, \quad \mathscr{L}_q((\hat{\mathsf{e}}_t)_\sharp\boldsymbol{\pi}|\mathfrak{m}) < \infty \quad \text{for every } t \in [0,1]. \tag{5.1.5}$$

Notice that $\boldsymbol{\pi}$ is a stretchable $\mathcal{T}_q$-test plan if and only if $\mathrm{Restr}_s^t(\boldsymbol{\pi}) \in \mathcal{T}_q$ for every $s, t \in [0, 1]$, $s < t$. Clearly the class of nonparametric $\mathcal{T}_q$-test plans depends on the full e.m.t.m. structure of $\mathbb{X}$; however, when there is no risk of ambiguity, we will simply write $\mathcal{T}_q$.

Definition 5.1.2 ($\mathcal{T}_q$-Negligible Sets of Rectifiable Arcs) Let P be a property concerning nonparametric arcs in $\mathrm{RA}(X)$. We say that P holds $\mathcal{T}_q$-a.e. (or for $\mathcal{T}_q$-almost every arc $\gamma \in \mathrm{RA}(X)$) if for any $\boldsymbol{\pi} \in \mathcal{T}_q$ the set

$$N := \{\gamma : P(\gamma) \text{ does not hold }\}$$

is contained in a $\boldsymbol{\pi}$-negligible Borel set.

Since $\mathcal{T}_q \subset \mathcal{B}_q$, it is clear that for every Borel set $\Gamma \subset \mathrm{RA}(X)$

$$\mathrm{Mod}_p(\gamma) = 0 \quad \Rightarrow \quad \mathrm{Cont}_p(\Gamma) = 0 \quad \Rightarrow \quad \Gamma \text{ is } \mathcal{T}_q\text{-negligible.} \tag{5.1.6}$$

Notice that we can revert the first implication in (5.1.6) e.g. when (X, τ, d) is a Souslin and complete e.m.t. space, see Theorem 4.3.2.

Lemma 5.1.3 *If a set $N \subset \mathrm{RA}_0(X)$ is $\boldsymbol{\pi}$-negligible for every $\boldsymbol{\pi} \in \mathcal{T}_q^*$ then N is $\mathcal{T}_q$-negligible.*

Proof Let us fix $\boldsymbol{\pi} \in \mathcal{T}_q$ and let $\pi_i = (\mathsf{e}_i)_\sharp \pi = h_i \mathfrak{m}$ with $h_i \in L^q(X, \mathfrak{m})$, h_i Borel nonnegative. We set

$$H_{i,k} := \{x \in X : h_i(x) \le k\}, \quad \mathrm{H}_k := \{\gamma \in \mathrm{RA}(X) : \ell(\gamma) \le k,\ \mathsf{e}_i(\gamma) \in H_{i,k}\ i = 0, 1\}.$$

Clearly

$$\lim_{k\to\infty} \boldsymbol{\pi}(\mathrm{RA}(X) \setminus \mathrm{H}_k) \le \lim_{k\to\infty} \big(\pi_0(X \setminus H_{0,k}) + \pi_1(X \setminus H_{1,k})\big) = 0.$$

We can also find an increasing sequence of compact sets $\mathrm{K}_n \subset \mathrm{RA}_0(X)$ such that the restriction of ℓ to K_n is continuous and strictly positive and $\boldsymbol{\pi}(\mathrm{RA}_0(X) \setminus \mathrm{K}_n) \le 2^{-n}$. It follows that $\boldsymbol{\pi}_n := \boldsymbol{\pi}|_{\mathrm{K}_n \cap \mathrm{H}_n}$ belongs to $\mathcal{T}_q^*$ and we can find a Borel set B_n with $\mathrm{RA}_0(X) \supset B_n \supset N$ such that $\boldsymbol{\pi}_n(B_n) = 0$. Setting $B := \cap_n B_n$ clearly $B \supset N$ and $\boldsymbol{\pi}_n(B) = 0$ so that $\boldsymbol{\pi}(B) = 0$ as well. □

Recall that a Borel function $g : X \to [0, +\infty]$ is an *upper gradient* [22] for $f : X \to \mathbb{R}$ if

$$\Big|\int_{\partial\gamma} f\Big| \le \int_\gamma g \quad \text{for every } \gamma \in \mathrm{RA}(X) \quad \text{where} \quad \int_{\partial\gamma} f := f(\gamma_1) - f(\gamma_0). \tag{5.1.7}$$

Definition 5.1.4 ($\mathcal{T}_q$-Weak Upper Gradients) Given $f : X \to \mathbb{R}$, a $\mathfrak{m}$-measurable function $g : X \to [0, \infty]$ is a $\mathcal{T}_q$-weak upper gradient (w.u.g.) of f if

$$\Big| \int_{\partial\gamma} f \Big| \le \int_\gamma g < \infty \qquad \text{for } \mathcal{T}_q\text{-almost every } \gamma \in \mathrm{RA}(X). \tag{5.1.8}$$

Remark 5.1.5 (Truncations) If $T : \mathbb{R} \to \mathbb{R}$ is a 1-Lipschitz map and g is a $\mathcal{T}_q$-w.u.g. of f, then g is a $\mathcal{T}_q$-w.u.g. of $T \circ f$ as well. Conversely, if g is a $\mathcal{T}_q$-w.u.g. of $f_k := -k \vee f \wedge k$ for every $k \in \mathbb{N}$, it is easy to see that g is a $\mathcal{T}_q$-w.u.g. of f. By this property, in the proof of many statements concerning $\mathcal{T}_q$-w.u.g. it will not be restrictive to assume f bounded.

The definition of weak upper gradient enjoys natural invariance properties w.r.t. modifications in $\mathfrak{m}$-negligible sets. We will also show that if g is $\mathfrak{m}$-measurable the integral in (5.1.8) is well defined for $\mathcal{T}_q$-a.e. arc γ.

Proposition 5.1.6 (Measurability and Invariance Under Modifications in $\mathfrak{m}$-Negligible Sets)

(a) *If $f, \tilde f : X \to \mathbb{R}$ differ in a $\mathfrak{m}$-negligible set then for $\mathcal{T}_q$-a.e. arc γ*

$$f(\gamma_0) = \tilde f(\gamma_0), \quad f(\gamma_1) = \tilde f(\gamma_1), \quad f \circ R_\gamma = \tilde f \circ R_\gamma \mathscr{L}^1\text{-a.e. in } (0, 1). \tag{5.1.9}$$

(b) *If g is $\mathfrak{m}$-measurable then the map $s \mapsto g(R_\gamma(s))$ is $\mathscr{L}^1$-measurable for $\mathcal{T}_q$-a.e. arc γ.*

(c) *Let $f, \tilde f : X \to \mathbb{R}$ and $g, \tilde g : X \to [0, \infty]$ be such that both $\{f \neq \tilde f\}$ and $\{g \neq \tilde g\}$ are $\mathfrak{m}$-negligible. If g is a $\mathcal{T}_q$-weak upper gradient of f then $\tilde g$ is a $\mathcal{T}_q$-weak upper gradient of $\tilde f$.*

Proof

(a) Let $N \supset \{f \neq \tilde f\}$ be a $\mathfrak{m}$-negligible Borel set and let $\boldsymbol{\pi} \in \mathcal{T}_q$ be a test plan. We have

$$\int \Big(\int_\gamma \chi_N\Big)\,\mathrm{d}\boldsymbol{\pi}(\gamma) = \mu_{\boldsymbol{\pi}}(N) = 0 \quad \text{since } \mu_{\boldsymbol{\pi}} \ll \mathfrak{m} \text{ and } \mathfrak{m}(N) = 0,$$

so that $\int_\gamma \chi_N = \ell(\gamma)\int_0^1 \chi_N(R_\gamma(s))\,\mathrm{d}s = 0$ for $\boldsymbol{\pi}$-a.e. γ. For any arc γ for which the integral is null $f(R(\gamma_s))$ coincides a.e. in $[0, 1]$ with $\tilde f(R(\gamma_s))$. The same argument shows the sets $\big\{\gamma : f(\gamma_t) \neq \tilde f(\gamma_t)\big\} \subset \{\gamma : \gamma_t \in N\}$, $t = 0, 1$ are $\boldsymbol{\pi}$-negligible because $(\mathrm{e}_t)_\sharp\boldsymbol{\pi} \ll \mathfrak{m}$, which implies that $\boldsymbol{\pi}(\{\gamma : \gamma_t \in N\}) = (\mathrm{e}_t)_\sharp\boldsymbol{\pi}(N) = 0$.

(b) If $\tilde g$ is a Borel modification of g the set $\{g \neq \tilde g\}$ is a $\mathfrak{m}$-negligible set; by the previous Claim (a), $g(R(\gamma_s))$ coincides $\mathscr{L}^1$-a.e. in $[0, 1]$ with the Borel map $\tilde g(R(\gamma_s))$ and it is therefore $\mathscr{L}^1$-measurable.

(c) follows immediately by Claim (a) as well, since for $\mathcal{T}_q$-a.e. arc γ $\int_{\partial\gamma} f = \int_{\partial\gamma} \tilde{f}$ and $\int_\gamma g = \int_\gamma \tilde{g}$.

□

Remark 5.1.7 (Local Lipschitz Constants of d*-Lipschitz Functions Are Weak Upper Gradients)* If $f \in \mathrm{Lip}_b(X, \tau, \mathsf{d})$ then the local Lipschitz constant lip f is an upper gradient and therefore it is also a $\mathcal{T}_q$-weak upper gradient. An analogous property holds for the d-slopes (notice that the topology τ does not play any role in the definition)

$$|\mathrm{D}^\pm f|(x) := \lim_{\mathsf{d}(y,x)\to 0} \frac{(f(y) - f(x))_\pm}{\mathsf{d}(y,x)}, \quad |\mathrm{D} f|(x) := \lim_{\mathsf{d}(y,x)\to 0} \frac{|f(y) - f(x)|}{\mathsf{d}(y,x)}$$

of an arbitrary d-Lipschitz functions $f \in \mathrm{B}_b(X)$: if $|\mathrm{D}f|$ is $\mathfrak{m}$-measurable (this property is always satisfied if, e.g., (X, τ) is Souslin, see [8, Lemma 2.6]) then it is a weak upper gradient of f.

It is easy to check that for every $\alpha, \beta \in \mathbb{R}$

$$\text{if } g_i \text{ is a } \mathcal{T}_q\text{-w.u.g. of } f_i, i = 0, 1, \text{ then } \quad |\alpha| g_0 + |\beta| g_1 \text{ is a } \mathcal{T}_q\text{-w.u.g. of } \alpha f_0 + \beta f_1. \tag{5.1.10}$$

In particular the set $S := \{(f, g) : g \text{ is a } \mathcal{T}_q\text{-w.u.g. of } f\}$ is convex.

If we know a priori the integrability of f and the L^p-summability of g then Definition 5.1.4 can be formulated in a slightly different way:

Lemma 5.1.8 *Let $f \in L^1(X, \mathfrak{m})$ and $g \in L^p(X, \mathfrak{m})$, $g \geq 0$. g is a $\mathcal{T}_q$-weak upper gradient of f if and only if*

$$\int \big(f(\gamma_1) - f(\gamma_0)\big)\,\mathrm{d}\boldsymbol{\pi}(\gamma) \leq \int \Big(\int_\gamma g\Big)\,\mathrm{d}\boldsymbol{\pi}(\gamma) \quad \textit{for every } \boldsymbol{\pi} \in \mathcal{T}_q^*. \tag{5.1.11}$$

Equivalently, setting $\pi_i := (\mathsf{e}_i)_\sharp \boldsymbol{\pi}$, $i = 0, 1$,

$$\int_X f\,\mathrm{d}(\pi_1 - \pi_0) \leq \int_X g\,\mathrm{d}\mu_{\boldsymbol{\pi}} \quad \textit{for every } \boldsymbol{\pi} \in \mathcal{T}_q^*. \tag{5.1.12}$$

Proof It is clear that (5.1.8) yields (5.1.11) simply by integration w.r.t. $\boldsymbol{\pi}$; notice that the integrals in (5.1.11) (and in (5.1.12)) are well defined since $\pi_i = h_i\mathfrak{m}$ and $\mu_{\boldsymbol{\pi}} = h\mathfrak{m}$ for functions $h_i \in L^\infty(X, \mathfrak{m})$ and $h \in L^q(X, \mathfrak{m})$.

Let us prove the converse implication. It is not restrictive to assume that f, g are Borel. By Lemma 5.1.3, if (5.1.8) is not true, there exists a nontrivial test plan $\boldsymbol{\pi} \in \mathscr{T}_q^*$ such that the Borel set $A := \Big\{\gamma \in \mathrm{RA}(X) : |\int_{\partial\gamma} f| > \int_\gamma g\Big\}$ satisfies $\boldsymbol{\pi}(A) > 0$ (notice that (5.1.8) is always satisfied on constant arcs). By possible reducing A we can also find $\theta > 0$ such that $A' := \Big\{\gamma \in \mathrm{RA}(X) : \int_{\partial\gamma} f \geq \theta + \int_\gamma g\Big\}$ satisfies

$\pi(A') > 0$. Thus defining $\pi' := \pi_{|A'}$, it is immediate to check that $\pi' \in \mathcal{T}_q^*$; a further integration with respect to π' of the previous inequality yields

$$\int \Big(f(\gamma_1) - f(\gamma_0)\Big)\,\mathrm{d}\pi'(\gamma) \ge \theta\pi(A') + \int \Big(\int_\gamma g\Big)\,\mathrm{d}\pi'(\gamma),$$

which contradicts (5.1.11). □

Definition 5.1.9 (Sobolev Regularity Along a Rectifiable Arc) We say that a map $f : X \to \mathbb{R}$ is *Sobolev* (resp. *strictly Sobolev*) along an arc γ if $f \circ R_\gamma$ coincides $\mathscr{L}^1$-a.e. in $[0, 1]$ (resp. $\mathscr{L}^1$-a.e. in $[0, 1]$ and in $\{0, 1\}$) with an absolutely continuous map $f_\gamma : [0, 1] \to \mathbb{R}$. In this case, we say that a map $g : X \to \mathbb{R}$ is a Sobolev upper gradient (S.u.g.) for f along γ if $\ell(\gamma) g \circ R_\gamma \in L^1(0, 1)$ and

$$\left|\frac{\mathrm{d}}{\mathrm{d}t} f_\gamma\right| \le \ell(\gamma) g \circ R_\gamma \quad \text{a.e. in } [0, 1]. \tag{5.1.13}$$

We can give an intrinsic formulation of the (strict) Sobolev regularity with Sobolev upper gradient g which does not involve the absolutely continuous representative.

Lemma 5.1.10 *Let us suppose that $\gamma \in \mathrm{RA}(X)$, $f, g : X \to \mathbb{R}$ such that $f \circ R_\gamma$, $\ell(\gamma) g \circ R_\gamma \in L^1(0, 1)$, and $\mathcal{C}$ (resp. $\mathcal{C}_c$) is a countable dense subset of $\mathrm{C}^1([0, 1])$ (resp. of $\mathrm{C}^1_c(0, 1)$).*

(a) *f is Sobolev along γ with Sobolev u.g. g if and only if*

$$\Big| - \int_0^1 \varphi'(t) f(R_\gamma(t))\,\mathrm{d}t\Big| \le \ell(\gamma) \int_0^1 |\varphi(t)|\, g(R_\gamma(t))\,\mathrm{d}t \tag{5.1.14}$$

for every $\varphi \in \mathcal{C}_c$.

(b) *f is strictly Sobolev along γ with Sobolev u.g. g if and only if*

$$\Big|\varphi(1) f(R_\gamma(1)) - \varphi(0) f(R_\gamma(0)) - \int_0^1 \varphi'(t) f(R_\gamma(t))\,\mathrm{d}t\Big| \le \ell(\gamma) \int_0^1 |\varphi(t)|\, g(R_\gamma(t))\,\mathrm{d}t \tag{5.1.15}$$

for every $\varphi \in \mathcal{C}$.

In particular, if f, g are Borel maps, the sets of curves $\gamma \in \mathrm{RA}(X)$ along which $\int_\gamma (|f| + g) < \infty$ and f is Sobolev (resp. strictly Sobolev) with S.u.g. g is Borel in $\mathrm{RA}(X)$.

Proof

(a) One implication is obvious. If (5.1.14) holds for every $\varphi \in \mathcal{C}_c$ then it can be extended to every $\varphi \in \mathrm{C}^1_c(0, 1)$. In particular we have

$$\Big| - \int_0^1 \varphi'(t) f(R_\gamma(t))\,\mathrm{d}t\Big| \le C \sup_{t\in[0,1]} |\varphi(t)| \quad \text{where} \quad C := \int_\gamma g$$

for every $\varphi \in \mathrm{C}_c^1(0,1)$, so that the distributional derivative of $f \circ R_\gamma$ can be represented by Radon measure $\mu \in \mathcal{M}((0,1))$ with finite total variation. Equation (5.1.14) also yields

$$\Big|\int_0^1 \varphi \,\mathrm{d}\mu\Big| \le \ell(\gamma)\int_0^1 |\varphi|\, g\,\mathrm{d}t \quad \text{for every } \varphi \in \mathrm{C}_c^1(0,1)$$

so that $\mu = h\mathscr{L}^1$ is absolutely continuous w.r.t. $\mathscr{L}^1$ with density satisfying $|h| \le \ell(\gamma) g \circ R_\gamma$ $\mathscr{L}^1$-a.e. It follows that $f \circ R_\gamma \in W^{1,1}(0,1)$ and its absolutely continuous representative f_γ satisfies (5.1.13).

(b) follows as in the previous claim (a); from (5.1.15) it is also not difficult to check that $f_\gamma(i) = f \circ R_\gamma(i)$ for $i = 0$ or $i = 1$.

Concerning the last statement, arguing as in the proof of Theorem 2.2.13(e), it is not difficult to show that every bounded (resp. nonnegative) function h and for every $\psi \in \mathrm{C}([0,1])$ (resp. nonnegative) the real maps

$$\gamma \mapsto \int_0^1 \psi(t) h(\gamma(t))\,\mathrm{d}t \quad \text{are Borel in } (\mathrm{C}([0,1];X), \tau_{\mathrm{C}}).$$

Since by Theorem 2.2.13(d) the map $\gamma \mapsto R_\gamma$ is Borel from $(\mathrm{RA}(X), \tau_{\mathrm{A}})$ to $(\mathrm{C}([0,1];X), \tau_{\mathrm{C}})$ and $\mathcal{C}$, $\mathcal{C}_c$ are countable, we deduce that the sets characterized by the family of inequalities (5.1.14) or (5.1.15) are Borel in $\mathrm{RA}(X)$.

□

Remark 5.1.11 (Sobolev Regularity Along $\mathcal{T}_q$-Almost Every Arc) By Proposition 5.1.6(a), it is easy to check that the properties to be Sobolev or strictly Sobolev along $\mathcal{T}_q$-almost every arc with S.u.g. g are invariant with respect to modification of f and g in $\mathfrak{m}$-negligible sets, and thus they make sense for Lebesgue classes. It is also not restrictive to consider only arcs $\gamma \in \mathrm{RA}_0(X)$.

In the next Theorem we prove that the existence of a $\mathcal{T}_q$-weak upper gradient yields strict Sobolev regularity along $\mathcal{T}_q$-almost every arc. This property is based on a preliminary lemma, which provides this property for stretchable plans in $\mathcal{T}_q^*$.

Lemma 5.1.12 *Assume that $\boldsymbol{\pi} \in \mathcal{T}_q^*$ is a stretchable plan and that $g : X \to [0,\infty]$ is a $\mathcal{T}_q$-weak upper gradient of a $\mathfrak{m}$-measurable function $f : X \to \mathbb{R}$. Then f is strictly Sobolev along $\boldsymbol{\pi}$-almost every arc with S.u.g. g, i.e.* (5.1.15) *or, equivalently,*

$$\Big|\frac{\mathrm{d}}{\mathrm{d}t} f_\gamma\Big| \le \ell(\gamma) g \circ R_\gamma \quad \textit{a.e. in } [0,1], \qquad f_\gamma(i) = f(R_\gamma(i)) \quad i \in \{0,1\}, \tag{5.1.16}$$

hold for $\boldsymbol{\pi}$-almost every $\gamma \in \mathrm{RA}(X)$.

Proof Arguing as in Proposition 5.1.6 it is not restrictive to assume that f, g are Borel function. Since $\boldsymbol{\pi} \in \mathcal{T}_q^*$ we know that there exists a compact set $\mathcal{K} \subset \mathrm{RA}(X)$ satisfying (5.1.4). The stretchable condition (5.1.5) and an obvious change of variables related to the maps $R_\gamma^{s\to t}$ of (5.1.1) yields for every $s < t$ in $[0,1]$ and for $\boldsymbol{\pi}$-almost every $\gamma \in \mathcal{K}$,

$$|f(R_\gamma(t)) - f(R_\gamma(s))| \le (t-s)\ell(\gamma)\int_0^1 g(R_\gamma^{s\to t}(r))\,\mathrm{d}r = \ell(\gamma)\int_s^t g(R_\gamma(r))\,\mathrm{d}r, \tag{5.1.17}$$

since $\mathrm{Restr}_s^t(\boldsymbol{\pi}) \in \mathcal{T}_q$. We apply Fubini's Theorem to the product measure $\mathscr{L}^2 \otimes \boldsymbol{\pi}$ in $(0,1)^2 \times \mathcal{K}$ and we use the fact that the maps characterizing the inequality (5.1.17) are jointly Borel with respect to $(s,t,\gamma) \in (0,1)^2 \times \mathcal{K}$: here we use the continuity of R from $\mathcal{K}$ to $\mathrm{BVC}_c([0,1];X)$ endowed with the topology τ_{C}. It follows that for $\boldsymbol{\pi}$-a.e. γ the function f satisfies

$$|f(R_\gamma(t)) - f(R_\gamma(s))| \le \ell(\gamma)\int_s^t g(R_\gamma(r))\,\mathrm{d}r \qquad \text{for } \mathscr{L}^2\text{-a.e. } (t,s) \in (0,1)^2.$$

An analogous argument shows that for $\boldsymbol{\pi}$-a.e. γ

$$\begin{cases} |f(R_\gamma(s)) - f(\gamma_0)| \le \ell(\gamma)\int_0^s g(R_\gamma(r))\,\mathrm{d}r \\ |f(\gamma_1) - f(R_\gamma(s))| \le \ell(\gamma)\int_s^1 g(R_\gamma(r))\,\mathrm{d}r \end{cases} \qquad \text{for } \mathscr{L}^1\text{-a.e. } s \in (0,1). \tag{5.1.18}$$

Since $g \circ R_\gamma \in L^1(0,1)$ for $\boldsymbol{\pi}$-a.e. $\gamma \in \mathcal{K}$ with $\ell(\gamma) > 0$,, by the next Lemma 5.1.13 it follows that $f \circ R_\gamma \in W^{1,1}(0,1)$ for $\boldsymbol{\pi}$-a.e. γ and (understanding the derivative of $f \circ R_\gamma$ as the distributional one)

$$\left|\frac{\mathrm{d}}{\mathrm{d}t}(f \circ R_\gamma)\right| \le \ell(\gamma) g \circ R_\gamma \quad \mathscr{L}^1\text{-a.e. in } (0,1), \quad \text{for } \boldsymbol{\pi}\text{-a.e. } \gamma. \tag{5.1.19}$$

We conclude that $f \circ R_\gamma \in W^{1,1}(0,1)$ for $\boldsymbol{\pi}$-a.e. γ, and therefore it admits an absolutely continuous representative f_γ for which (5.1.16) holds; moreover, by (5.1.18) , it is immediate to check that $f(\gamma(t)) = f_\gamma(t)$ for $t \in \{0,1\}$ and $\boldsymbol{\pi}$-a.e. γ. □

Lemma 5.1.13 *Let $f : (0,1) \to \mathbb{R}$ and $g \in L^q(0,1)$ nonnegative satisfy*

$$\big|f(t) - f(s)\big| \le \left|\int_s^t g(r)\,\mathrm{d}r\right| \quad \textit{for } \mathscr{L}^2\textit{-a.e. } (s,t) \in (0,1)\times(0,1). \tag{5.1.20}$$

Then $f \in W^{1,q}(0,1)$ and $|f'| \le g$ $\mathscr{L}^1$-a.e. in $(0,1)$.

We refer to [7, Lemma 2.1] for the proof.

We can considerably refine Lemma 5.1.12, by removing the assumption that $\boldsymbol{\pi}$ is stretchable and by considering arbitrary nonparametric dynamic plans in $\mathcal{T}_q$.

Theorem 5.1.14 *Assume that $g : X \to [0, \infty]$ is a $\mathcal{T}_q$-weak upper gradient of a $\mathfrak{m}$-measurable function $f : X \to \mathbb{R}$. Then f is strictly Sobolev and satisfies* (5.1.16) *along $\mathcal{T}_q$-almost every arc.*

Proof By Lemma 5.1.3 it is sufficient to prove the property for every $\boldsymbol{\pi} \in \mathcal{T}_q^*$, so that we can also assume that there exists a compact set $\mathcal{K} \subset \mathrm{RA}(X)$ satisfying (5.1.4).

For every $r \in [0, 1/3]$ and $s \in [2/3, 1]$ we consider the rescaled plans

$$\boldsymbol{\pi}_r^+ := (\mathrm{Restr}_r^1)_\sharp \boldsymbol{\pi}, \quad \boldsymbol{\pi}_s^- := (\mathrm{Restr}_0^s)_\sharp(\boldsymbol{\pi}), \tag{5.1.21}$$

which form two continuous (thus Borel) collections depending on $r \in [0, 1/3]$, $s \in [2/3, 1]$. We then set

$$\boldsymbol{\pi}^+ := 3\int_0^{1/3} \boldsymbol{\pi}_r^+ \,\mathrm{d}r, \quad \boldsymbol{\pi}^- := 3\int_{2/3}^1 \boldsymbol{\pi}_s^- \,\mathrm{d}s. \tag{5.1.22}$$

Notice that we can equivalently characterize $\boldsymbol{\pi}^+, \boldsymbol{\pi}^-$ as the push forward measures of

$$\boldsymbol{\sigma}^+ := 3\mathscr{L}^1|_{(0,1/3)} \otimes \boldsymbol{\pi} \text{ and } \boldsymbol{\sigma}^- := 3\mathscr{L}^1|_{(2/3,1)} \otimes \boldsymbol{\pi},$$

through the continuous maps $(r, \gamma) \mapsto \mathrm{Restr}_r^1(\gamma)$ and $(s, \gamma) \mapsto \mathrm{Restr}_0^s(\gamma)$ respectively:

$$\boldsymbol{\pi}^+ = (\mathrm{Restr}_\cdot^1)_\sharp\big(3\mathscr{L}^1|_{(0,1/3)} \otimes \boldsymbol{\pi}\big), \quad \boldsymbol{\pi}^- = (\mathrm{Restr}_0^\cdot)_\sharp\big(3\mathscr{L}^1|_{(2/3,1)} \otimes \boldsymbol{\pi}\big). \tag{5.1.23}$$

Let us check that $\boldsymbol{\pi}^\pm$ belong to $\mathcal{T}_q^*$ and are stretchable. We only consider $\boldsymbol{\pi}^+$, since the argument for $\boldsymbol{\pi}^-$ is completely analogous. Recalling Lemma 4.2.3, for every nonnegative $f \in \mathrm{C}_b(X)$ we have

$$\begin{aligned}\int\int_\gamma f \,\mathrm{d}\boldsymbol{\pi}^+(\gamma) &\overset{(5.1.22)}{=} 3\int_0^{1/3} \Big(\int\int_\gamma f \,\mathrm{d}\boldsymbol{\pi}_r^+(\gamma)\Big)\,\mathrm{d}r \\ &\overset{(5.1.2)}{=} 3\int_0^{1/3} \Big(\int \ell(\gamma)\int_r^1 f(R_\gamma(s))\,\mathrm{d}s\,\mathrm{d}\boldsymbol{\pi}(\gamma)\Big)\,\mathrm{d}r \\ &\le 3\int_0^{1/3} \Big(\int\int_\gamma f\,\mathrm{d}\boldsymbol{\pi}(\gamma)\Big)\,\mathrm{d}r \overset{(4.2.9)}{\le} \|f\|_{L^p}\,\mathrm{Bar}_q(\boldsymbol{\pi}),\end{aligned} \tag{5.1.24}$$

which shows that $\mathrm{Bar}_q(\boldsymbol{\pi}^+) \le \mathrm{Bar}_q(\boldsymbol{\pi})$.

Let us now prove that $(\hat{\mathsf{e}}_s)_\sharp \boldsymbol{\pi}^+ \ll \mathfrak{m}$ with density in $L^q(X,\mathfrak{m})$ for every $s \in [0,1]$; setting $\ell_o := \min_{\mathcal{K}} \ell > 0$, if $s < 1$ we have

$$\begin{aligned}
\int f \,\mathrm{d}(\hat{\mathsf{e}}_s)_\sharp \boldsymbol{\pi}^+ &= \int f(R_\gamma(s))\,\mathrm{d}\boldsymbol{\pi}^+(\gamma) \overset{(5.1.22)}{=} 3\int_0^{1/3}\Big(\int f(R_\gamma(s))\,\mathrm{d}\boldsymbol{\pi}_r^+(\gamma)\Big)\,\mathrm{d}r\\
&\overset{(5.1.23)}{=} 3\int_0^{1/3}\Big(\int f(R_\gamma^{r\to 1}(s))\mathrm{d}\boldsymbol{\pi}(\gamma)\Big)\,\mathrm{d}r\\
&= 3\int\Big(\int_0^{1/3} f(R_\gamma(r(1-s)+s))\,\mathrm{d}r\Big)\mathrm{d}\boldsymbol{\pi}(\gamma)\\
&= \frac{3}{1-s}\int\Big(\int_s^{s+(1-s)/3} f(R_\gamma(\theta))\,\mathrm{d}\theta\Big)\,\mathrm{d}\boldsymbol{\pi}(\gamma)\\
&\le \frac{3}{\ell_o(1-s)}\int\int_\gamma f\,\mathrm{d}\boldsymbol{\pi}(\gamma) \le \frac{3}{\ell_o(1-s)}\|f\|_{L^p}\,\mathrm{Bar}_q(\boldsymbol{\pi}).
\end{aligned} \tag{5.1.25}$$

On the other hand, if $s = 1$, we can use the fact that $(\mathsf{e}_1)_\sharp \boldsymbol{\pi}_r = (\mathsf{e}_1)_\sharp \boldsymbol{\pi}$

$$\begin{aligned}
\int f(\gamma_1)\,\mathrm{d}\boldsymbol{\pi}^+(\gamma) &= 3\int_0^{1/3}\Big(\int f(\gamma(1))\,\mathrm{d}\boldsymbol{\pi}_r^+(\gamma)\Big)\,\mathrm{d}r = 3\int_0^{1/3}\Big(\int f(\gamma(1))\,\mathrm{d}\boldsymbol{\pi}(\gamma)\Big)\,\mathrm{d}r\\
&= \int f\,\mathrm{d}(\mathsf{e}_1)_\sharp \boldsymbol{\pi},
\end{aligned}$$

so that $(\mathsf{e}_1)_\sharp \boldsymbol{\pi}^+ = (\mathsf{e}_1)_\sharp \boldsymbol{\pi}$ which has an L^q density w.r.t. $\mathfrak{m}$.

Let us now select a Borel representative of f. Applying Lemma 5.1.12 we know that f is Sobolev along $\boldsymbol{\pi}^+$ and $\boldsymbol{\pi}^-$-almost every arc. Recalling the representation result (5.1.23) and applying Fubini's Theorem, we can find a $\boldsymbol{\pi}$-negligible set $N \subset \mathcal{K}$ such that for every $\gamma \in \mathcal{K} \setminus N$ the map f is Sobolev along the arcs $\mathrm{Restr}_r^1(\gamma)$ and $\mathrm{Restr}_0^s(\gamma)$ for $\mathscr{L}^1$-a.e. $r \in [0,1/3]$ and $\mathscr{L}^1$-a.e. $s \in [2/3,1]$ and (5.1.16) holds. Choosing arbitrarily $r \in [0,1/3]$ and $s \in [2/3,1]$ so that such a property holds, since the absolutely continuous representative f_γ should coincide along the curve $t \mapsto R_\gamma(t)$ in the interval $[r,s]$, one immediately sees that f is Sobolev along γ and (5.1.16) holds as well. We conclude that f is Sobolev along $\boldsymbol{\pi}$-a.e. arc and since $\boldsymbol{\pi}$ is arbitrary in $\mathcal{T}_q^*$ we get the thesis. □

Remark 5.1.15 (Equivalent Formulation) By a similar argument we obtain an equivalent formulation of the weak upper gradient property when f is Sobolev along $\mathcal{T}_q$-almost every arc: a function g satisfying $\int_\gamma g < \infty$ for $\mathcal{T}_q$-almost every arc γ is a $\mathcal{T}_q$-weak upper gradient of f if and only if (5.1.15) holds for every φ in a dense subset of $\mathrm{C}^1([0,1])$ and $\mathcal{T}_q$-almost every arc γ, or, equivalently, the function f_γ of Definition 5.1.9 satisfies (5.1.16) $\mathcal{T}_q$-almost everywhere.

5.1.2 The Link with Mod_p-Weak Upper Gradients

In this section we will show that the definition of $\mathcal{T}_q$-weak upper gradient can be equivalently stated in terms of Mod_p, as in the Newtonian approach to metric Sobolev spaces. Part of the results stated here could also be derived as a consequence of the identification Theorem of Sect. 5.2, so we will just sketch the main ideas.

First of all we can associate a q-barycentric entropy to plans in $\mathcal{T}_q$: we consider the measure $\tilde{\mu}_{\boldsymbol{\pi}} \in \mathcal{M}_+(X)$ defined by

$$\int_X f\,\mathrm{d}\tilde{\mu}_{\boldsymbol{\pi}} := \int \Big(f(\gamma_0) + f(\gamma_1) + \int_\gamma f\Big)\,\mathrm{d}\boldsymbol{\pi}(\gamma) = \int \Big(\int f\,\mathrm{d}\tilde{\mathrm{M}}\gamma\Big)\,\mathrm{d}\boldsymbol{\pi}(\gamma) \tag{5.1.26}$$

where $\tilde{\mathrm{M}} : \mathrm{RA}(X) \to \mathcal{M}_+(X)$ has been defined in (4.1.12). We then set

$$\frac{1}{q}\widetilde{\mathrm{Bar}}_q^q(\boldsymbol{\pi}) := \mathscr{L}^q(\tilde{\mu}_{\boldsymbol{\pi}}|\mathfrak{m}) = \frac{1}{q}\int_X h^q\,\mathrm{d}\mathfrak{m} \quad \text{if } \tilde{\mu}_{\boldsymbol{\pi}} = h\mathfrak{m}. \tag{5.1.27}$$

and it is easy to check that

$$\boldsymbol{\pi} \in \mathcal{T}_q \text{ if and only if } \widetilde{\mathrm{Bar}}_q(\boldsymbol{\pi}) < \infty. \tag{5.1.28}$$

It is clear that $\widetilde{\mathrm{Bar}}_q(\boldsymbol{\pi}) \geq \mathrm{Bar}_q(\boldsymbol{\pi})$ for every dynamic plan $\boldsymbol{\pi}$. Arguing as in the proof of Lemma 4.2.5 one can also see that for every $k \geq 0$

$$\text{if } (X, \mathsf{d}) \text{ is complete then the set } \Big\{\boldsymbol{\pi} \in \mathcal{M}_+(\mathrm{RA}(X)) : \widetilde{\mathrm{Bar}}_q(\boldsymbol{\pi}) \leq k\Big\} \text{ is compact.} \tag{5.1.29}$$

By duality we obtain the corresponding notion of content

$$\frac{1}{p}\widetilde{\mathrm{Cont}}_p(\Gamma) := \sup\Big\{\boldsymbol{\pi}(\Gamma) - \frac{1}{q}\widetilde{\mathrm{Bar}}_q^q(\boldsymbol{\pi}) : \boldsymbol{\pi} \in \mathcal{T}_q\Big\}, \tag{5.1.30}$$

and we can obtain an important characterization of $\widetilde{\mathrm{Mod}}_p$, as for Theorems 4.3.1 and 4.3.2:

Theorem 5.1.16

(a) *If* Γ *is a compact subset of* $\mathrm{RA}(X)$ *then*

$$\widetilde{\mathrm{Cont}}_p^p(\Gamma) = \widetilde{\mathrm{Mod}}_p(\Gamma) = \widetilde{\mathrm{Mod}}_{p,c}(\Gamma). \tag{5.1.31}$$

(b) $\widetilde{\mathrm{Mod}}_p$ *is a* $\mathscr{K}(\mathrm{RA}(X), \tau_{\mathrm{A}})$*-Choquet capacity in* $\mathrm{RA}(X)$.

(c) *For every universally measurable* $\Gamma \subset \mathrm{RA}(X)$

$$\widetilde{\mathrm{Cont}}_p(\Gamma) = \sup\left\{\widetilde{\mathrm{Cont}}_p(\mathrm{K}) : \mathrm{K} \subset \Gamma,\ \mathrm{K}\ \textit{compact}\right\}. \tag{5.1.32}$$

(d) *If* (X, d) *is complete and* (X, τ) *is Souslin then every* $\mathscr{B}(\mathrm{RA}(X), \tau_{\mathrm{A}})$*-analytic set* Γ *is* $\widetilde{\mathrm{Mod}}_p$*-capacitable and satisfies* $\widetilde{\mathrm{Mod}}_p(\Gamma) = \widetilde{\mathrm{Cont}}_p^p(\Gamma)$*. In particular* Γ *is* $\widetilde{\mathrm{Mod}}_p$*-negligible if and only if it is* $\mathcal{T}_q$*-negligible.*

We leave the proof to the reader: Claim (a) is based on the same min-max argument of Theorem 4.3.1 (replacing integration w.r.t. ν_γ with integration w.r.t. $\tilde{\nu}_\gamma$), Claims (b–d) can be obtained by arguing as in Theorem 4.3.2. In fact, the proofs would be slightly easier, since compactness of sublevels of $\widetilde{\mathrm{Bar}}_q$ and tightness of $\widetilde{\mathrm{Mod}}_p$ behave better than the corresponding properties for Bar_q and Mod_p. It would also be possible to derive the proofs by a general duality between Mod_p and a corresponding notion of content in $\mathcal{M}_+(X)$, see [2, §5].

Let us now observe that if we consider only Sobolev regularity along arcs, we can improve Theorem 5.1.14.

Proposition 5.1.17

(a) *If* $g \in L^p(X, \mathfrak{m})$*,* $g \geq 0$*, is a* $\mathcal{T}_q$*-weak upper gradient of a* $\mathfrak{m}$*-measurable function* $f : X \to \mathbb{R}$*, then* f *is Sobolev with S.u.g.* g *along* $\mathcal{B}_q$*-almost every arc;* (5.1.13) *holds for* $\mathcal{B}_q$*-almost every* $\gamma \in \mathrm{RA}(X)$.
(b) *If moreover* (X, d) *is complete and* (X, τ) *is Souslin, then* f *is Sobolev with S.u.g.* g *along* Mod_p*-almost every arc and* (5.1.13) *holds* Mod_p*-a.e.*

Proof

(a) It is not restrictive to assume that f and g are Borel. Let us show that for every plan $\boldsymbol{\pi} \in \mathcal{B}_q$ f is Sobolev along $\boldsymbol{\pi}$-a.e. arc γ and (5.1.13) holds $\boldsymbol{\pi}$-a.e. Since $\boldsymbol{\pi}$ is Radon and both the properties trivially hold along constant arcs, it is not restrictive to assume that $\boldsymbol{\pi}$ it is concentrated on a compact set $\mathrm{K} \subset \mathrm{RA}_0(X)$ where ℓ is continuous. In particular the map $T : (r, \gamma) \mapsto \mathrm{Restr}_{1-r}^r(\gamma)$ is continuous in $[0, 1/3] \times \mathrm{K}$. Arguing as in the proof of Theorem 5.1.14 we define

$$\boldsymbol{\pi}_r := (\mathrm{Restr}_{1-r}^r)_\sharp \boldsymbol{\pi}, \quad \tilde{\boldsymbol{\pi}} := 3\int_0^{1/3} \boldsymbol{\pi}_r \,\mathrm{d}r = T_\sharp(3\mathscr{L}^1|_{(0,1/3)} \otimes \boldsymbol{\pi}), \tag{5.1.33}$$

and by calculations similar to (5.1.24) and (5.1.25) we can check that $\tilde{\boldsymbol{\pi}} \in \mathcal{T}_q$. By Theorem 5.1.14 we deduce that f is Sobolev along $\tilde{\boldsymbol{\pi}}$-a.e. arc and (5.1.13) holds for $\tilde{\boldsymbol{\pi}}$-a.e. γ. Applying Fubini's Theorem we can find a $\boldsymbol{\pi}$-negligible Borel set $\mathrm{N} \subset \mathrm{RA}(X)$ such that for every $\gamma \in \mathrm{RA}(X) \setminus \mathrm{N}$ f is Sobolev with S.u.g. g along the arcs $\mathrm{Restr}_{1-r}^r(\gamma)$ for $\mathscr{L}^1$-a.e. $r \in (0, 1/3)$. For every $\gamma \in \mathrm{RA}(X) \setminus \mathrm{N}$ we can thus find a vanishing sequence $r_n \downarrow 0$ such that f is Sobolev and (5.1.13) holds along $\mathrm{Restr}_{1-r_n}^{r_n}(\gamma)$. We can thus pass to the limit and obtain the same properties along γ.

(b) As in the previous Claim, it is not restrictive to assume f, g Borel; by Remark 5.1.5 we can also suppose that f is bounded. Let us consider a countable dense subset $\mathcal{C}_c$ of $C^1_c(0,1)$ and let us define the sets

$$A_0 := \Big\{\gamma \in \mathrm{RA}_0(X) : \int_\gamma g < \infty\Big\}, \tag{5.1.34}$$

$$B_0 := \Big\{\gamma \in A_0 : \Big|\int_0^1 \varphi'(t) f(R_\gamma(t))\,\mathrm{d}t\Big| \le \ell(\gamma)\int_0^1 |\varphi(t)|\, g(R_\gamma(t))\,\mathrm{d}t \quad \text{for every } \varphi \in \mathcal{C}_c\Big\}. \tag{5.1.35}$$

By Theorem 2.2.13(e) A_0 is a Borel set; Proposition 4.1.2(b) shows that $\mathrm{Mod}_p(\mathrm{RA}_0(X) \setminus A_0) = 0$. By Lemma 5.1.10 for every arc $\gamma \in A_0$, f is Sobolev along γ with S.u.g. g if and only if $\gamma \in B_0$. Lemma 5.1.10 also shows that $A_0 \setminus B_0$ is Borel. Since by Claim (a) we know that $\mathrm{Cont}_p(A_0 \setminus B_0) = 0$, we get $\mathrm{Mod}_p(A_0 \setminus B_0) = 0$ by Theorem 4.3.2(d).

□

According to the Definition 5.1.9, Proposition 5.1.17 ensures that for Mod_p-a.e. arc γ a function f with $\mathcal{T}_q$-w.u.g. in $L^p(X, \mathfrak{m})$ coincides $\mathscr{L}^1$-a.e. with an absolutely continuous function f_γ. We can in fact prove a much better result, which establishes a strong connection with the theory of Newtonian Sobolev spaces.

Theorem 5.1.18 (Good Representative) *Let us suppose that (X, d) is complete and (X, τ) is Souslin. Every $\mathfrak{m}$-measurable function f with a $\mathcal{T}_q$-w.u.g. $g \in L^p(X, \mathfrak{m})$ admits a Borel $\mathfrak{m}$-representative $\tilde f$ such that $\tilde f \circ R_\gamma$ is absolutely continuous with S.u.g. g along* Mod_p*-a.e. arc γ (and a fortiori along $\mathcal{T}_q$-a.e. arc).*

Proof As usual, it is not restrictive to assume that f, g are Borel maps, f bounded. We will also denote by f_γ the absolutely continuous representative of $f \circ R_\gamma$ whenever f is Sobolev along γ.

Claim 1 There exists $h \in \mathcal{L}^p_+(X, \mathfrak{m})$, such that

$$f \text{ is Sobolev with S.u.g. } g \text{ along all the arcs of } H := \Big\{\gamma \in \mathrm{RA}_0(X) : \int_\gamma h < \infty\Big\}, \tag{5.1.36}$$

$$f \text{ is } \underline{strictly} \text{ Sobolev with S.u.g. } g \text{ along all the arcs of}$$
$$H_0 := \Big\{\gamma \in \mathrm{RA}_0(X) : h(\gamma_0) + h(\gamma_1) + \int_\gamma h < \infty\Big\}. \tag{5.1.37}$$

Notice that $H_0 \subset H$, $\mathrm{Mod}_p(\mathrm{RA}_0(X) \setminus H) = 0$, and $\widetilde{\mathrm{Mod}}_p(\mathrm{RA}_0(X) \setminus H_0) = 0$. By Proposition 5.1.17 and Proposition 4.1.2(b) we can find a Borel function $h' \in \mathcal{L}^p_+(X, \mathfrak{m})$ such that f is Sobolev with S.u.g. g along all the arcs of $H' := \Big\{\gamma \in \mathrm{RA}_0(X) : \int_\gamma h' < \infty\Big\}$. Notice that $\mathrm{Mod}_p(\mathrm{RA}_0(X) \setminus H') = 0$.

In order to get (5.1.37) we argue as in the proof of Proposition 5.1.17(b): we fix a countable set $\mathcal{C}$ dense in $C^1([0,1])$ and we consider the sets

$$A := \Big\{\gamma \in \mathrm{RA}(X) : \int_\gamma g < \infty\Big\}, \tag{5.1.38}$$

$$B := \Big\{\gamma \in A : \Big|\varphi(1)f(R_\gamma(1)) - \varphi(0)f(R_\gamma(0)) - \int_0^1 \varphi'(t)f(R_\gamma(t))\,\mathrm{d}t\Big|$$
$$\le \ell(\gamma)\int_0^1 |\varphi(t)|\,g(R_\gamma(t))\,\mathrm{d}t \quad \text{for every } \varphi \in \mathcal{C}\Big\}. \tag{5.1.39}$$

By Theorem 2.2.13(e) A is a Borel set; Proposition 4.1.2(b) shows that $\widetilde{\mathrm{Mod}}_p(\mathrm{RA}(X) \setminus A) = 0$. By Lemma 5.1.10 for every arc $\gamma \in A$, f is Sobolev along γ and (5.1.16) holds if and only if $\gamma \in B$, so that $\widetilde{\mathrm{Cont}}_p(A \setminus B) = 0$. Lemma 5.1.10 also shows that B is Borel, so that $\widetilde{\mathrm{Mod}}_p(\mathrm{RA}(X) \setminus B) = 0$ by Theorem 5.1.16(d). We can eventually apply Proposition 4.1.2 to find $h_0' \in \mathcal{L}^p_+(X, \mathfrak{m})$ such that f is Sobolev with S.u.g. g along all the arcs of $H_0' := \big\{\gamma \in \mathrm{RA}_0(X) : h_0'(\gamma_0) + h_0'(\gamma_1) + \int_\gamma h_0' < \infty\big\}$ We can eventually set $h := h' + h_0'$ and define the sets H and H_0 accordingly.

Claim 2 If $\gamma, \gamma' \in H$ and $R_\gamma(r) = R_{\gamma'}(r')$ for some $r, r' \in [0,1]$ then $f_\gamma(r) = f_\gamma(r')$.

Let us argue by contradiction assuming that there exist $\gamma, \gamma' \in H$ and $r, r' \in [0,1]$ such that $R_\gamma(r) = R_{\gamma'}(r') = x$ but $f_\gamma(r) \neq f_{\gamma'}(r')$. Up to a possible inversion of the orientation of γ or γ' it is not restrictive to assume that $r > 0$ and $r' < 1$. We can then consider the curve γ'' obtained by gluing $\gamma_- = \mathrm{Restr}_0^r(\gamma)$ and $\gamma_+ = \mathrm{Restr}_{r'}^1(\gamma')$, with $\ell(\gamma'') = r\ell(\gamma) + (1-r')\ell(\gamma')$. Clearly $\gamma'' \in \mathrm{RA}(X)$ and $\int_{\gamma''} h = \int_{\gamma_-} h + \int_{\gamma_+} h < \infty$, so that $\gamma'' \in H$ as well. Moreover, if $r'' = r\ell(\gamma)/\ell(\gamma'')$ we have $R_{\gamma''}^{0\to r''}(t) = R_\gamma^{0\to r}(t)$ and $R_{\gamma''}^{r''\to 1}(t) = R_{\gamma'}^{r'\to 1}(t)$ for every $t \in [0,1]$. It follows that $f_{\gamma''}(t) = f_\gamma(rt/r'')$ for $t \in [0, r'']$ and $f_{\gamma''}(t) = f_{\gamma'}(r' + (1-r')(t - r'')/(1-r''))$ so that $\lim_{t\uparrow r_2} f_{\gamma''}(t) = f_\gamma(r) \neq \lim_{t\downarrow r_2} f_{\gamma''}(t) = f_{\gamma'}(r')$, which conflicts with the fact that $f_{\gamma''}$ is absolutely continuous.

Claim 3 Let us set

$$\tilde f(x) := \begin{cases} f_\gamma(r) & \text{if } x = R_\gamma(r) \text{ for some } \gamma \in H \text{ and } r \in [0,1], \\ f(x) & \text{otherwise.} \end{cases} \tag{5.1.40}$$

Then $\tilde f$ is well defined, $\tilde f(R_\gamma) \equiv f_\gamma$ for every $\gamma \in H$, and $\tilde f(x) = f(x)$ in $\{x \in X : h(x) < \infty\}$. In particular $\{\tilde f \neq f\}$ is $\mathfrak{m}$-negligible and $\tilde f = f_\gamma$ along Mod_p*-a.e. arc (and a fortiori along $\mathcal{B}_q$ and $\mathcal{T}_q$-a.e. arc).*

The facts that $\tilde f$ is well defined and $\tilde f(R_\gamma) \equiv f_\gamma$ for every $\gamma \in H$ follow directly from the previous claim. Let us now argue by contradiction and let us suppose that there exists $x \in X$ with $\tilde f(x) \neq f(x)$ and $h(x) < \infty$. By definition of $\tilde f$ there exists an arc $\gamma \in H$ and $r \in [0,1]$ such that $R_\gamma(r) = x$. Since $\gamma \in H$ we know that $\int_\gamma h < \infty$: we can thus find $s \in [0,1] \setminus r$ such that $h(R_\gamma(s)) < \infty$. Assuming that $r < s$ (otherwise we switch the order of r and s), we can consider the arc $\gamma' := \mathrm{Restr}_r^s(\gamma)$ which satisfies $\int_{\gamma'} h \le \int_\gamma h < \infty$ and $h(R_{\gamma'}(0)) = h(R_\gamma(r)) = h(x) < \infty$ and $h(R_{\gamma'}(1)) = h(R_\gamma(s)) < \infty$. We deduce that $\gamma' \in H_0$ so that f is *strictly* Sobolev along γ' and therefore $\tilde{f}(x) = \tilde{f}(R_{\gamma'}(0)) = f_{\gamma'}(R_{\gamma'}(0)) = f(R_{\gamma'}(0)) = f(x)$, a contradiction.

□

Let us apply the previous representation Theorem to prove the equivalence of the notion of $\mathcal{T}_q$-w.u.g. with the "Newtonian" one introduced in [57].

Definition 5.1.19 (Newtonian Weak Upper Gradient) Let $f \in \mathcal{L}^p(X, \mathfrak{m})$. We say that f belongs to the Newtonian space $N^{1,p}(\mathbb{X})$ if f is absolutely continuous along Mod_p-a.e. arc $\gamma \in \mathrm{RA}_0(X)$ and there exists a nonnegative $g \in L^p(X, \mathfrak{m})$ such that

$$\Big|\int_{\partial\gamma} f\Big| \le \int_\gamma g \quad \text{for } \mathrm{Mod}_p\text{-a.e. arc } \gamma \in \mathrm{RA}_0(X). \tag{5.1.41}$$

In this case, we say that g is a $N^{1,p}$-weak upper gradient of f.

Functions with Mod_p-weak upper gradient have the important Beppo-Levi property of being absolutely continuous along Mod_p-a.e. arc γ. Because of the implication (5.1.6), functions with Mod_p-weak upper gradient have also $\mathcal{T}_q$-weak upper gradient. A priori there is an important difference between the two definitions, since Definition 5.1.19 is not invariant w.r.t. modifications of f in a $\mathfrak{m}$-negligible set. However, as an application of Theorem 5.1.18, we can show that these two notions are essentially equivalent modulo the choice of a representative in the equivalence class:

Corollary 5.1.20 *Let us suppose that $\mathbb{X}$ is a complete Souslin e.m.t.m. space. A function $f \in L^p(X, \mathfrak{m})$ admits a $\mathcal{T}_q$-weak upper gradient $g \in L^p(X, \mathfrak{m})$ if and only if there is a Borel representative $\tilde{f} : X \to \mathbb{R}$ with $\mathfrak{m}(\{\tilde{f} \neq f\}) = 0$ which belongs to the Newtonian space $N^{1,p}(\mathbb{X})$. Equivalently, $\tilde{f}$ is absolutely continuous along* Mod_p*-a.e. arc and g satisfies* (5.1.41) Mod_p*-a.e. In particular, the class of $\mathcal{T}_q$-w.u.g. for f coincides with the class of $N^{1,p}$-w.u.g. for a suitable Borel representative $\tilde{f}$ of f.*

5.1.3 Minimal $\mathcal{T}_q$-Weak Upper Gradient and the Sobolev Space $W^{1,p}(\mathbb{X}, \mathcal{T}_q)$

We want now to characterize the minimal $\mathcal{T}_q$-w.u.g. of a function and the corresponding notion of Sobolev space. We first prove two important properties. The first one directly involves the characterization of Theorem 5.1.14.

Proposition 5.1.21 (Locality) *Let $f : X \to \mathbb{R}$ be $\mathfrak{m}$-measurable and let g_1, g_2 be weak upper gradients of f w.r.t. $\mathcal{T}_q$. Then* $\min\{g_1, g_2\}$ *is a $\mathcal{T}_q$-weak upper gradient of f.*

Proof We know from Theorem 5.1.14 that f is Sobolev along $\mathcal{T}_q$-almost every arc. Then, the claim is a direct consequence of Remark 5.1.15 and (5.1.16). □

Another important property of weak upper gradients is their stability w.r.t. weak L^p convergence.

Theorem 5.1.22 (Stability w.r.t. Weak Convergence) *Assume that $f_n \in L^1(X, \mathfrak{m})$ and that $g_n \in L^p(X, \mathfrak{m})$ are $\mathcal{T}_q$-weak upper gradients of f_n. If $f_n \rightharpoonup f$ weakly in $L^1(X, \mathfrak{m})$ and $g_n \rightharpoonup g$ weakly in $L^p(X, \mathfrak{m})$ as $n \to \infty$, then f is Sobolev along $\mathcal{T}_q$-a.e. arc and g is a $\mathcal{T}_q$-weak upper gradient of f.*

Proof We can apply Lemma 5.1.8: we fix a test plan $\boldsymbol{\pi} \in \mathcal{T}_q^*$ and set $h_0, h_1 \in L^\infty(X, \mathfrak{m})$, $h \in L^q(X, \mathfrak{m})$ such that

$$(\mathsf{e}_0)_\sharp \boldsymbol{\pi} = \pi_0 = h_0 \mathfrak{m}, \quad (\mathsf{e}_1)_\sharp \boldsymbol{\pi} = \pi_1 = h_1 \mathfrak{m}, \quad \mu_{\boldsymbol{\pi}} = h\mathfrak{m}.$$

Since g_n is a $\mathcal{T}_q$-weak upper gradient for f_n we know that

$$\int_X f_n \,\mathrm{d}(\pi_1 - \pi_0) = \int_X f_n(h_1 - h_0)\,\mathrm{d}\mathfrak{m} \le \int_X g_n \,\mathrm{d}\mu_{\boldsymbol{\pi}} = \int_X g_n h \,\mathrm{d}\mathfrak{m}.$$

Passing to the limit by weak convergence in L^1 and L^p we immediately get

$$\int_X f \,\mathrm{d}(\pi_1 - \pi_0) \le \int_X g \,\mathrm{d}\mu_{\boldsymbol{\pi}}. \tag{5.1.42}$$

Since $\boldsymbol{\pi} \in \mathcal{T}_q$ is arbitrary, we conclude. □

We can now formalize the notion of $\mathcal{T}_q$-minimal weak upper gradient. For the sake of simplicity, here we will consider only the case of functions with $\mathcal{T}_q$-w.u.g. in $L^p(X, \mathfrak{m})$.

Definition 5.1.23 (Minimal $\mathcal{T}_q$-Weak Upper Gradient) Let $f \in L^1(X, \mathfrak{m})$ be a $\mathfrak{m}$-measurable function with a $\mathcal{T}_q$-weak upper gradient in $L^p(X, \mathfrak{m})$. The $\mathcal{T}_q$-minimal weak upper gradient $|\mathrm{D}f|_{w,\mathcal{T}_q}$ of f is the $\mathcal{T}_q$-weak upper gradient characterized, up to $\mathfrak{m}$-negligible sets, by the property

$$|\mathrm{D}f|_{w,\mathcal{T}_q} \le g \qquad \mathfrak{m}\text{-a.e. in } X, \text{ for every } \mathcal{T}_q - \text{weak upper gradient } g \text{ of } f. \tag{5.1.43}$$

Uniqueness of the minimal weak upper gradient is obvious. For existence, let us consider a minimizing sequence $(g_n)_{n\in\mathbb{N}} \subset L^p(X, \mathfrak{m})$ for the problem

$$\inf\left\{\int_X g^p \,\mathrm{d}\mathfrak{m} : \ g \text{ is a } \mathcal{T}_q\text{-weak upper gradient of } f\right\}.$$

We immediately see, thanks to Theorem 5.1.22, that we can assume with no loss of generality that $g_n \rightharpoonup g_\infty$ in $L^p(X, \mathfrak{m})$ and g_∞ is the $\mathcal{T}_q$-weak upper gradient of f of minimal L^p-norm. This minimality, in conjunction with Proposition 5.1.21, gives (5.1.43) for $|\mathrm{D}f|_{w,\mathcal{T}_q} := g_\infty$.

Definition 5.1.24 (The Weak $(\mathcal{T}_q, p)$-Energy and the Sobolev Space $W^{1,p}(\mathbb{X}, \mathcal{T}_q)$) Let $f \in L^1(X, \mathfrak{m})$ with a $\mathcal{T}_q$-weak upper gradient $g \in L^p(X, \mathfrak{m})$. The weak $(\mathcal{T}_q, p)$-energy of f is defined by

$$\mathsf{wCE}_{p,\mathcal{T}_q}(f) = \mathsf{wCE}_p(f) := \int_X |\mathrm{D}f|^p_{w,\mathcal{T}_q}\,\mathrm{d}\mathfrak{m}. \tag{5.1.44}$$

If, moreover, $f \in L^p(X, \mathfrak{m})$ then we say that f belongs to the space $W^{1,p}(\mathbb{X}, \mathcal{T}_q)$. $W^{1,p}(\mathbb{X}, \mathcal{T}_q)$ is a Banach space endowed with the norm

$$\|f\|^p_{W^{1,p}(\mathbb{X},\mathcal{T}_q)} := \int_X \Big(f^p + |\mathrm{D}f|^p_{w,\mathcal{T}_q}\Big)\,\mathrm{d}\mathfrak{m} = \|f\|^p_{L^p(X,\mathfrak{m})} + \mathsf{wCE}_{p,\mathcal{T}_q}(f). \tag{5.1.45}$$

Remark 5.1.25 (The $\mathcal{T}_q$-Notation) Even if we will mainly use minimal w.u.g. induced by $\mathcal{T}_q$-test plan, we will keep the explicit occurrence of $\mathcal{T}_q$ in the notation $|\mathrm{D}f|_{w,\mathcal{T}_q}$ and $\mathsf{wCE}_{p,\mathcal{T}_q}$ in order to distinguish these notions from other definitions of weak upper gradients based on different class of test plan (also on parametric arcs), which usually share the symbol $|\mathrm{D}f|_w$. We will use the shorter notation wCE_p only when no risk of confusion will be possible.

By using the same approach, the construction of the minimal p-weak upper gradient $|\mathrm{D}f|_{w,N^{1,p}}$ can also be performed for functions in the Newtonian space $N^{1,p}(\mathbb{X})$, and gives raise to the (semi)norm

$$\|f\|^p_{N^{1,p}(\mathbb{X})} := \int_X \Big(|f|^p + |\mathrm{D}f|^p_{w,N^{1,p}}\Big)\,\mathrm{d}\mathfrak{m}. \tag{5.1.46}$$

Taking Corollary 5.1.20 into account we easily have:

Corollary 5.1.26 (The Link with the Newtonian Space $N^{1,p}(\mathbb{X})$) *Let us suppose that $\mathbb{X}$ is a complete Souslin e.m.t.m. space. (The Lebesgue equivalence class of) every function $f \in N^{1,p}(\mathbb{X})$ belongs to $W^{1,p}(\mathbb{X}, \mathcal{T}_q)$. Conversely, every function $f \in W^{1,p}(\mathbb{X}, \mathcal{T}_q)$ has an equivalent representative $\tilde f$ in $N^{1,p}(\mathbb{X})$ with*

$$|\mathrm{D}f|_{w,\mathcal{T}_q} = |\mathrm{D}\tilde f|_{w,N^{1,p}} \quad a.e., \qquad \|f\|_{W^{1,p}(\mathbb{X},\mathcal{T}_q)} = \|\tilde f\|_{N^{1,p}(\mathbb{X})}. \tag{5.1.47}$$

It is easy to check using (5.1.10) and Theorem 5.1.22 that the weak Cheeger energy $\mathsf{wCE}_{p,\mathcal{T}_q}$ is a convex, p-homogeneous, weakly lower-semicontinuous functional in $L^1(X, \mathfrak{m})$. It is also easy to state a first comparison with the strong Cheeger energy

CE_p (the corresponding inequalities for $\mathsf{CE}_{p,\mathscr{A}}$ and $|\mathrm{D}f|_{\star,\mathscr{A}}$ follow trivially by (3.2.40) and (3.2.41)).

Lemma 5.1.27 *Every function $f \in H^{1,p}(\mathbb{X})$ belongs to $W^{1,p}(\mathbb{X}, \mathcal{T}_q)$ and*

$$\mathsf{CE}_p(f) \geq \mathsf{wCE}_{p,\mathcal{T}_q}(f), \quad |\mathrm{D}f|_\star \geq |\mathrm{D}f|_{w,\mathcal{T}_q} \quad \mathfrak{m}\text{-a.e. in } X. \tag{5.1.48}$$

Proof We already notice that for a Lipschitz function $f \in \mathrm{Lip}_b(X, \tau, \mathsf{d})$ lip f is a $\mathcal{T}_q$-w.u.g. so that

$$\mathsf{pCE}_p(f) \geq \mathsf{wCE}_{p,\mathcal{T}_q}(f). \tag{5.1.49}$$

It is then sufficient to take an optimal sequence $f_n \in \mathrm{Lip}_b(X, \tau, \mathsf{d})$ as in (3.1.9) and to apply the stability Theorem 5.1.22. □

Proposition 5.1.28 (Chain Rule for Minimal Weak Upper Gradients) *If $f \in L^1(X, \mathfrak{m})$ has a $\mathcal{T}_q$-weak upper gradient in $L^p(X, \mathfrak{m})$, the following properties hold:*

(a) for any $\mathscr{L}^1$-negligible Borel set $N \subset \mathbb{R}$ it holds $|\mathrm{D}f|_{w,\mathcal{T}_q} = 0$ $\mathfrak{m}$-a.e. on $f^{-1}(N)$.

(b) $|\mathrm{D}\phi(f)|_{w,\mathcal{T}_q} = \phi'(f)|\mathrm{D}f|_{w,\mathcal{T}_q}$ $\mathfrak{m}$-a.e. in X, with the convention $0 \cdot \infty = 0$, for any nondecreasing function ϕ, Lipschitz on an interval containing the image of f.

Proof We use the equivalent formulation of Remark 5.1.15 and the well-known fact that both (a) and (b) are true when $X = \mathbb{R}$ (endowed with the Euclidean distance and the Lebesgue measure) and f is absolutely continuous. We can prove (a) setting

$$G(x) := \begin{cases} |\mathrm{D}f|_{w,\mathcal{T}_q}(x) & \text{if } f(x) \in \mathbb{R} \setminus N, \\ 0 & \text{if } f(x) \in N, \end{cases}$$

and noticing that the validity of (a) for real-valued absolutely continuous maps gives that G is $\mathcal{T}_q$-weak upper gradient of f. Then, the minimality of $|\mathrm{D}f|_{w,\mathcal{T}_q}$ gives $|\mathrm{D}f|_{w,\mathcal{T}_q} \leq G$ $\mathfrak{m}$-a.e. in X.

By a similar argument based on (5.1.16) we can prove that $|\mathrm{D}\phi(f)|_{w,\mathcal{T}_q} \leq \phi'(f)|\mathrm{d}f|_{w,\mathcal{T}_q}$ $\mathfrak{m}$-a.e. in X. Then, the same subadditivity argument of Theorem 3.1.12(c) provides the equality $\mathfrak{m}$-a.e. in X. □

5.1.4 Invariance Properties of Weak Sobolev Spaces

In this section we will state a few useful results on the behaviour of weak Sobolev spaces with respect to some basic operations.

Restriction

Lemma 5.1.29 *Let* $\mathbb{X} = (X, \tau, \mathsf{d}, \mathfrak{m})$ *be an e.m.t.m. space and let* $Y \subset X$ *be a* d-*closed set such that* $\mathfrak{m}(X \setminus Y) = 0$. *Then every dynamic plan* $\boldsymbol{\pi} \in \mathcal{B}_q$ *is concentrated on* $\mathrm{RA}(Y)$.

Proof Let $\boldsymbol{\pi} \in \mathcal{M}_+(\mathrm{RA}(X))$ with $\mathrm{Bar}_q(\boldsymbol{\pi}) < \infty$. Setting $Z := X \setminus Y$ we have

$$\int \int_\gamma \chi_Z \, \mathrm{d}\boldsymbol{\pi}(\gamma) = \int_X \chi_Z \, \mathrm{d}\mu_{\boldsymbol{\pi}} = 0$$

since $\boldsymbol{\mu}_{\boldsymbol{\pi}} \ll \mathfrak{m}$ and $\mathfrak{m}(Z) = 0$. We deduce that for $\boldsymbol{\pi}$-a.e. γ $\int_\gamma \chi_Z = 0$, i.e. $\mathscr{L}^1(\{t \in [0, 1] : R_\gamma(t) \in Z\}) = 0$. Since Z is d-open, it follows that $R_\gamma([0, 1]) \subset Y$, i.e. $\boldsymbol{\pi}$-a.e. γ belongs to $\mathrm{RA}(Y)$. □

Corollary 5.1.30 (Invariance of $W^{1,p}$ by Restriction) *Let* $Y \subset X$ *be a* d-*closed set such that* $\mathfrak{m}(X \setminus Y) = 0$. *Setting* $\mathbb{Y} := (Y, \tau, \mathsf{d}, \mathfrak{m})$, *we have* $W^{1,p}(\mathbb{X}, \mathcal{T}_q) = W^{1,p}(\mathbb{Y}; \mathcal{T}_q)$ *and for every Sobolev function* f *the minimal* $\mathcal{T}_q(\mathbb{X})$-*weak upper gradient coincides with the minimal* $\mathcal{T}_q(\mathbb{Y})$-*weak upper gradient.*

Measure Preserving Isometric Embeddings

Let $\mathbb{X} = (X, \tau, \mathsf{d}, \mathfrak{m})$ and $\mathbb{X}' = (X', \tau', \mathsf{d}', \mathfrak{m}')$ be e.m.t.m. spaces and let suppose that $\iota : X \to X'$ is a measure-preserving embedding according to Definition 2.1.28. We will call $\mathcal{T}_q{}' = \mathcal{T}_q(\mathbb{X}')$ the class of nonparametric test plans in $\mathcal{M}_+(\mathrm{RA}(X'))$.

Starting from ι we can define a continuous injective map $J : \mathrm{C}([0, 1]; X) \to \mathrm{C}([0, 1]; X')$ by setting $J(\gamma) := \iota \circ \gamma$. Thanks to the isometric property of ι, $J(\mathrm{BVC}([0, 1]; X) \subset \mathrm{BVC}([0, 1]; X')$ and clearly J preserves equivalence classes of curves, so that J induces a continuous injective map from $\mathrm{RA}(X)$ to $\mathrm{RA}(X')$ satisfying

$$\int_{J\gamma} f' = \int_\gamma f' \circ \iota, \quad \ell(J\gamma) = \ell(\gamma) \quad \text{for every } \gamma \in \mathrm{RA}(X),\ f' \in \mathrm{B}_b(X'). \tag{5.1.50}$$

It is interesting to notice that

$$\iota \text{ is surjective} \quad \Rightarrow \quad J \text{ is surjective.} \tag{5.1.51}$$

In fact, given an arc $\gamma' \in \mathrm{RA}(X')$ we can consider the curve $R := \iota^{-1} \circ R_{\gamma'}$ which satisfies

$$\mathsf{d}(R(s), R(t)) = \mathsf{d}'(R_{\gamma'}(s), R_{\gamma'}(t)) = \ell(\gamma')|t - s|$$

so that $R \in \mathrm{Lip}_c([0,1];(X,\mathsf{d})) \subset \mathrm{BVC}([0,1];(X,\mathsf{d}))$ and $\gamma = \mathsf{q}(R) \in \mathrm{RA}(X)$ with $J\gamma = \gamma'$.

Lemma 5.1.31 *For every dynamic plan $\boldsymbol{\pi} \in \mathcal{M}_+(\mathrm{RA}(X))$ the push forward $\boldsymbol{\pi}' := J_\sharp\boldsymbol{\pi}$ is a dynamic plan in $\mathcal{M}_+(\mathrm{RA}(X'))$ satisfying*

$$\mu_{\boldsymbol{\pi}'} = \iota_\sharp\mu_{\boldsymbol{\pi}}, \quad \int \ell(\gamma)\,\mathrm{d}\boldsymbol{\pi}(\gamma) = \int \ell(\gamma')\,\mathrm{d}\boldsymbol{\pi}'(\gamma'), \quad (\mathsf{e}_i)_\sharp\boldsymbol{\pi}' = \iota_\sharp((\mathsf{e}_i)_\sharp\boldsymbol{\pi}) \quad i = 0, 1. \tag{5.1.52}$$

In particular

$$\mathrm{Bar}_q(\boldsymbol{\pi}') = \mathrm{Bar}_q(\boldsymbol{\pi}) \tag{5.1.53}$$

and $\boldsymbol{\pi}'$ belongs to $\mathcal{T}_q{}'$ if and only if $\boldsymbol{\pi}$ belongs to $\mathcal{T}_q$.

Proof For every nonnegative $f' \in \mathrm{C}_b(X')$ we have by (5.1.50)

$$\int f'\,\mathrm{d}\mu_{\boldsymbol{\pi}'} = \int\int_{\gamma'} f'\,\mathrm{d}\boldsymbol{\pi}'(\gamma') = \int\int_{J\gamma} f'\,\mathrm{d}\boldsymbol{\pi}(\gamma) \overset{(5.1.50)}{=} \int\int_{\gamma} f'\circ\iota\,\mathrm{d}\boldsymbol{\pi}(\gamma) = \int f'\circ\iota\,\mathrm{d}\mu_{\boldsymbol{\pi}},$$

which shows the first identity of (5.1.52). The second follows easily by choosing $f \equiv 1$ and the third identity is a consequence of the relation $\mathsf{e}_i \circ J = \iota\circ\mathsf{e}_i$, $i = 0, 1$. Since ι is injective, (5.1.53) is a consequence of the general properties of relative entropy functionals

$$\mathrm{Bar}_q^q(\boldsymbol{\pi}') = \mathscr{L}^q(\mu_{\boldsymbol{\pi}'}|\mathfrak{m}') = \mathscr{L}^q(\iota_\sharp\mu_{\boldsymbol{\pi}'}|\iota_\sharp\mathfrak{m}') = \mathscr{L}^q(\mu_{\boldsymbol{\pi}}|\mathfrak{m}) = \mathrm{Bar}_q^q(\boldsymbol{\pi}). \tag{5.1.54}$$

A similar argument shows that $\mathscr{L}^q((\mathsf{e}_i)_\sharp\boldsymbol{\pi}'|\mathfrak{m}') = \mathscr{L}^q((\mathsf{e}_i)_\sharp\boldsymbol{\pi}|\mathfrak{m})$. □

A simple but important application of the previous two Lemma yields the following result.

Theorem 5.1.32 *Let $\iota : X \to X'$ be a measure-preserving isometric imbedding of $\mathbb{X}$ into $\mathbb{X}'$. For every $f' \in W^{1,p}(\mathbb{X}', \mathcal{T}_q')$ the function $f := \iota^* f'$ belongs to $W^{1,p}(\mathbb{X}, \mathcal{T}_q)$ and*

$$|\mathrm{D}f|_{w,\mathcal{T}_q} \le \iota^*(|\mathrm{D}f'|_{w,\mathcal{T}_q'}) \quad \mathfrak{m}\text{-a.e. in } X. \tag{5.1.55}$$

If moreover ι is surjective or (X,d) is complete then ι^* is an isomorphism between $W^{1,p}(\mathbb{X}', \mathcal{T}_q{}')$ and $W^{1,p}(\mathbb{X}, \mathcal{T}_q)$ whose inverse is ι_* and

$$|\mathrm{D}f|_{w,\mathcal{T}_q} = \iota^*(|\mathrm{D}f'|_{w,\mathcal{T}_q'}) \quad \mathfrak{m}\text{-a.e. in } X. \tag{5.1.56}$$

Proof Let $g' \in L^p(X', \mathfrak{m}')$ be a $\mathcal{T}_q{}'$-weak upper gradient of f' in X'. We want to show that $g := \iota^* g' \in L^p(X, \mathfrak{m})$ is a $\mathcal{T}_q$-weak upper gradient for f: we use the equivalent characterization (5.1.12) of Lemma 5.1.8.

For every plan $\boldsymbol{\pi} \in \mathcal{T}_q$ Lemma 5.1.31 shows that $\boldsymbol{\pi}' := J_\sharp \boldsymbol{\pi} \in \mathcal{T}_q{}'$ with $\mu_{\boldsymbol{\pi}'} = \iota_\sharp \mu_{\boldsymbol{\pi}}$ and $(\mathsf{e}_i)_\sharp \boldsymbol{\pi}' = \iota_\sharp (\mathsf{e}_i)_\sharp \boldsymbol{\pi}$. We then obtain

$$\begin{aligned}\int_X f \,\mathrm{d}\pi_0 - \int_X f \,\mathrm{d}\pi_1 &= \int_X f' \circ \iota \,\mathrm{d}\pi_0 - \int_X f' \circ \iota \,\mathrm{d}\pi_1 \\ &= \int_{X'} f' \,\mathrm{d}(\iota_\sharp \pi_0) - \int_{X'} f' \,\mathrm{d}(\iota_\sharp \pi_1) \\ &= \int_{X'} f' \,\mathrm{d}\pi_0' - \int_{X'} f' \,\mathrm{d}\pi_1' \le \int_{X'} g' \,\mathrm{d}\mu_{\boldsymbol{\pi}'} \\ &= \int_X g' \,\mathrm{d}(\iota_\sharp \mu_{\boldsymbol{\pi}}) = \int g' \circ \iota \,\mathrm{d}\mu_{\boldsymbol{\pi}} = \int g \,\mathrm{d}\mu_{\boldsymbol{\pi}}.\end{aligned}$$

If ι is surjective, then J is surjective by (5.1.51), so that the very same argument shows that any $\mathcal{T}_q$-weak upper gradient for $f \in L^p(X, \mathfrak{m})$ yields a weak $\mathcal{T}_q{}'$ weak upper gradient $g' := \iota_* g$ for $\iota_* f$ in $L^p(X', \mathfrak{m}')$ with $\|g'\|_{L^q(X',\mathfrak{m}')} = \|g\|_{L^q(X,\mathfrak{m})}$ thus showing (5.1.56).

When (X, d) is complete then $(\iota(X), \mathsf{d}')$ is complete (and therefore d'-closed) in X', so that by Corollary 5.1.30 $W^{1,p}(\iota(X), \tau', \mathsf{d}', \mathfrak{m}'; \mathcal{T}_q{}') = W^{1,p}(X', \tau', \mathsf{d}', \mathfrak{m}'; \mathcal{T}_q{}')$ with equality of minimal $\mathcal{T}_q{}'$-weak upper gradients. On the other hand, $\iota : X \to \iota(X)$ is a measure preserving surjective embedding and we can apply the previous statement. □

Length Distances and Conformal Invariance

We refer to the definitions and notation of Sect. 2.3.

Lemma 5.1.33 *Let $\mathbb{X} = (X, \tau, \mathsf{d}, \mathfrak{m})$ be an e.m.t.m. space and let $\delta : X \times X \to [0, +\infty]$ an extended distance such that*

$$(X, \tau, \delta) \text{ is an extended metric-topological space}, \quad \mathsf{d} \le \delta \le \mathsf{d}_\ell \quad \text{in } X \times X. \tag{5.1.57}$$

Then $W^{1,p}(X, \tau, \mathsf{d}, \mathfrak{m}; \mathcal{T}_q) = W^{1,p}(X, \tau, \delta, \mathfrak{m}; \mathcal{T}_q)$ and the corresponding minimal weak $\mathcal{T}_q$-upper gradients coincide.

Proof We know that the class of rectifiable arcs $\mathrm{RA}(X, \mathsf{d})$ and $\mathrm{RA}(X, \delta)$ coincide, since $\mathsf{d}_\ell = \delta_\ell$, with the same length. Therefore, the corresponding classes of dynamic plans in $\mathcal{T}_q$ coincide. □

By the previous result, we can always replace d with d'_ℓ, d''_ℓ or d'''_ℓ in the definition of the Sobolev spaces. We can also use d_ℓ whenever d_ℓ is τ-continuous or when (X, τ) is compact.

Remark 5.1.34 The (easy) proof of the previous Lemma shows that the definition of the Sobolev space $W^{1,p}(X, \tau, \delta, \mathfrak{m}; \mathcal{T}_q)$ can be extended to a slightly more general setting: in fact, the condition that (X, τ, δ) is an e.t.m. space can be relaxed by asking that there exists an extended distance $\mathsf{d} : X \times X \to [0, +\infty]$ such that

$$(X, \tau, \mathsf{d}) \text{ is an e.t.m. space} \quad \text{and} \quad \mathsf{d} \le \delta \le \mathsf{d}_\ell. \tag{5.1.58}$$

We now discuss the case of a conformal distance d_g induced by a continuous function $g \in C_b(X)$ with $\inf_X g > 0$.

Proposition 5.1.35 *Let* $\mathbb{X} = (X, \tau, \mathsf{d}, \mathfrak{m})$ *be an e.m.t.m. space, let* $g \in C_b(X)$ *with* $0 < m_g \le g \le M_g < \infty$, *and let* $\delta : X \times X \to [0, +\infty]$ *be an extended distance such that*

$$\mathbb{X}' = (X, \tau, \delta, \mathfrak{m}) \text{ is an e.m.t.m. space}, \quad \mathsf{d}'_g \le \delta \le \mathsf{d}_g \quad \text{in } X \times X. \tag{5.1.59}$$

Then $\mathcal{T}_q(\mathbb{X})$ *coincides with* $\mathcal{T}'_q := \mathcal{T}_q(\mathbb{X}')$, *a function* $f \in L^p(X, \mathfrak{m})$ *belongs to the Sobolev space* $W^{1,p}(\mathbb{X}'; \mathcal{T}'_q)$ *if and only if* $f \in W^{1,p}(\mathbb{X}, \mathcal{T}_q)$, *and the corresponding minimal* $\mathcal{T}_q$*-weak upper gradients in* $\mathbb{X}$ *and in* $\mathbb{X}'$ *(which we call* $|\mathrm{D} f|_{w,\mathbb{X}}$ *and* $|\mathrm{D} f|_{w,\mathbb{X}'}$ *respectively) satisfy*

$$|\mathrm{D} f|_{w,\mathbb{X}} = g|\mathrm{D} f|_{w,\mathbb{X}'}. \tag{5.1.60}$$

Proof Denoting by $\int_{\gamma'}$ the integration of a function f along an arc γ with respect to the δ arc-length, we can easily check that

$$\int_{\gamma'} f = \int_\gamma g f. \tag{5.1.61}$$

It follows that if $\boldsymbol{\pi} \in \mathcal{T}_q$

$$\int \int_{\gamma'} f \,\mathrm{d}\boldsymbol{\pi}(\gamma) = \int \int_\gamma g f \,\mathrm{d}\boldsymbol{\pi}(\gamma) = \int g f \,\mathrm{d}\mu_{\boldsymbol{\pi}} = \int f \,\mathrm{d}\mu'_{\boldsymbol{\pi}}, \quad \mu'_{\boldsymbol{\pi}} := g\mu_{\boldsymbol{\pi}}.$$

We deduce that

$$m_g^q \mathscr{L}^q(\mu_{\boldsymbol{\pi}}|\mathfrak{m}) \le \mathscr{L}^q(\mu'_{\boldsymbol{\pi}}|\mathfrak{m}) \le M_g^q \mathscr{L}^q(\mu_{\boldsymbol{\pi}}|\mathfrak{m})$$

so that $\mathcal{T}_q = \mathcal{T}'_q$. If $h \in L^p(X, \mathfrak{m})$ is a weak $\mathcal{T}_q$-upper gradient for f in $\mathbb{X}$ we have

$$\int f \,\mathrm{d}(\pi_0 - \pi_1) \le \int h \,\mathrm{d}\mu_{\boldsymbol{\pi}} = \int g(g^{-1}h)\,\mathrm{d}\mu_{\boldsymbol{\pi}} = \int g^{-1}h\,\mathrm{d}\mu'_{\boldsymbol{\pi}},$$

which shows that $g^{-1}h$ is a $\mathcal{T}'_q$-weak upper gradient for f in $\mathbb{X}'$. □

5.1.5 The Approach by Parametric Dynamic Plans

Let us give a brief account of the definition of the Sobolev space $W^{1,p}$ by parametric dynamic plans [2, 8], i.e. Radon measures on suitable subsets of $\mathrm{C}([0,1];(X,\tau))$, and their relations with the notions we introduced in the previous Sections.

We first define the space $\mathrm{AC}^q([0,1];X)$ as the collection of curves $\gamma \in \mathrm{BVC}([0,1];(X,\mathsf{d}))$ such that V_γ is absolutely continuous with derivative $|\dot\gamma| := V'_\gamma \in L^q(0,1)$. The q-energy of a curve γ is defined by

$$\mathrm{E}_q(\gamma) := \int_0^1 |\dot\gamma|^q\,\mathrm{d}t \quad \text{if } \gamma \in \mathrm{AC}^q([0,1];X) \qquad \mathrm{E}_q(\gamma) := +\infty \text{ otherwise;} \tag{5.1.62}$$

it defines a τ_{C}-lower semicontinuous map. It follows in particular that $\mathrm{AC}^q([0,1];X)$ is a F_σ (thus Borel) subset of $\mathrm{C}([0,1];(X,\tau))$.

Recall that $\mathsf{e}_t : \mathrm{C}([0,1];(X,\tau)) \to X$ is the evaluation map $\mathsf{e}_t(\gamma) = \gamma(t)$.

Definition 5.1.36 (Parametric q-Test Plan, [7, 8]) We denote by T_q the collection of all the Radon probability measures $\boldsymbol{\sigma}$ on $\mathrm{C}([0,1];(X,\tau))$ satisfying the following two properties:

(T1) there exists $M_{\boldsymbol{\sigma}} > 0$ such that

$$(\mathsf{e}_t)_\sharp\boldsymbol{\sigma} \le M_{\boldsymbol{\sigma}}\,\mathfrak{m} \quad \text{for every } t \in [0,1]. \tag{5.1.63}$$

(T2) $\boldsymbol{\sigma}$ is concentrated on $\mathrm{AC}^q([0,1];X)$, i.e. $\boldsymbol{\sigma}\big(\mathrm{C}([0,1];(X,\tau)\setminus \mathrm{AC}^q([0,1];X)\big) = 0$;

We will call T^*_q the subset of dynamic plans in T_q with finite q-energy:

$$\mathcal{E}_q(\boldsymbol{\sigma}) := \int \mathrm{E}_q(\gamma)\,\mathrm{d}\boldsymbol{\sigma}(\gamma) < \infty. \tag{5.1.64}$$

We will say that a set $\Sigma \subset \mathrm{C}([0,1];(X,\tau))$ is T_q-negligible (resp. T^*_q-negligible) if $\boldsymbol{\sigma}(\Sigma) = 0$ for every $\boldsymbol{\sigma} \in \mathrm{T}_q$ (resp. $\boldsymbol{\sigma} \in \mathrm{T}^*_q$).

As usual, we will say that a property P on curves of $\mathrm{C}([0,1];(X,\tau))$ holds T_q-a.e. if the set where P does not hold is T_q-negligible.

Notice that if a set Σ is T_q^*-negligible then it is also T_q negligible: it is sufficient to approximate every plan $\boldsymbol{\sigma} \in \mathrm{T}_q$ by an increasing sequence of plans satisfying (5.1.64).

Starting from the notion of T_q-exceptional sets, we can introduce the corresponding definition of T_q-weak upper gradient and Sobolev space.

Definition 5.1.37 **(T_q-Weak Upper Gradient)** We say that a function $f \in L^p(X, \mathfrak{m})$ belongs to the Sobolev space $W^{1,p}(\mathbb{X}, \mathrm{T}_q)$ if there exists a function $g \in \mathcal{L}_+^p(X, \mathfrak{m})$ such that

$$|f(\gamma(1)) - f(\gamma(0))| \le \int_0^1 g(\gamma(t))|\dot{\gamma}|(t)\,\mathrm{d}t \tag{5.1.65}$$

for T_q-a.e. $\gamma \in \mathrm{AC}^{\mathrm{q}}([0,1]; X)$. Every function g with the stated property is called a T_q-w.u.g. of f.

The properties of Sobolev functions in $W^{1,p}(\mathbb{X}, \mathrm{T}_q)$ can be studied by arguments similar to the ones we presented in Sects. 5.1.1 and 5.1.3, obtaining corresponding results adapted to the parametric T_q-setting: we refer to [7, 8] for the precise statements and proofs.

However, by adapting the arguments of [2], it is possible to prove directly that the notions of $\mathcal{T}_q$ and T_q weak upper gradient coincide, obtaining the equivalence of the corresponding Sobolev spaces $W^{1,p}(\mathbb{X}, \mathcal{T}_q)$ and $W^{1,p}(\mathbb{X}, \mathrm{T}_q)$.

First of all, it is not difficult to check that for every $f \in L^p(X, \mathfrak{m})$ and $g \in \mathcal{L}_+^p(X, \mathfrak{m})$

$$g \text{ is a } \mathcal{T}_q\text{-w.u.g. of } f \quad \Rightarrow \quad g \text{ is a } \mathrm{T}_q\text{-w.u.g. of } f, \tag{5.1.66}$$

since for every parametric dynamic plan $\boldsymbol{\sigma} \in \mathrm{T}_q^*$ the corresponding nonparametric version $\boldsymbol{\pi} := \mathfrak{q}_\sharp\boldsymbol{\sigma}$ belongs to $\mathcal{T}_q$; recall that we denoted by $\mathfrak{q} : \mathrm{C}([0,1]; (X,\tau)) \to \mathrm{A}(X,\tau)$ the quotient map. In fact

$$(\mathsf{e}_i)_\sharp\boldsymbol{\pi} = (\mathsf{e}_i)_\sharp\boldsymbol{\sigma} \le M_{\boldsymbol{\sigma}}\mathfrak{m} \quad i = 0, 1,$$

and for every bounded Borel function $f : X \to \mathbb{R}$

$$\begin{aligned}
\int\int_\gamma f\,\mathrm{d}\boldsymbol{\pi}(\gamma) &= \int\int_{\mathfrak{q}(\eta)} f\,\mathrm{d}\boldsymbol{\sigma}(\eta) = \int\int_0^1 f(\eta(t))|\dot{\eta}|(t)\,\mathrm{d}\boldsymbol{\sigma}(\eta)\\
&\le \int\Big(\int_0^1 f^p(\eta(t))\,\mathrm{d}t\Big)^{1/p}\mathrm{E}_q^{1/q}(\eta)\,\mathrm{d}\boldsymbol{\sigma}(\eta)\\
&\le \mathcal{E}_q^{1/q}\Big(\int\int_0^1 f^p(\eta(t))\,\mathrm{d}t\,\mathrm{d}\boldsymbol{\sigma}(\eta)\Big)^{1/p}\\
&\le \mathcal{E}_q^{1/q}\Big(\int_0^1\int_0^1 f^p(\mathsf{e}_t(\eta))\,\mathrm{d}\boldsymbol{\sigma}(\eta)\,\mathrm{d}t\Big)^{1/p} \le (M_{\boldsymbol{\sigma}}\,\mathcal{E}_q)^{1/q}\Big(\int_X f^p\,\mathrm{d}\mathfrak{m}\Big)^{1/p}.
\end{aligned}$$

We can deduce that for every $\Gamma \subset \mathrm{RA}(X)$

$$\Gamma \text{ is } \mathcal{T}_q\text{-negligible} \quad \Rightarrow \quad \mathsf{q}^{-1}(\Gamma) \text{ is } \mathrm{T}_q\text{-negligible} \tag{5.1.67}$$

and therefore we get (5.1.66). In order to prove the converse property we introduce the notion of parametric barycenter of a Radon measure $\boldsymbol{\sigma} \in \mathcal{M}_+(\mathrm{C}([0,1];(X,\tau)))$: it is the image measure $\varrho_{\boldsymbol{\sigma}} := \mathsf{e}_\sharp(\boldsymbol{\sigma} \otimes \mathscr{L}^1) \in \mathcal{M}_+(X)$, which satisfies

$$\int_X f\,\mathrm{d}\varrho_{\boldsymbol{\sigma}} = \int\int_0^1 f(\mathsf{e}_t(\gamma))\,\mathrm{d}t\,\mathrm{d}\boldsymbol{\sigma}(\gamma) \quad \text{for every } f \in \mathrm{B}_b(X). \tag{5.1.68}$$

We say that $\boldsymbol{\sigma}$ has *parametric barycenter in* $L^q(X,\mathfrak{m})$ if $\varrho_{\boldsymbol{\sigma}} = h_{\boldsymbol{\sigma}}\mathfrak{m} \ll \mathfrak{m}$ for a density $h_{\boldsymbol{\sigma}} \in L^q(X,\mathfrak{m})$.

The proof of the converse implication of (5.1.66) is based on the following two technical Lemmata.

Lemma 5.1.38 *Let us suppose that* $g \in \mathcal{L}^p_+(X,\mathfrak{m})$ *is a* T_q*-w.u.g. of* $f \in L^p(X,\mathfrak{m})$ *and let* $\boldsymbol{\sigma} \in \mathcal{M}_+(\mathrm{C}([0,1];(X,\tau)))$ *be a dynamic plan satisfying* (5.1.64) *and*

$$(\mathsf{e}_i)_\sharp\boldsymbol{\sigma} \le M\mathfrak{m} \ll \mathfrak{m} \quad i=0,1, \qquad \varrho_{\boldsymbol{\sigma}} \le M\mathfrak{m} \qquad \text{for a constant } M>0. \tag{5.1.69}$$

Then (5.1.65) *holds for* $\boldsymbol{\sigma}$*-a.e.* γ.

Proof The argument is similar (but simpler) than the one used for the proof of Theorem 5.1.14. For $0 \le r < s \le 1$ we consider the Borel maps D^+_r, D^-_s : $\mathrm{C}([0,1],X) \times [0,1] \to \mathrm{C}([0,1];(X,\tau))$ defined by

$$D^+[\gamma,r](t) := \gamma((r+t)\wedge 1), \quad D^-[\gamma,s](t) := \gamma((t-s)\vee 0).$$

We then set $\lambda := 3\mathscr{L}^1 \llcorner (1/3,2/3)$ and which can also be characterized as

$$\boldsymbol{\sigma}^+ = (D^+)_\sharp(\boldsymbol{\sigma}\otimes\lambda), \quad \boldsymbol{\sigma}^- = (D^-)_\sharp(\boldsymbol{\sigma}\otimes\lambda).$$

We easily get for every $t \ge 2/3$ $(\mathsf{e}_t)_\sharp\boldsymbol{\sigma}^+ = (\mathsf{e}_1)_\sharp\boldsymbol{\sigma} \le M\mathfrak{m}$, whereas for every $t \in [0,2/3)$ and every nonnegative Borel $f : X \to \mathbb{R}$

$$\begin{aligned}
\int f(\mathsf{e}_t(\gamma))\,\mathrm{d}\boldsymbol{\sigma}^+(\gamma) &= 3\int_{1/3}^{2/3}\Big(\int f(\gamma((t+r)\wedge 1))\,\mathrm{d}\boldsymbol{\sigma}(\gamma)\Big)\,\mathrm{d}r \\
&= 3\int_{1/3}^{(1-t)\wedge 2/3}\Big(\int f(\gamma(r+t))\,\mathrm{d}\boldsymbol{\sigma}(\gamma)\Big)\,\mathrm{d}r \\
&\quad + 3(1/3-t)_+ \cdot \int f(\gamma(1))\,\mathrm{d}\boldsymbol{\sigma}(\gamma) \\
&\le 3\int_X f\,\mathrm{d}\varrho_{\boldsymbol{\sigma}} + \int_X f\,\mathrm{d}(\mathsf{e}_1)_\sharp\boldsymbol{\sigma} \le 4M\int_X f\,\mathrm{d}\mathfrak{m}
\end{aligned}$$

so that $(\mathsf{e}_t)_\sharp\boldsymbol{\sigma}^+ \le 4M\mathfrak{m}$ for every $t \in [0,1]$. An analogous calculation holds for $\boldsymbol{\sigma}^-$, so that both satisfy (2.2.28). Since $\mathrm{E}_q(D^\pm(\gamma, r)) \le \mathrm{E}_q(\gamma)$ for every $r \in [0,1]$ we also get (5.1.64). We deduce that $\boldsymbol{\sigma}^+, \boldsymbol{\sigma}^-$ belong to T_q so that (5.1.65) holds for $\boldsymbol{\sigma}^+$ and $\boldsymbol{\sigma}^-$-a.e. curve γ. Applying Fubini's theorem, we can find a common Borel and $\boldsymbol{\sigma}$-negligible set $N \subset \mathrm{C}([0,1]; (X,\tau))$ such that for every $\gamma \in \mathrm{AC}^q([0,1]; X) \setminus N$

$$
\begin{aligned}
|f(\gamma(1-s)) - f(\gamma(0)| &= |f(D^-[\gamma, s](1)) - f(D^-[\gamma, s](0))| \\
&\le \int_0^1 g(\gamma((t-s)\vee 0))|\dot\gamma((t-s)\vee 0)|\,\mathrm{d}t \\
&= \int_0^{1-s} g(\gamma(t))|\dot\gamma|(t)\,\mathrm{d}t \quad \text{for a.e. } s \in (1/2, 3/2)
\end{aligned}
$$

and similarly

$$
\begin{aligned}
|f(\gamma(1)) - f(\gamma(r)| &= |f(D^+[\gamma, r](1)) - f(D^+[\gamma, r](0))| \\
&\le \int_0^1 g(\gamma((t+r)\wedge 1))|\dot\gamma((t-s)\wedge 1)|\,\mathrm{d}t \\
&= \int_r^1 g(\gamma(t))|\dot\gamma|(t)\,\mathrm{d}t \quad \text{for a.e. } r \in (1/2, 3/2)
\end{aligned}
$$

For every $\gamma \in \mathrm{AC}^q([0,1]; X)$ we can thus find a common value $r = 1-s \in (1/2, 3/2)$ such that the previous inequality hold, obtaining

$$
\begin{aligned}
|f(\gamma(1)) - f(\gamma(0)| &\le \int_0^{1-s} g(\gamma(t))|\dot\gamma|(t)\,\mathrm{d}t + \int_r^1 g(\gamma(t))|\dot\gamma|(t)\,\mathrm{d}t \\
&= \int_0^1 g(\gamma(t))|\dot\gamma|(t)\,\mathrm{d}t,
\end{aligned}
$$

which yields (5.1.65). □

The second lemma is a reparametrization technique taken from [2, Theorem 8.5].

Lemma 5.1.39 *For every nonparametric dynamic plan* $\boldsymbol{\pi} \in \mathcal{T}_q^*$ *there exists a parametric dynamic plan* $\boldsymbol{\sigma}$ *satisfying* (5.1.69) *such that*

$$
\boldsymbol{\pi} \ll \mathsf{q}_\sharp\boldsymbol{\sigma}. \tag{5.1.70}
$$

Combining Lemmas 5.1.38 and 5.1.39 we obtain the following result, which shows the equivalence of the parametric and nonparametric approaches.

Corollary 5.1.40 *For every $f \in L^p(X, \mathfrak{m})$ and $g \in \mathcal{L}^p_+(X, \mathfrak{m})$*

$$g \text{ is a } \mathcal{T}_q\text{-w.u.g. of } f \quad \Longleftrightarrow \quad g \text{ is a } \mathrm{T}_q\text{-w.u.g. of } f. \tag{5.1.71}$$

In particular $W^{1,p}(\mathbb{X}, \mathcal{T}_q) = W^{1,p}(\mathbb{X}, \mathrm{T}_q)$.

Proof We have only to prove the converse implication of (5.1.66). Let $g \in \mathcal{L}^p_+(X, \mathfrak{m})$ be a T_q-w.u.g. and let $\boldsymbol{\pi} \in \mathcal{T}^*_q$. By Lemma 5.1.39 there exists a parametric dynamic plan $\boldsymbol{\sigma}$ satisfying (5.1.69) such that $\boldsymbol{\pi} \ll \mathfrak{q}_\sharp\boldsymbol{\sigma}$. By Lemma 5.1.38 we know that (5.1.65) holds for $\boldsymbol{\sigma}$-a.e. curve, i.e.

$$|f(\gamma_1) - f(\gamma_0)| \le \int_\gamma g \quad \text{for } \mathfrak{q}_\sharp\boldsymbol{\sigma}\text{-a.e. } \gamma \in \mathrm{RA}(X).$$

Since $\boldsymbol{\pi} \ll \mathfrak{q}_\sharp\boldsymbol{\sigma}$ we deduce that (5.1.11) holds as well, so that we can apply Lemma 5.1.8. □

Remark 5.1.41 As for Cheeger's energy and the relaxed gradient, if no additional assumption on $(X, \tau, \mathsf{d}, \mathfrak{m})$ is made, it is well possible that the weak upper gradient is trivial. We will discuss this issue in the next Theorem 5.2.9.

5.1.6 Notes

Sections 5.1.1 and **5.1.3** contain new definitions of weak upper gradient and weak Sobolev spaces based on the class of $\mathcal{T}_q$-weak upper gradients. This approach has some useful characteristics:

- it involves measures on nonparametric arcs; notice that the notion of upper gradient is inherently invariant w.r.t. parametrization, so arcs provide a natural setting;
- it is invariant w.r.t. modification on $\mathfrak{m}$-negligible sets;
- it seems quite close to the class $\mathcal{B}_q$: one has only to add the control of the initial and final points of the arcs
- the corresponding Modulus $\widetilde{\mathrm{Mod}}_p$ is strictly related to Mod_p, so that via the selection of a "good representative" the Sobolev class $W^{1,p}(\mathbb{X}, \mathcal{T}_q)$ coincides with $N^{1,p}(\mathbb{X})$;
- it is directly connected with the dual of the Cheeger energy.

Of course, the study of the properties of the $\mathcal{T}_q$ w.u.g. retains many ideas of the corresponding analysis based on Radon measures on parametric curves [7, 8] as the stability, the Sobolev property along $\mathcal{T}^*_q$-a.e. arc, the chain rule. The rescaling technique of Theorem 5.1.14 has been also used in [2].
It is worth noticing that Corollary 5.1.26 could also be derived as a consequence of Theorem 5.2.7, as in [7, 8]. Here we followed the more direct approach of [2], which shows the closer link between $W^{1,p}$ and $N^{1,p}$.

Section 5.1.2 combines various methods introduced by [2]: apart from some topological aspects, Theorem 5.1.16 is a particular case of the identity between Modulus and Content at the level of collection of Radon measures, Proposition 5.1.17 uses the invariance of the Sobolev property by restriction and Theorem 5.1.18 is strongly inspired by [2, Theorem 10.3].

Section 5.1.4 contains natural invariance properties of weak Sobolev spaces: the most important one is (5.1.56) of Theorem 5.1.32, which will play a crucial role in the final part of the proof of the identification Theorem 5.2.7.

Section 5.1.5 contains a brief discussion of the equivalence between the non-parametric and parametric approaches to weak upper gradients and weak Sobolev spaces. It uses some of the arguments of [2] to show that the two approaches lead to equivalent definitions.

5.2 *Identification of Sobolev Spaces*

In this Section we will prove the main identification Theorem for the Sobolev spaces $H^{1,p}(\mathbb{X}, \mathscr{A})$ and $W^{1,p}(\mathbb{X}, \mathcal{T}_q)$ when (X, d) is complete. As a first step we study a dual characterization of the weak $(\mathcal{T}_q, p)$-energy.

5.2.1 Dual Cheeger Energies

For every $\mu_0, \mu_1 \in \mathcal{M}_+(X)$ we will introduce the (possibly empty) set

$$\Pi(\mu_0, \mu_1) := \Big\{\boldsymbol{\pi} \in \mathcal{M}_+(\mathrm{RA}(X)), \quad (\mathsf{e}_i)_\sharp\boldsymbol{\pi} = \mu_i\Big\}, \tag{5.2.1}$$

and we define the cost functional

$$\begin{aligned}\mathscr{D}_q(\mu_0, \mu_1) &:= \inf\Big\{\mathrm{Bar}_q^q(\boldsymbol{\pi}) : \boldsymbol{\pi} \in \Pi(\mu_0, \mu_1)\Big\},\\ \mathscr{D}_q(\mu_0, \mu_1) &= +\infty \quad \text{if } \Pi(\mu_0, \mu_1) = \emptyset.\end{aligned} \tag{5.2.2}$$

Notice that $\Pi(\mu_0, \mu_1)$ is surely empty if $\mu_0(X) \neq \mu_1(X)$.

Let us check that if $\mathscr{D}_q(\mu_0, \mu_1) < +\infty$ and (X, d) is complete, then the infimum in (5.2.2) is attained. Notice that $\Pi(\mu_0, \mu_1)$ is a closed convex subset of $\mathcal{M}_+(\mathrm{RA}(X))$.

Lemma 5.2.1 *Let us suppose that (X, d) is complete. For every $\mu_0, \mu_1 \in \mathcal{M}_+(X)$, if $\mathscr{D}_q(\mu_0, \mu_1) < \infty$ then there exists a minimizer $\boldsymbol{\pi}_{\min} \in \Pi(\mu_0, \mu_1)$ which realizes the infimum in (5.2.2). The set $\Pi_o(\mu_0, \mu_1)$ of optimal plans is a compact convex subset of $\mathcal{M}_+(\mathrm{RA}(X))$ and for every $\boldsymbol{\pi} \in \Pi_o(\mu_0, \mu_1)$ the induced measure $\mu_{\boldsymbol{\pi}}$ is uniquely determined and is independent of the choice of the minimizer.*

Proof Let $\boldsymbol{\pi}' \in \Pi(\mu_0, \mu_1)$ with $\mathrm{Bar}_q\, \boldsymbol{\pi}' = E < \infty$ and define $\mathcal{K} := \big\{\boldsymbol{\pi} \in \Pi(\mu_0, \mu_1) : \mathrm{Bar}_q(\boldsymbol{\pi}) \le E\big\}$. We can apply Lemma 4.2.5: for every $\boldsymbol{\pi} \in \Pi(\mu_0, \mu_1)$ $\boldsymbol{\pi}(\mathrm{RA}(X)) = \mu_0(X)$ so that condition (T1) is satisfied. Concerning (T2) it is sufficient to we use the tightness of μ_0 to find compact sets $H_\varepsilon \subset X$ such that $\mu_0(X \setminus H_\varepsilon) \le \varepsilon$; clearly

$$\boldsymbol{\pi}(\{\gamma : \mathsf{e}(\gamma) \cap H_\varepsilon = \emptyset\}) \le \boldsymbol{\pi}(\{\gamma : \mathsf{e}_0(\gamma) \cap H_\varepsilon = \emptyset\}) = \mu_0(X \setminus H_\varepsilon) \le \varepsilon.$$

Since the functional Bar_q is lower semicontinuous with respect to weak convergence, we conclude that the minimum is attained. The convexity and the compactness of $\Pi_o(\mu_0, \mu_1)$ are also immediate; the uniqueness of $\mu_{\boldsymbol{\pi}}$ when $\boldsymbol{\pi}$ varies in $\Pi_o(\mu_0, \mu_1)$ depends on the strict convexity of the $L^q(X, \mathfrak{m})$-norm and on the convexity of $\Pi_o(\mu_0, \mu_1)$. □

We want to compare $\mathscr{D}_q$ with the dual of the pre-Cheeger energy:

$$\frac{1}{q}\mathsf{pCE}_p^*(\mu) := \sup\Big\{\int_X f\,\mathrm{d}\mu - \frac{1}{p}\mathsf{pCE}_p(f) : f \in \mathrm{Lip}_b(X, \tau, \mathsf{d})\Big\}, \quad \mu = \mu_0 - \mu_1 \in \mathcal{M}(X). \tag{5.2.3}$$

Notice that by Lemma A.7 we have the equivalent representation

$$\Big(\mathsf{pCE}_p^*(\mu)\Big)^{1/q} = \sup\Big\{\int_X f\,\mathrm{d}\mu : f \in \mathrm{Lip}_b(X, \tau, \mathsf{d}),\ \mathsf{pCE}_p(f) \le 1\Big\}. \tag{5.2.4}$$

Whenever $\mu = h\mathfrak{m}$ with $h \in L^q(X, \mathfrak{m})$, we can also consider the dual of the Cheeger energy

$$\frac{1}{q}\mathsf{CE}_p^*(h) := \sup\Big\{\int_X f\,h\,\mathrm{d}\mathfrak{m} - \frac{1}{p}\mathsf{CE}_p(f) : f \in H^{1,p}(\mathbb{X})\Big\}, \tag{5.2.5}$$

and of the weak $(\mathcal{T}_q, p)$-energy wCE_p (defined by a formula analogous to (5.2.5)) that we will denote by wCE_p^*. An obvious necessary condition for the finiteness of pCE_p^* and of CE_p^* is given by

$$\mathsf{pCE}_p^*(\mu) < +\infty \quad \Rightarrow \quad \mu(X) = 0; \qquad \mathsf{CE}_p^*(h) < +\infty \quad \Rightarrow \quad \int_X h\,\mathrm{d}\mathfrak{m} = 0. \tag{5.2.6}$$

Since $\mathsf{wCE}_p(f) \le \mathsf{CE}_p(f)$ for every $f \in L^p(X, \mathfrak{m})$ and $\mathsf{CE}_p(f) \le \mathsf{pCE}_p(f)$ for every $f \in \mathrm{Lip}_b(X, \tau, \mathsf{d})$, it is clear that

$$\mathsf{wCE}_p^*(h) \ge \mathsf{CE}_p^*(h) \ge \mathsf{pCE}_p^*(h\mathfrak{m}) \quad \text{for every } h \in L^p(X, \mathfrak{m}). \tag{5.2.7}$$

Lemma 5.2.2 *For every $\mu_0, \mu_1 \in \mathcal{M}_+(X)$ we have*

$$\mathscr{D}_q(\mu_0, \mu_1) \ge \mathsf{pCE}_p^*(\mu_0 - \mu_1). \tag{5.2.8}$$

If moreover $\mu_i = h_i \mathfrak{m}$ with $h_i \in L^p(X, \mathfrak{m})$, $h_i \ge 0$, then

$$\mathscr{D}_q(h_0\mathfrak{m}, h_1\mathfrak{m}) \ge \mathsf{wCE}_p^*(h) \quad h = h_0 - h_1. \tag{5.2.9}$$

Proof We observe that for every $\boldsymbol{\pi} \in \Pi(\mu_0, \mu_1)$ we have

$$\frac{1}{q}\operatorname{Bar}_q^q(\boldsymbol{\pi}) = \sup_{g \in \mathcal{L}_+^p(X,\mathfrak{m})} \iint_\gamma g \, \mathrm{d}\boldsymbol{\pi} - \frac{1}{p}\int g^p \, \mathrm{d}\mathfrak{m}. \tag{5.2.10}$$

Restricting the supremum to the functions $g := \operatorname{lip} f$ for some $f \in \operatorname{Lip}_b(X, \mathsf{d}, \mathfrak{m})$ and observing that in this case for every $\gamma \in \mathrm{RA}(X)$,

$$f(\gamma_0) - f(\gamma_1) \le \int_\gamma g \tag{5.2.11}$$

we get

$$\begin{aligned}\iint_\gamma g \, \mathrm{d}\boldsymbol{\pi} - \frac{1}{p}\int g^p \, \mathrm{d}\mathfrak{m} &\ge \int \big(f(\gamma_0) - f(\gamma_1)\big) \, \mathrm{d}\boldsymbol{\pi} - \frac{1}{p}\mathsf{pCE}_p(f) \\ &= \int_X f \, \mathrm{d}(\mu_0 - \mu_1) - \frac{1}{p}\mathsf{pCE}_p(f)\end{aligned}$$

so that (5.2.8) follows by taking the supremum w.r.t. f and the infimum w.r.t. $\boldsymbol{\pi}$.

When $\mu_i = h_i\mathfrak{m}$ with $h_i \in L^p(X, \mathfrak{m})$ nonnegative, any dynamic plan $\boldsymbol{\pi} \in \Pi(\mu_0, \mu_1)$ with $\operatorname{Bar}_q(\boldsymbol{\pi}) < \infty$ belongs to $\mathcal{T}_q$. Restricting the supremum of (5.2.10) to (the Borel representative of) functions $g = |\mathrm{D}f|_{w,\mathcal{T}_q}$ for some $f \in W^{1,p}(\mathbb{X}, \mathcal{T}_q)$ it follows that (5.2.11) holds for $\mathcal{T}_q$-a.e. curve, in particular for $\boldsymbol{\pi}$-a.e. curve γ. We can then perform the same integration with respect to $\boldsymbol{\pi}$ and obtain (5.2.9). □

5.2.2 $H = W$

The Compact Case

Let us first consider the case when (X, τ) is compact. For every strictly positive function $g \in \mathrm{C}_b(X)$ (we will still use the notation $\mathrm{C}_b(X)$ even if the subscript $_b$ is redundant, being X compact) we will denote by d_g the conformal distance we studied in Sect. 2.3.2 and by $\mathsf{K}_{\mathsf{d}_g}$ the Kantorovich-Rubinstein distance induced by d_g, see Sect. 2.1.4. Notice that (X, τ, d_g) is a geodesic e.m.t. space thanks to Theorem 2.3.2.

Theorem 5.2.3 *Let us suppose that (X, τ) is compact; then for every $\mu_0, \mu_1 \in \mathcal{M}_+(X)$ with $\mu_0(X) = \mu_1(X)$ we have*

$$\mathscr{D}_q(\mu_0, \mu_1) = \sup\Big\{\mathsf{K}_{\mathsf{d}_g}(\mu_0, \mu_1) - \frac{1}{p}\int_X g^p\,\mathrm{dm} : g \in \mathrm{C}_b(X),\ g > 0\Big\}. \tag{5.2.12}$$

Proof Let us introduce the convex set

$$\mathcal{C} := \Big\{(g, \psi_0, \psi_1) \in \big(\mathrm{C}_b(X)\big)^3 : g(x) > 0 \quad \text{for every } x \in X\Big\} \tag{5.2.13}$$

and the dual representation of the convex set $\Pi(\mu_0, \mu_1)$ given by two Lagrange multipliers $\psi_0, \psi_1 \in \mathrm{C}_b(X)$: $\boldsymbol{\pi} \in \Pi(\mu_0, \mu_1)$ if and only if (here $\gamma_i = \mathsf{e}_i(\gamma)$, $i = 0, 1$)

$$\sup_{\psi_0, \psi_1 \in \mathrm{C}_b(X)} \int_X \psi_0\,\mathrm{d}\mu_0 - \int \psi_0(\gamma_0)\,\mathrm{d}\boldsymbol{\pi}(\gamma) - \Big(\int_X \psi_1\,\mathrm{d}\mu_1 - \int \psi_1(\gamma_1)\,\mathrm{d}\boldsymbol{\pi}(\gamma)\Big) < +\infty; \tag{5.2.14}$$

Notice that whenever the supremum in (5.2.14) is finite, it vanishes. We first observe that

$$\frac{1}{q}\mathscr{D}_q(\mu_0, \mu_1) = \inf_{\boldsymbol{\pi} \in \mathcal{M}_+(\mathrm{RA}(X))} \sup_{(g, \psi_0, \psi_1) \in \mathcal{C}} \mathcal{L}((g, \psi_0, \psi_1); \boldsymbol{\pi})$$

where the Lagrangian function $\mathcal{L}$ is given by

$$\begin{aligned}\mathcal{L}((g, \psi_0, \psi_1); \boldsymbol{\pi}) := &\int \Big(\int_\gamma g + \psi_1(\gamma_1) - \psi_0(\gamma_0)\Big)\,\mathrm{d}\boldsymbol{\pi}(\gamma) \\ &+ \int_X \psi_0\,\mathrm{d}\mu_0 - \int_X \psi_1\,\mathrm{d}\mu_1 - \frac{1}{p}\int_X g^p\,\mathrm{dm},\end{aligned} \tag{5.2.15}$$

and it is clearly convex w.r.t. $\boldsymbol{\pi}$ and concave w.r.t. (g, ψ_0, ψ_1). We want to apply Von Neumann Theorem A.8 and to invert the order of inf and sup.

Selecting $g_\star \equiv 1,\ \psi_{1,\star} \equiv 1,\ \psi_{0,\star} \equiv 0$ we see that for every $C \geq 0$ the sublevel

$$\mathcal{K}_C := \Big\{\boldsymbol{\pi} \in \mathcal{M}_+(\mathrm{RA}(X)) : \mathcal{L}((g_\star, \psi_{0,\star}, \psi_{1,\star}); \boldsymbol{\pi}) \leq C\Big\} \tag{5.2.16}$$

is not empty (it contains the null plan) and compact, since for every $\boldsymbol{\pi} \in \mathcal{K}_C$ we have

$$\boldsymbol{\pi}(\mathrm{RA}(X)) + \int \ell(\gamma)\,\mathrm{d}\boldsymbol{\pi} \leq C + \frac{1}{p}\mathrm{m}(X) + \mu_1(X), \tag{5.2.17}$$

so that $\mathcal{K}_C$ is equi-tight, thanks to Theorem 2.2.13(g) (here we use the compactness of (X, τ)).

We therefore obtain

$$\mathscr{D}_q(\mu_0, \mu_1) = \sup_{(g,\psi_0,\psi_1)\in\mathcal{C}} \inf_{\boldsymbol{\pi}\in\mathcal{M}_+(\mathrm{RA}(X))} \mathcal{L}((g, \psi_0, \psi_1); \boldsymbol{\pi}). \tag{5.2.18}$$

We can introduce the conformal (extended) distance generated by g

$$\mathsf{d}_g(x_0, x_1) := \inf\Big\{ \int_\gamma g : \gamma \in \mathrm{RA}(X),\ \gamma_0 = x_0,\ \gamma_1 = x_1 \Big\} \tag{5.2.19}$$

observing that if the triple (g, ψ_0, ψ_1) does not belong to the subset of $\mathcal{C}$

$$\Sigma := \Big\{(g, \psi_0, \psi_1) \in \mathrm{C}_b(X)^3 : g > 0,\ \psi_0(x_0) - \psi_1(x_1) \le \mathsf{d}_g(x_0, x_1) \quad \text{for every } x_0, x_1 \in X\Big\} \tag{5.2.20}$$

we would have

$$\inf_{\boldsymbol{\pi}} \mathcal{L}((g, \psi_0, \psi_1); \boldsymbol{\pi}) = -\infty.$$

On the other hand, if $(g, \psi_0, \psi_1) \in \Sigma$ the infimum in (5.2.18) is attained at $\boldsymbol{\pi} = 0$ so that

$$\inf_{\boldsymbol{\pi}} \mathcal{L}((g, \psi_0, \psi_1); \boldsymbol{\pi}) = \int_X \psi_0 \,\mathrm{d}\mu_0 - \int_X \psi_1 \,\mathrm{d}\mu_1 - \frac{1}{p}\int_X g^p \,\mathrm{dm}$$

and therefore (5.2.18) reads

$$\mathscr{D}_q(\mu_0, \mu_1) = \sup\Big\{ \int_X \psi_0 \,\mathrm{d}\mu_0 - \int_X \psi_1 \,\mathrm{d}\mu_1 - \frac{1}{p}\int_X g^p \,\mathrm{dm} : (g, \psi_0, \psi_1) \in \Sigma,\ g > 0\Big\}, \tag{5.2.21}$$

which coincides with (5.2.12) thanks to (2.3.28). □

Theorem 5.2.4 *Let us suppose that (X, τ) is compact; then for every $\mu_0, \mu_1 \in \mathcal{M}_+(X)$ we have*

$$\mathscr{D}_q(\mu_0, \mu_1) = \mathsf{pCE}_p^*(\mu_0 - \mu_1). \tag{5.2.22}$$

Proof Combining (5.2.12) with (2.3.29) we easily get

$$\mathscr{D}_q(\mu_0, \mu_1) = \sup\Big\{ \int_X \varphi \,\mathrm{d}(\mu_0 - \mu_1) - \frac{1}{p}\int_X g^p \,\mathrm{dm} : \\ g \in \mathrm{C}_b(X),\ g > 0,\ \varphi \in \mathrm{Lip}_b(X, \tau, \mathsf{d}),\ \mathrm{lip}_{\mathsf{d}}\,\varphi \le g\Big\}, \tag{5.2.23}$$

so that

$$\mathscr{D}_q(\mu_0,\mu_1) \overset{(5.2.23)}{\leq} \sup_{\varphi\in \mathrm{Lip}_b(X,\tau,\mathsf{d})} \int \varphi\,\mathrm{d}(\mu_0-\mu_1) - \frac{1}{p}\int \mathrm{lip}^p(\varphi)\,\mathrm{d}\mathfrak{m} = \mathsf{pCE}_p^*(\mu_0-\mu_1).$$

Since we already proved that $\mathscr{D}_q(\mu_0,\mu_1) \geq \mathsf{pCE}_p^*(\mu_0-\mu_1)$ we conclude. □

Corollary 5.2.5 *Let us suppose that (X,τ) is compact. For every $h \in L^q(X,\mathfrak{m})$ with $\int_X h\,\mathrm{d}\mathfrak{m} = 0$ we have*

$$\mathscr{D}_q(h_+\mathfrak{m}, h_-\mathfrak{m}) = \mathsf{CE}_p^*(h) = \mathsf{wCE}_p^*(h) = \mathsf{pCE}_p^*(h\mathfrak{m}). \tag{5.2.24}$$

Proof Combining (5.2.9) and (5.2.7) we know that for every $h \in L^q(X,\mathfrak{m})$

$$\mathscr{D}_q(h_+\mathfrak{m}, h_-\mathfrak{m}) \geq \mathsf{wCE}_p^*(h) \geq \mathsf{CE}_p^*(h) \geq \mathsf{pCE}_p^*(h\mathfrak{m}).$$

Equality then follows by Theorem 5.2.4. □

By Fenchel-Moreau duality we can now recover for every $f \in L^p(X,\mathfrak{m})$

$$\begin{aligned}\frac{1}{p}\mathsf{CE}_p(f) &= \sup_{h\in L^q(X,\mathfrak{m})} \int_X hf\,\mathrm{d}\mathfrak{m} - \frac{1}{q}\mathsf{CE}_p^*(h)\\ &= \sup_{h\in L^q(X,\mathfrak{m})} \int_X hf\,\mathrm{d}\mathfrak{m} - \frac{1}{q}\mathsf{wCE}_p^*(h)\\ &= \frac{1}{p}\mathsf{wCE}_p(f),\end{aligned}$$

and we obtain the identification of the strong and weak Cheeger energy and of the Sobolev spaces, including the case of a compatible algebra $\mathscr{A}$, thanks to Theorem 3.2.7.

Corollary 5.2.6 *Let us suppose that (X,τ) is compact. Then for every algebra $\mathscr{A}$ compatible with $\mathbb{X}$ we have*

$$H^{1,p}(\mathbb{X},\mathscr{A}) = H^{1,p}(\mathbb{X}) = W^{1,p}(\mathbb{X},\mathcal{T}_q) \tag{5.2.25}$$

with equality of norms; in particular

$$\mathsf{CE}_{p,\mathscr{A}}(f) = \mathsf{CE}_p(f) = \mathsf{wCE}_{p,\mathcal{T}_q}(f) \quad \textit{for every } f \in L^p(X,\mathfrak{m}), \tag{5.2.26}$$

and for every $f \in W^{1,p}(\mathbb{X},\mathcal{T}_q)$

$$|\mathrm{D}f|_{\star,\mathscr{A}} = |\mathrm{D}f|_\star = |\mathrm{D}f|_{w,\mathcal{T}_q} \quad \mathfrak{m}\textit{-a.e. in } X. \tag{5.2.27}$$

The Complete Case

Let us now extend the previous result to the case when (X, d) is complete, by removing the compactness assumption.

Theorem 5.2.7 *Let us suppose that* (X, d) *is complete and let* $\mathscr{A}$ *be an algebra compatible with* $\mathbb{X}$*. Then the same conclusions* (5.2.25), (5.2.26) *and* (5.2.27) *hold.*

Proof Let us consider the Gelfand compactification $\hat{\mathbb{X}} = (\hat{X}, \hat{\tau}, \hat{\mathsf{d}}, \hat{\mathfrak{m}})$ of Theorem 2.1.34 induced by $\mathscr{A}$. Since (X, d) is complete, we can apply Theorem 5.1.32 and we obtain that ι_* induces an isomorphism of $W^{1,p}(\mathbb{X}, \mathfrak{T}_q)$ onto $W^{1,p}(\hat{\mathbb{X}}, \hat{\mathfrak{T}}_q)$ with

$$|\mathrm{D}f|_{w,\mathfrak{T}_q} = \iota^*(|\mathrm{D}\hat{f}|_{w,\hat{\mathfrak{T}}_q}), \quad \hat{f} := \iota_* f. \tag{5.2.28}$$

Since $(\hat{X}, \hat{\tau})$ is compact, by Corollary 5.2.6 we know that $\hat{f} \in H^{1,p}(\hat{X}, \hat{\mathscr{A}})$ with

$$|\mathrm{D}\hat{f}|_{\star,\hat{\mathscr{A}}} = |\mathrm{D}\hat{f}|_{w,\hat{\mathfrak{T}}_q} \quad \hat{\mathfrak{m}} - \text{a.e.} \tag{5.2.29}$$

Finally, applying Lemma 3.1.14 we obtain that $f = \iota^* \hat{f}$ belongs to $H^{1,p}(\mathbb{X}, \mathscr{A})$ with

$$|\mathrm{D}f|_{\star,\mathscr{A}} \le \iota^*\big(|\mathrm{D}\hat{f}|_{\star,\hat{\mathscr{A}}}\big). \tag{5.2.30}$$

Combining the previous inequalities we obtain

$$|\mathrm{D}f|_{\star,\mathscr{A}} \le |\mathrm{D}f|_{w,\mathfrak{T}_q} \quad \mathfrak{m}\text{-a.e.} \tag{5.2.31}$$

Recalling (5.1.48) we conclude. □

We can also extend to the complete case the dual characterizations of Theorem 5.2.4 and Corollary 5.2.5.

Theorem 5.2.8 *Let us suppose that* (X, d) *is complete. Then for every* $\mu_0, \mu_1 \in \mathcal{M}_+(X)$ *we have*

$$\mathscr{D}_q(\mu_0, \mu_1) = \mathsf{pCE}_p^*(\mu_0 - \mu_1), \tag{5.2.32}$$

and whenever $\mu_i = h_i \mathfrak{m}$ *with* $h_i \in L^q(X, \mathfrak{m})$ *and* $h = h_0 - h_1$

$$\mathscr{D}_q(h_0\mathfrak{m}, h_1\mathfrak{m}) = \mathsf{CE}_p^*(h) = \mathsf{wCE}_p^*(h) = \mathsf{pCE}_p^*(h\mathfrak{m}). \tag{5.2.33}$$

Proof It is sufficient to prove that $\mathscr{D}_q(\mu_0, \mu_1) \le \mathsf{pCE}_p^*(\mu_0 - \mu_1)$. Keeping the same notation of the previous proof and using the compactification $\hat{\mathbb{X}}$ induced by the canonical algebra $\mathscr{A} = \mathrm{Lip}_b(X, \tau, \mathsf{d})$, we consider the Radon measures

$\hat{\mu}_i := \iota_\sharp \mu_i \in \mathcal{M}_+(\hat{X})$. It is easy to check that for every plan $\boldsymbol{\pi} \in \Pi(\mu_0, \mu_1)$ the push forward $J_\sharp \boldsymbol{\pi}$ (where $J(\gamma) = \iota \circ \gamma$) belongs to $\Pi(\hat{\mu}_0, \hat{\mu}_1)$ in $\mathrm{RA}(\hat{X})$, so that $\mathcal{D}_q(\mu_0, \mu_1) \geq \mathcal{D}_q(\hat{\mu}_0, \hat{\mu}_1)$. On the other hand, by Lemma 5.1.29 and the completeness of $(\iota(X), \hat{\mathsf{d}})$, every plan $\hat{\boldsymbol{\pi}} \in \Pi(\hat{\mu}_0, \hat{\mu}_1)$ is concentrated on curves in $J(\mathrm{RA}(X))$ so that $\mathcal{D}_q(\mu_0, \mu_1) = \mathcal{D}_q(\hat{\mu}_0, \hat{\mu}_1)$. Recalling that for every $f \in \mathrm{Lip}_b(X, \tau, \mathsf{d})$ we have $\hat{f} = \Gamma(f) \in \mathrm{Lip}_b(\hat{X}, \hat{\tau}, \hat{\mathsf{d}})$ with $\mathrm{lip}_{\hat{\mathsf{d}}} \hat{f}(\iota(x)) \geq \mathrm{lip}_{\mathsf{d}} f(x)$ we get

$$\begin{aligned}
\mathcal{D}_q(\mu_0, \mu_1) &= \mathcal{D}_q(\hat{\mu}_0, \hat{\mu}_1) \\
&= \sup_{\hat{f} \in \mathrm{Lip}_b(\hat{X}, \hat{\tau}, \hat{\mathsf{d}})} \int_{\hat{X}} \hat{f} \, \mathrm{d}(\hat{\mu}_0 - \hat{\mu}_1) - \frac{1}{p} \int_{\hat{X}} \big(\mathrm{Lip}_{\hat{\mathsf{d}}} \hat{f}(\hat{x})\big)^p \, \mathrm{d}\hat{\mathfrak{m}}(\hat{x}) \\
&= \sup_{\hat{f} \in \mathrm{Lip}_b(\hat{X}, \hat{\tau}, \hat{\mathsf{d}})} \int_X \hat{f} \circ \iota \, \mathrm{d}(\mu_0 - \mu_1) - \frac{1}{p} \int_X \big(\mathrm{Lip}_{\hat{\mathsf{d}}} \hat{f}(\iota(x))\big)^p \, \mathrm{d}\mathfrak{m}(x) \\
&\leq \sup_{f \in \mathrm{Lip}_b(X, \tau, \mathsf{d})} \int_X f \, \mathrm{d}(\mu_0 - \mu_1) - \frac{1}{p} \int_X \big(\mathrm{Lip}_{\mathsf{d}} f(x)\big)^p \, \mathrm{d}\mathfrak{m}(x) \\
&= \mathsf{pCE}_p^*(\mu_0 - \mu_1).
\end{aligned}$$

□

As a consequence of the above result, we can prove a simple characterization of nontriviality for the Cheeger energy.

Theorem 5.2.9 *Let us suppose that (X, d) is complete. The following properties are equivalent:*

(a) *The Cheeger energy is trivial: $\mathsf{CE}_p(f) \equiv 0$ for every $f \in L^p(X, \mathfrak{m})$.*
(b) *The Cheeger energy $\mathsf{CE}_p(f)$ is finite for every $f \in L^p(X, \mathfrak{m})$.*
(c) $\mathrm{RA}_0(X)$ *is $\mathcal{T}_q$-negligible.*
(d) $\mathrm{RA}_0(X)$ *is $\mathcal{B}_q$-negligible (equivalently, if (X, τ) is Souslin,* $\mathrm{RA}_0(X)$ *is* Mod_p*-negligible).*

Proof The implication $(a) \Rightarrow (b)$ is obvious.

$\boldsymbol{(b) \Rightarrow (a)}$ If the Cheeger energy is always finite then the Sobolev norm of $H^{1,p}(\mathbb{X}) \subset L^p(X, \mathfrak{m})$ is equivalent to the L^p-norm [20, Corollary 2.8], so that there exists a constant $C > 0$ such that

$$\mathsf{CE}_p(f) \leq C \|f\|^p_{L^p(X,\mathfrak{m})} \quad \text{for every } f \in L^p(X, \mathfrak{m}). \tag{5.2.34}$$

Let us show that (5.2.34) implies $\mathsf{CE}_p(f) \equiv 0$ for every $f \in H^{1,p}(\mathbb{X})$. We consider the 2 periodic Lipschitz function $\phi : \mathbb{R} \to \mathbb{R}$ satisfying $\phi(r) = |r|$ for $r \in [-1, 1]$ and we set

$$\phi_n(r) := \phi(nr), \quad f_n(x) := \phi_n(f(x)).$$

Thanks to the locality of the minimal relaxed gradient we have $|\mathrm{D}f_n|_\star(x) = n|\mathrm{D}f|_\star(x)$ so that

$$\mathsf{CE}_p(f) = \frac{1}{n^p}\mathsf{CE}_p(f_n) \le \frac{C}{n^p}\|f_n\|^p_{L^p(X,\mathfrak{m})} \le \frac{C}{n^p}\mathfrak{m}(X) \to 0 \quad \text{as } n \to \infty.$$

(c) ⇒ (a) If (c) holds then for every nonvanishing $h \in L^p(X, \mathfrak{m})$ the class $\Pi(h_+\mathfrak{m}, h_-\mathfrak{m})$ is empty so that $\mathsf{CE}^*_p(h) = \mathscr{D}_q(h_+\mathfrak{m}, h_-\mathfrak{m}) = +\infty$. By duality we obtain $\mathsf{CE}_p \equiv 0$.

(a) ⇒ (c) Let $\boldsymbol{\pi} \in \mathfrak{T}_q$ with $\boldsymbol{\pi}(\mathrm{RA}(X)) > 0$ and let $(\mathsf{d}_i)_{i\in I}$ be a directed family of continuous semidistances as in (2.1.24a,b,c,d). It is not restrictive to assume that $\boldsymbol{\pi}$ is concentrated on a compact set Γ on which ℓ is continuous. The image $K = \mathsf{e}(\Gamma)$ is compact in (X, τ): for every $i \in I$ we can find a countable set $K_i \subset K$ such that for every $x \in K$ $\inf_{y\in K_i} \mathsf{d}_i(x, y) = 0$. For every $y \in K_i$ the minimal relaxed gradient of the function $x \mapsto \mathsf{d}_i(x, y)$ vanishes, so that there exists a $\boldsymbol{\pi}$-negligible set $N_i \subset \Gamma$ such that the function $t \mapsto \mathsf{d}_i(R_\gamma(t), y)$ is constant for every $y \in K_i$ and $\gamma \in \Gamma \setminus N_i$. By continuity we deduce that $t \mapsto \mathsf{d}_i(R_\gamma(t), y)$ is constant for every $y \in K$ so that $\mathsf{d}_i(R_\gamma(t), \gamma_0) = 0$ for every $\gamma \in \Gamma \setminus N_i$; by integration we obtain

$$\int \Big(\int_\gamma \mathsf{d}_i(x, \gamma_0)\Big)\,\mathrm{d}\boldsymbol{\pi}(\gamma) = 0. \tag{5.2.35}$$

On the other hand, for every $\gamma \in \Gamma$ Beppo Levi's Monotone Convergence Theorem yields

$$\lim_{i\in I} \int_\gamma \mathsf{d}_i(x, \gamma_0) = \int_\gamma \mathsf{d}(x, \gamma_0).$$

A further application of the same theorem thanks to the fact that the function $\gamma \mapsto \int_\gamma \mathsf{d}_i(x, \gamma_0)$ is continuous on Γ with respect to the τ_{A} topology yields

$$0 = \lim_{i\in I} \int \Big(\int_\gamma \mathsf{d}_i(x, \gamma_0)\Big)\,\mathrm{d}\boldsymbol{\pi}(\gamma) = \int \Big(\int_\gamma \mathsf{d}(x, \gamma_0)\Big)\,\mathrm{d}\boldsymbol{\pi}(\gamma) \tag{5.2.36}$$

which shows that $\boldsymbol{\pi}$-a.e. γ is constant, a contradiction.

(d) ⇔ (c) The implication $(d) \Rightarrow (c)$ is obvious. In order to prove the converse one, we argue by contradiction and we suppose that there exists a plan $\boldsymbol{\pi} \in \mathfrak{B}_q$ with $\boldsymbol{\pi}(\mathrm{RA}_0(X)) > 0$. We can then argue as in the proof of Proposition 5.1.17 and define a new plan $\tilde{\boldsymbol{\pi}} \in \mathfrak{T}_q$ according to (5.1.33). It is clear that $\tilde{\boldsymbol{\pi}}(\mathrm{RA}_0(X)) > 0$ as well.

□

5.2.3 Notes

The representation Theorems 5.2.3, 5.2.4, and 5.2.8 are new. The proof of Theorem $H = W$ has been given in [57] in the case of doubling, p-Poincaré spaces [16, Theorem 5.1] and in [7, 8] for general spaces by a completely different method: it relies on three basic ingredients:

- the properties of the L^2-gradient flow of the Cheeger energy (in particular the comparison principle),
- the estimate of the Wasserstein velocity of the evolution curve, by means of a suitable version of the Kuwada's Lemma,
- the representation of the solution as the evaluation at time t of a dynamic plan concentrated on curves with finite q-energy,
- the derivation of the Shannon-Reny entropy along the flow, by using the weak upper gradients of the solutions.

It is curious that the refined estimates of the Hopf-Lax flow play a crucial role in the second step.

A different proof of Theorem 5.2.9 in the context of Newtonian spaces can be found in [45, Prop. 7.1.33].

5.3 *Examples and Applications*

5.3.1 Refined Invariance of the (Strong) Cheeger Energy

Invariance w.r.t. the Algebra $\mathscr{A}$

Theorem 5.3.1 (Invariance of the Cheeger Energy w.r.t. $\mathscr{A}$) *For every e.m.t.m. space $\mathbb{X}$ and every compatible algebra $\mathscr{A}$ the Sobolev space $H^{1,p}(\mathbb{X}, \mathscr{A})$ is independent of the compatible algebra $\mathscr{A}$ and coincides with $H^{1,p}(\mathbb{X})$.*

Proof It is sufficient to combine Theorem 5.2.7 with Corollary 3.1.16. □

We can rephrase the previous statement as a density result: if $\mathscr{A}$ is a compatible algebra for $\mathbb{X}$,

$$\begin{gathered}\text{for every } f \in H^{1,p}(\mathbb{X}) \text{ there exists a sequence } f_n \in \mathscr{A} \text{ such that} \\ f_n \to f, \quad \operatorname{lip} f_n \to |\mathrm{D}f|_\star \quad \text{strongly in } L^p(X, \mathfrak{m}).\end{gathered} \tag{5.3.1}$$

We can also slightly relax the assumption that $\mathscr{A}$ is unital.

Proposition 5.3.2 *Let $\mathbb{X} = (X, \tau, \mathsf{d}, \mathfrak{m})$ be an e.m.t.m. space and let $\mathscr{A} \subset \mathrm{Lip}_b(X, \tau, \mathsf{d})$ be an algebra of functions satisfying* (2.1.52) *(we do not assume that $\mathbb{1} \in \mathscr{A}$). If there exists a sequence of compact sets $K_n \subset X$ and functions $f_n \in \mathscr{A}$*

such that

$$f_n(x) \ge 1 \quad \textit{for every } x \in K_n, \qquad \lim_{n\to\infty} \int_{X\setminus K_n} \Big(1 + |\operatorname{lip} f_n(x)|^p\Big)\, \mathrm{d}\mathfrak{m}(x) = 0 \tag{5.3.2}$$

then $\mathscr{A}$ *satisfies* (5.3.1).

Since $\mathfrak{m}$ is tight, (5.3.2) is surely satisfied if for every compact $K \subset X$ there exists a function $f \in \mathscr{A}$ such that

$$f(x) \ge 1 \quad \text{for every } x \in K, \quad \operatorname{Lip}(f, X) \le C \quad \text{for a constant } C \text{ independent of } K. \tag{5.3.3}$$

Proof of Proposition 5.3.2 Let K_n, f_n be satisfying (5.3.2) and let $[-c_n, c_n] \supset f_n(X)$. Choosing a sequence $i \mapsto \varepsilon_i \downarrow 0$ we can consider the polynomial $P_{n,i} = 2P_{\varepsilon_i}^{c_n,-1/2,1/2}$ where $P_\varepsilon^{c,\alpha,\beta}$ is given by Corollary 2.1.24. Notice that $P_{n,i}(0) = 0$ so that the functions $h_{n,i} := P_{n,i} \circ f_n$ belong to $\mathscr{A}$. It is easy to check that

$$\lim_{i\to\infty} h_{n,i} = f_n' := (-1 \vee 2f_n \wedge 1), \quad \operatorname{lip} h_{n,i} \le 2 \operatorname{lip} f_n, \tag{5.3.4}$$

$$\lim_{i\to\infty} \operatorname{lip} h_{n,i}(x) = \lim_{i\to\infty} P'_{n,i}(f_n(x)) \operatorname{lip} f_n(x) = 0 \quad \text{for every } x \text{ in a neighborhood of } K_n. \tag{5.3.5}$$

By Lebesgue Dominated Convergence Theorem we get

$$\begin{aligned} &\lim_{i\to\infty} \int_{K_n} |h_{n,i} - 1|^p \,\mathrm{d}\mathfrak{m} = 0, \qquad \lim_{i\to\infty} \int_{K_n} |\operatorname{lip} h_{n,i}|^p \,\mathrm{d}\mathfrak{m} = 0, \\ &\int_{X\setminus K_n} \Big(|h_{n,i} - 1|^p + |\operatorname{lip} h_{n,i}|^p\Big) \,\mathrm{d}\mathfrak{m} \le 2^p \int_{X\setminus K_n} \Big(1 + |\operatorname{lip} f_n|^p\Big) \,\mathrm{d}\mathfrak{m}. \end{aligned} \tag{5.3.6}$$

We can now introduce the algebra $\tilde{\mathscr{A}} = \mathscr{A} \oplus \{a\mathbb{1}\} = \{\tilde f = f + c\mathbb{1} : f \in \mathscr{A},\ c \in \mathbb{R}\}$ which is clearly unital and compatible with $\mathbb{X}$ according to definition 2.1.17. Applying (5.3.1) to $\tilde{\mathscr{A}}$, for every $f \in H^{1,p}(\mathbb{X})$ we can find a sequence $\tilde f_k = f_k + a_k\mathbb{1} \in \tilde{\mathscr{A}}$, $k \in \mathbb{N}$, such that

$$\tilde f_k \to f, \quad \operatorname{lip} \tilde f_k \to |\mathrm{D}f|_\star \quad \text{strongly in } L^p(X, \mathfrak{m}) \quad \text{as } k \to \infty.$$

For every $k > 0$, by (5.3.6) and (5.3.2) we can find $i = i(k)$ and $n = n(k)$ sufficiently big, such that $u_k := h_{n,i} \in \mathscr{A}$ such that

$$-1 \le u_k \le 1, \quad a_k^p \int_X \Big(|u_k - 1|^p + |\operatorname{lip} u_k|^p\Big) \,\mathrm{d}\mathfrak{m} \le 1/k^p \tag{5.3.7}$$

We can then consider $f'_k := f_k + a_k u_k \in \mathscr{A}$, observing that

$$\|f'_k - \tilde{f}_k\|_p \le a_k \|u_k - \mathbb{1}\|_{L^p(X,\mathfrak{m})} \le 1/k,$$
$$\|\operatorname{lip} f'_k\|_{L^p(X,\mathfrak{m})} \le \|\operatorname{lip} \tilde{f}_k\|_{L^p(X,\mathfrak{m})} + a_k \|\operatorname{lip} u_k\|_{L^p(X,\mathfrak{m})} \le \|\operatorname{lip} \tilde{f}_k\|_{L^p(X,\mathfrak{m})} + 1/k.$$

We conclude that the sequence $(f'_k)_{k\in\mathbb{N}} \subset \mathscr{A}$ satisfies

$$\lim_{k\to\infty} \|f'_k - f\|_{L^p(X,\mathfrak{m})} = 0, \qquad \lim_{k\to\infty} \int_X |\operatorname{lip} f'_k|^p \,\mathrm{d}\mathfrak{m} = \mathsf{CE}_p(f).$$

□

Let us show two simple examples of applications of Proposition 5.3.2 and condition (5.3.3):

1. If d is τ-continuous, one can always consider the algebra

$$\mathrm{Lip}_{bs}(X,\tau,\mathsf{d}) := \Big\{ f \in \mathrm{Lip}(X,\tau,\mathsf{d}) : f \text{ has } \mathsf{d}\text{-bounded support} \Big\} \tag{5.3.8}$$

2. If (X, d) is proper (i.e. every closed bounded set is compact, and thus τ is the topology induced by d) then the algebra

$$\mathrm{Lip}_c(X,\tau,\mathsf{d}) := \Big\{ f \in \mathrm{Lip}(X,\tau,\mathsf{d}) : f \text{ has compact support} \Big\} \tag{5.3.9}$$

   satisfies (5.3.1).

Invariance w.r.t. Measure-Preserving Embeddings

Let us now consider the invariance of the strong Cheeger energy w.r.t. measure preserving embeddings. Thanks to the previous Theorem 5.3.1 it is sufficient to consider the case of the canonical algebra.

Theorem 5.3.3 (Invariance of the (Strong) Cheeger Energy w.r.t. Measure Preserving Embeddings) *Let $\mathbb{X} = (X, \tau, \mathsf{d}, \mathfrak{m})$ and $\mathbb{X}' = (X', \tau', \mathsf{d}', \mathfrak{m}')$ be two e.m.t.m. spaces and let $\iota : X \to X'$ be a measure preserving embedding of $\mathbb{X}$ into $\mathbb{X}'$ according to Definition 2.1.28. Then ι^* is an isomorphism between $H^{1,p}(\mathbb{X}')$ onto $H^{1,p}(\mathbb{X})$ and*

$$\textit{for every } f = \iota^* f' \in H^{1,p}(\mathbb{X}) \quad |\mathrm{D} f|_\star = \iota^*\big(|\mathrm{D} f|_\star'\big). \tag{5.3.10}$$

Proof Let $\bar{\mathbb{X}}$ and $\bar{\mathbb{X}}'$ be the completion of $\mathbb{X}$ and $\mathbb{X}'$ (where X and X' can be identified as d and d' dense subsets of $\bar{X}$ and $\bar{X}'$ respectively, see Remark 2.1.37). Since $\iota : X \to X'$ is an isometry and X is d-dense in $\bar{X}$, ι can be extended to an isometric embedding $\bar{\iota}$ of $\bar{X}$ into $\bar{X}'$. Using property (C4) of Corollary 2.1.36 one can check that $\bar{\iota}$ is also continuous from $(\bar{X}, \bar{\tau})$ to $(\bar{X}', \bar{\tau}')$ and since $\bar{X}\setminus X$ and $\bar{X}'\setminus X'$ are $\bar{\mathfrak{m}}$ and $\bar{\mathfrak{m}}'$ negligible subsets respectively, we also see that $\bar{\iota}$ is measure-preserving. We conclude that ι is a measure-preserving imbedding of $\bar{X}$ into $\bar{X}'$.

Since the Cheeger energy is invariant w.r.t. completion by Corollary 3.1.16, the above argument shows that it is not restrictive to assume that $\mathbb{X}$ and $\mathbb{X}'$ are complete. By Theorem 5.2.7 out thesis follows by the property for the spaces $W^{1,p}(\mathbb{X}, \mathcal{T}_q)$ and $W^{1,p}(\mathbb{X}', \mathcal{T}_q')$ and the corresponding weak upper gradients, proved in Theorem 5.1.32. □

Recalling the examples of Example 2.1.30, we obtain two useful properties:

Corollary 5.3.4 (Invariance w.r.t. the Topology) *Let* $\mathbb{X} = (X, \tau, \mathsf{d}, \mathfrak{m})$ *be an e.m.t.m. space and let* τ' *be a coarser topology such that* (X, τ', d) *is an e.m.t. space. Then* $H^{1,p}(X, \tau, \mathsf{d}, \mathfrak{m})$ *is isomorphic to* $H^{1,p}(X, \tau', \mathsf{d}, \mathfrak{m})$ *with equal minimal relaxed gradients.*

Corollary 5.3.5 (Restriction) *Let* $\mathbb{X} = (X, \tau, \mathsf{d}, \mathfrak{m})$ *be an e.m.t.m. space and let* $Y \subset X$ *be a* $\mathfrak{m}$*-measurable subset of* X *with* $\mathfrak{m}(X \setminus Y) = 0$*. If* $\mathbb{Y}$ *is the associated e.m.t.m. space according to Example 2.1.30(d),* $H^{1,p}(\mathbb{X})$ *is isomorphic to* $H^{1,p}(\mathbb{Y})$ *with equal minimal relaxed gradients. In particular,* $H^{1,p}(X, \tau, \mathsf{d}, \mathfrak{m})$ *is always isomorphic to* $H^{1,p}(\mathrm{supp}(\mathfrak{m}), \tau, \mathsf{d}, \mathfrak{m})$.

Invariance w.r.t. the Length and the Conformal Constructions

Thanks to Theorem 5.2.7, we can extend the results of Lemma 5.1.33 and Proposition 5.1.35 to the Cheeger energy.

Corollary 5.3.6 *Let* $\mathbb{X} = (X, \tau, \mathsf{d}, \mathfrak{m})$ *be a* complete *e.m.t.m. space and let* $\delta : X \times X \to [0, +\infty]$ *be an extended distance such that*

$$(X, \tau, \delta) \text{ is an extended metric-topological space,} \quad \mathsf{d} \le \delta \le \mathsf{d}_\ell \quad \text{in } X \times X. \tag{5.3.11}$$

Then $H^{1,p}(X, \tau, \mathsf{d}, \mathfrak{m}) = H^{1,p}(X, \tau, \delta, \mathfrak{m})$ *and the corresponding minimal relaxed gradients coincide.*

Corollary 5.3.7 *Let* $\mathbb{X} = (X, \tau, \mathsf{d}, \mathfrak{m})$ *be a* complete *e.m.t.m. space, let* $g \in \mathrm{C}_b(X)$ *with* $0 < m_g \le g \le M_g < \infty$*, and let* $\delta : X \times X \to [0, +\infty]$ *be an extended distance such that*

$$\mathbb{X}' = (X, \tau, \delta, \mathfrak{m}) \text{ is an e.m.t.m. space,} \quad \mathsf{d}_g' \le \delta \le \mathsf{d}_g \quad \text{in } X \times X. \tag{5.3.12}$$

A function $f \in L^p(X, \mathfrak{m})$ *belongs to the Sobolev space* $H^{1,p}(\mathbb{X}')$ *if and only if* $f \in H^{1,p}(\mathbb{X})$*, and the corresponding minimal relaxed gradients in* $\mathbb{X}$ *and in* $\mathbb{X}'$ *(which we call* $|\mathrm{D}f|_{\star,\mathbb{X}}$ *and* $|\mathrm{D}f|_{\star,\mathbb{X}'}$ *respectively) satisfy*

$$|\mathrm{D}f|_{\star,\mathbb{X}} = g|\mathrm{D}f|_{\star,\mathbb{X}'}. \tag{5.3.13}$$

5.3.2 Examples

Example 5.3.8 (Sobolev Spaces in $\mathbb{R}^d$ or in a Finsler-Riemannian Manifold) Let us consider the space $X := \mathbb{R}^d$ with the usual topology τ, the distance d induced by a norm $\|\cdot\|$ with dual norm $\|\cdot\|_*$, and a finite positive Borel measure $\mathfrak{m}$.

Being $(\mathbb{R}^d, \|\cdot\|)$ complete the weak and strong Sobolev spaces coincide. By Proposition 5.3.2 we can choose

$$\mathscr{A} := \mathrm{C}_c^\infty(\mathbb{R}^d), \quad \operatorname{lip} f(x) = \|\mathrm{D}f(x)\|_* \quad \text{for every } f \in \mathscr{A}. \tag{5.3.14}$$

We thus obtain

$$\begin{aligned} H^{1,p}(\mathbb{R}^d, \tau, \|\cdot\|, \mathfrak{m}) = \Big\{ & f \in L^p(\mathbb{R}^d, \mathfrak{m}) : \exists\, f_n \in \mathrm{C}_c^\infty(\mathbb{R}^d) \\ & f_n \to f \text{ in } L^p(X, \mathbb{R}^d),\ \sup_n \int_{\mathbb{R}^d} \|\mathrm{D}f_n\|_*^p \,\mathrm{d}\mathfrak{m} < \infty \Big\}. \end{aligned} \tag{5.3.15}$$

It is not difficult to check that this space is always reflexive (see also [1] and Corollary 5.3.11) and it is an Hilbert space if $\|\cdot\|$ is induced by a scalar product and $p = 2$, since pCE_2 is a quadratic form on $\mathscr{A}$. In this case we obtain the Sobolev space introduced by [19]. At least when the gradient operator is closable in $L^p(X, \mathfrak{m})$, the present metric approach also coincides with the definition of weighted Sobolev spaces given in [43] (a proof of the equivalence under doubling and Poincaré assumptions has been given in [16, Appendix 2]).

A completely analogous approach can be used in a complete Finsler or Riemannian manifold.

Example 5.3.9 (Sobolev Space on a Separable Banach Space) Let $(B, \|\cdot\|)$ be a separable Banach space endowed with the strong topology τ_s and the distance d induced by the norm. Let $\mathfrak{m}$ be a finite positive Borel measure in B and $\mathbb{B} := (B, \tau_s, \mathsf{d}, \mathfrak{m})$. We can consider the algebra $\mathscr{A} = \mathrm{Cyl}(B)$ of smooth cylindrical functions (see Example 2.1.19) so that

$$\operatorname{lip} f(x) = \|\mathrm{D}f(x)\|_* \quad \text{for every } f \in \mathscr{A} \tag{5.3.16}$$

and therefore

$$\begin{aligned} H^{1,p}(\mathbb{B}) = \Big\{ & f \in L^p(H, \mathfrak{m}) : \exists\, f_n \in \mathscr{A} \\ & f_n \to f \text{ in } L^p(B, \mathfrak{m}),\ \sup_n \int_B \|\mathrm{D}f_n\|_*^p \,\mathrm{d}\mathfrak{m} < \infty \Big\}. \end{aligned} \tag{5.3.17}$$

We can give an equivalent intrinsic characterization in terms of vector valued Sobolev differentials, in the case B is also reflexive. Some of the results below could be extended to the case when B has the Radon-Nikodym property [26]. If

$\boldsymbol{h} : B \to B'$ is a Borel map (recall the definition given in Sects. 2.1.1 and A.5 in the Appendix) it is not difficult to check that for every $\gamma \in \mathrm{RA}(B)$ we have $t \mapsto \langle \boldsymbol{h}(R_\gamma(t)), R'_\gamma(t)\rangle$ is Lebesgue-measurable. If $\int_\gamma \|\boldsymbol{h}\|_* < \infty$ we can thus consider the curvilinear integral

$$\int_\gamma \langle \boldsymbol{h}, \dot{\gamma}\rangle := \int_0^1 \langle \boldsymbol{h}(R_\gamma(t)), R'_\gamma(t)\rangle \,\mathrm{d}t \tag{5.3.18}$$

We will denote by $L^p(B, \mathfrak{m}; B')$ the Bochner space of Borel $\mathfrak{m}$-measurable maps $\boldsymbol{h} : B \to B'$ such that

$$\int_B \|\boldsymbol{h}(x)\|_*^p \,\mathrm{d}\mathfrak{m}(x) < \infty, \tag{5.3.19}$$

which is the dual of the Bochner space $L^q(B, \mathfrak{m}; B)$ [27, Theorem 8.20.3].

Given a function $f \in L^p(B, \mathfrak{m})$ we say that a Borel map $\boldsymbol{g} \in L^p(B, \mathfrak{m}; B')$ is a $\mathcal{T}_q$-weak gradient of f if

$$f(\gamma_1) - f(\gamma_0) = \int_\gamma \langle \boldsymbol{g}, \dot{\gamma}\rangle \quad \text{for } \mathcal{T}_q\text{-a.e. } \gamma \in \mathrm{RA}(B). \tag{5.3.20}$$

Notice that the integral in (5.3.20) is well defined since the fact that $\|\boldsymbol{g}\|_* \in L^q(B; \mathfrak{m})$ yields $\int_\gamma \|\boldsymbol{g}\|_* < \infty$ for $\mathcal{T}_q$-a.e.$\gamma \in \mathrm{RA}(B)$. Arguing as in Proposition 5.1.6, we can show that the class of weak gradients is invariant w.r.t. modifications in a $\mathfrak{m}$-negligible subset. We will use the symbols

$$\begin{aligned} \mathrm{WG}_p(f) &:= \Big\{\boldsymbol{g} \in L^p(B, \mathfrak{m}; B') : \boldsymbol{g} \text{ is a weak gradient of } f\Big\}, \\ \mathrm{WG}_p &:= \Big\{(f, \boldsymbol{g}) \in L^p(B, \mathfrak{m}) \times L^p(B, \mathfrak{m}; B') : \boldsymbol{g} \in \mathrm{WG}_p(f)\Big\}. \end{aligned} \tag{5.3.21}$$

Every curve $\gamma \in \mathrm{RA}(B)$ induces a vector measure $\boldsymbol{\nu}_\gamma \in \mathcal{M}(B; B)$ defined by

$$\int_B f \,\mathrm{d}\boldsymbol{\nu}_\gamma := \int_0^1 f(R_\gamma(t))R'_\gamma(t)\,\mathrm{d}t \quad \text{for every } f \in \mathrm{B}_b(B),$$

whose total variation is bounded by ν_γ: $|\boldsymbol{\nu}_\gamma| \le \nu_\gamma$. If $\boldsymbol{\pi} \in \mathcal{M}_+(\mathrm{RA}(B))$ is a dynamic plan we can then consider the vector measure

$$\boldsymbol{\mu}_\pi := \int_B \boldsymbol{\nu}_\gamma \,\mathrm{d}\boldsymbol{\pi}(\gamma), \quad |\boldsymbol{\mu}_\pi| \le \mu_\pi.$$

If $\boldsymbol{\pi} \in \mathcal{B}_q\mathfrak{m}$ then there exists a function $\boldsymbol{h}_\pi \in L^q(B, \mathfrak{m}; B)$ such that

$$\boldsymbol{\mu}_\pi = \boldsymbol{h}_\pi \mathfrak{m}, \qquad \int_B f \, \mathrm{d}\boldsymbol{\mu}_\pi = \int_B f(x)\boldsymbol{h}_\pi(x) \, \mathrm{d}\mathfrak{m}. \tag{5.3.22}$$

Theorem 5.3.10 *Let us suppose that B is a separable and reflexive Banach space and let $f \in L^p(B, \mathfrak{m})$.*

(a) *If $f \in \mathrm{C}^1(B) \cap \mathrm{Lip}(B)$ then $\mathrm{D}f \in \mathrm{WG}_p(f)$.*
(b) *A function $\boldsymbol{g} \in L^p(B, \mathfrak{m}; B')$ belongs to $\mathrm{WG}_p(f)$ if and only if for every $\boldsymbol{\pi} \in \mathcal{B}_q$ with $\boldsymbol{\mu}_\pi = \boldsymbol{h}_\pi \mathfrak{m}$*

$$\int_B f \, \mathrm{d}(\pi_1 - \pi_0) = \int_B \langle \boldsymbol{g}(x), \boldsymbol{h}_\pi(x)\rangle \, \mathrm{d}\mathfrak{m}. \tag{5.3.23}$$

(c) *The set WG_p is a (weakly) closed linear space of $L^p(B, \mathfrak{m}) \times L^p(B, \mathfrak{m}; B')$.*
(d) *If $(f, \boldsymbol{g}) \in \mathrm{WG}_p$ then $g := \|\boldsymbol{g}\|_*$ is a $\mathcal{T}_q$-weak upper gradient of f. Conversely, if g is a $(p, \mathrm{Cyl}(B))$-relaxed gradient of f then there exists $\boldsymbol{g} \in \mathrm{WG}_p(f)$ such that $\|\boldsymbol{g}\|_* \le g$.*
(e) *A function f belongs to the Sobolev space $H^{1,p}(\mathbb{B})$ if and only if there exists a weak gradient $\boldsymbol{g} \in L^p(B, \mathfrak{m}; B')$. In this case $\mathrm{WG}_p(f)$ has a unique element of minimal norm $\mathrm{D}_\mathfrak{m} f$,*

$$|\mathrm{D}f|_\star = \|\mathrm{D}_\mathfrak{m} f\|_* \quad \mathfrak{m}\text{-a.e.}, \qquad \mathsf{CE}_p(f) = \int_B \|\mathrm{D}_\mathfrak{m} f\|_*^p \, \mathrm{d}\mathfrak{m}, \tag{5.3.24}$$

and there exists a sequence $f_n \in \mathrm{Cyl}(B)$ such that

$$\lim_{n\to\infty} \|f_n - f\|_{L^p(B,\mathfrak{m})} + \|\mathrm{D}f_n - \mathrm{D}_\mathfrak{m} f\|_{L^p(B,\mathfrak{m};B')} = 0. \tag{5.3.25}$$

Proof

(a) is an obvious consequence of the chain rule of f along a Lipschitz curve.

(b) follows by the same argument of the proof of Lemma 5.1.8.

(c) is an immediate consequence of (5.3.23).

(d) The first statement is a consequence of (5.3.23), which yields

$$\int_B f \, \mathrm{d}(\pi_1 - \pi_0) \le \int_B \|\boldsymbol{g}\|_* \|\boldsymbol{h}_\pi\| \, \mathrm{d}\mathfrak{m} = \int_B \|\boldsymbol{g}\|_* \, \mathrm{d}(|\boldsymbol{\mu}_\pi|) \le \int_B \|\boldsymbol{g}\|_* \, \mathrm{d}\mu_\pi$$

so that $\|\boldsymbol{g}\|_*$ is a $\mathcal{T}_q$-weak upper gradient by Lemma 5.1.8.

Conversely, let g be a $(p, \mathrm{Cyl}(B))$-relaxed gradient of f. By definition, there exists a sequence f_n of cylindrical functions such that $f_n \to f$ in $L^p(B, \mathfrak{m})$ and $\mathrm{lip}\, f_n \rightharpoonup \tilde{g}$ in $L^p(B, \mathfrak{m})$ with $\tilde{g} \le g$. Since f_n are cylindrical, $\mathrm{lip}\, f_n(x) = \|\mathrm{D}f_n(x)\|_*$; since $L^p(B, \mathfrak{m}; B')$ is reflexive, there exists a subsequence (still denoted by f_n) such that $\mathrm{D} f_n \rightharpoonup \boldsymbol{g}$ in $L^p(B, \mathfrak{m}; B')$. Thanks to claim (c), $(f, \boldsymbol{g})$ belongs to WG_p and the weak lower semicontinuity of continuous convex functionals in a reflexive space yields for every Borel set $A \subset B$

$$\int_A \|\boldsymbol{g}\|_* \,\mathrm{d}\mathfrak{m} \le \liminf_{n\to\infty} \int_A \|\mathrm{D} f_n\|_* \,\mathrm{d}\mathfrak{m} = \int_A \tilde{g} \,\mathrm{d}\mathfrak{m} \le \int_A g \,\mathrm{d}\mathfrak{m}$$

so that $\|\boldsymbol{g}\|_* \le g$ $\mathfrak{m}$-a.e.

(e) The first statement follows by Claim (d) and the identification Theorem 5.2.7 between $H^{1,p}(\mathbb{B})$ and $W^{1,p}(\mathbb{B}, \mathcal{T}_q)$. Claim (d) and the strict convexity of the $L^p(B, \mathfrak{m}; B')$ norm yields (5.3.24). The proof of Claim (d) also shows that there exists a sequence $f_n \in \mathrm{Cyl}(B)$ such that $(f_n, \mathrm{D} f_n)$ weakly converges to $(f, \mathrm{D}_\mathfrak{m} f)$ in $L^p(B, \mathfrak{m}) \times L^p(B, \mathfrak{m}; B')$. We can now apply Mazur Theorem.

□

Corollary 5.3.11 *If B is a reflexive Banach space then $H^{1,p}(\mathbb{B})$ is reflexive. If moreover B is an Hilbert space then $H^{1,2}(\mathbb{B})$ is an Hilbert space.*

Proof The proof that $H^{1,p}(\mathbb{B})$ is reflexive is standard: we first notice that WG_p is a weakly closed subset of the reflexive space $L^p(B, \mathfrak{m}) \times L^p(B, \mathfrak{m}; B')$ and the projection on the first component $\mathrm{p} : (f, \boldsymbol{g}) \to f$ is a continuous and surjective map from WG_p onto $H^{1,p}(\mathbb{B})$ satisfying $\|f\|_{H^{1,p}(\mathbb{B})} = \min\big\{\|(f, \boldsymbol{g})\|_{\mathrm{WG}_p} : \mathrm{p}(f, \boldsymbol{g}) = f\big\}$. If L is a bounded linear functional on $H^{1,p}(\mathbb{B})$ then $L \circ \mathrm{p}$ belongs to WG_p'. If f_n is a bounded sequence in $H^{1,p}(\mathbb{B})$ then there exists a subsequence $k \mapsto f_{n(k)}$ and limits $(f, \boldsymbol{g}) \in \mathrm{WG}_p$ such that $(f_{n(k)}, \mathrm{D}_\mathfrak{m} f_{n(k)}) \rightharpoonup (f, \boldsymbol{g})$ in $L^p(B, \mathfrak{m}) \times L^p(B, \mathfrak{m}; B')$. It follows that

$$\lim_{k\to\infty} L(f_{n(k)}) = \lim_{k\to\infty} L \circ \mathrm{p}(f_{n(k)}, \mathrm{D}_\mathfrak{m} f_{n(k)})) = \lim_{k\to\infty} L \circ \mathrm{p}(f, \boldsymbol{g}) = L(f).$$

□

Remark 5.3.12 The same conclusion of the previous Corollary holds even if X is a *closed subset* of a reflexive and separable Banach (or Hilbert) space $\mathbb{B}$ endowed with the induced length distance d_ℓ (and, e.g., the strong topology τ_s). In this case we have $W^{1,p}(X, \tau_s, \mathsf{d}_\ell, \mathfrak{m}) = W^{1,p}(X, \tau_s, \mathsf{d}, \mathfrak{m})$ by Lemma 5.1.33 (see also Remark 5.1.34), $W^{1,p}(X, \tau_s, \mathsf{d}, \mathfrak{m}) = H^{1,p}(X, \tau_s, \mathsf{d}, \mathfrak{m})$ by Theorem 5.2.7, and eventually $H^{1,p}(X, \tau_s, \mathsf{d}, \mathfrak{m}) = H^{1,p}(\mathbb{B}, \tau_s, \mathsf{d}, \mathfrak{m})$ by Corollary 5.3.5. We can then apply Corollary 5.3.11.

Remark 5.3.13 If we consider the closed subspace

$$\mathrm{WG}_{p,o} := \{0\} \times \mathrm{WG}_p(0) = \Big\{(0, \boldsymbol{g}) : \boldsymbol{g} \in L^p(B, \mathfrak{m}; B') : \int_B \langle \boldsymbol{g}, \boldsymbol{h}_\pi \rangle \,\mathrm{d}\mathfrak{m} = 0 \quad \text{for every } \pi \in \mathcal{T}_q\Big\} \tag{5.3.26}$$

it would not be difficult to see that $H^{1,p}(\mathbb{B})$ is isomorphic to the quotient space $\mathrm{WG}_p/\mathrm{WG}_{p,o}$. The operator $f \mapsto \mathrm{D} f$ from $\mathrm{Cyl}(B)$ to $L^p(B, \mathfrak{m}; B')$ is closable if and only if $\mathrm{WG}_{p,o} = 0$. As typical example one can consider the case of an Hilbert space H endowed with a nondegenerate Gaussian measure, see e.g. [13, 23].

Example 5.3.14 (Wiener Space) Let $(X, \|\cdot\|)$ be a separable Banach space endowed with its strong topology τ and let $\mathfrak{m}$ be a centered non-degenerate Radon Gaussian measure. For every bounded linear functional $v \in X'$ let us set

$$R_{\mathfrak{m}}(v) := \int_X |\langle v, x\rangle|^2 \,\mathrm{d}\mathfrak{m}(x) \tag{5.3.27}$$

$R_{\mathfrak{m}}$ is a nondegenerate continuous quadratic form on X', whose dual characterizes the Cameron-Martin space $H(\mathfrak{m})$ as the subset of X where the functional

$$|x|_{H(\mathfrak{m})} = \sup\{\langle v, x\rangle : v \in X',\ R_{\mathfrak{m}}(v) \le 1\}, \tag{5.3.28}$$

is finite, and thus defines a Hilbertian norm. We also set $\mathsf{d}(x, y) := |x - y|_{H(\mathfrak{m})}$. As we have seen in Example 2.1.12, $\mathbb{X} = (X, \tau, \mathsf{d}, \mathfrak{m})$ is an Polish e.m.t.m. space. By using the algebra $\mathscr{A} = \mathrm{Cyl}(X)$ of smooth cylindrical functions it is not difficult to see that for every $f \in \mathscr{A}$ we have $\mathrm{D}f(x) \in X'$ and

$$\mathrm{lip}_{\mathsf{d}} f(x) = \sup_{v \in H(\mathfrak{m}),\, |v| \le 1} \langle \mathrm{D}f(x), v\rangle = (R_{\mathfrak{m}}(\mathrm{D}f(x)))^{1/2} = |\mathrm{D}f(x)|_{H(\mathfrak{m})'} \tag{5.3.29}$$

so that the metric Sobolev space $H^{1,p}(\mathbb{X})$ coincides with the usual Sobolev space $W^{1,p}(\mathfrak{m})$ [17] defined as the completion of the cylindrical functions with respect to the norm

$$\|f\|^p_{W^{1,p}(\mathfrak{m})} := \int_X \left(|f(x)|^p + |\mathrm{D}f(x)|^p_{H(\mathfrak{m})'}\right) \mathrm{d}\mathfrak{m}(x)$$

5.3.3 Distinguished Representations of Metric Sobolev Spaces

We have already seen that the strong Cheeger energy is invariant w.r.t. completion of the underlying space. We can now use Theorem 5.3.3 to obtain isomorphic realizations of the Sobolev space $H^{1,p}(\mathbb{X})$ with special e.m.t.m. space $\mathbb{X}$. Let us first fix the property we are interested in.

Definition 5.3.15 (Isomorphic Representations of Sobolev Spaces) Let $\mathbb{X}$, $\mathbb{X}'$ be two e.m.t.m. spaces. We say that $H^{1,p}(\mathbb{X}')$ is an isomorphic representation of $H^{1,p}(\mathbb{X})$ if there exists a linear isomorphism $\iota^* : H^{1,p}(\mathbb{X}') \to H^{1,p}(\mathbb{X})$ satisfying (5.3.10) induced by a measure preserving embedding $\iota : X \to X'$ from $\mathbb{X}$ into $\mathbb{X}'$.

All the statements below refers to an arbitrary e.m.t.m. space $\mathbb{X} = (X, \tau, \mathsf{d}, \mathfrak{m})$ and to the strong Sobolev space $H^{1,p}(\mathbb{X})$. Starting from a complete space, they also provide equivalent representations for the weak Sobolev space $W^{1,p}(\mathbb{X}, \mathcal{T}_q)$ thanks to Theorem 5.2.7.

A first example has already been used in the proof of Theorem 5.2.7. It is sufficient to use the compactification Theorem 2.1.34.

Corollary 5.3.16 (Compact Representation) *Every Sobolev space $H^{1,p}(\mathbb{X})$ admits an isomorphic representation $H^{1,p}(\hat{\mathbb{X}})$ where $\hat{X}$ is a compact e.m.t.m. space.*

Corollary 5.3.17 *Suppose that (X, τ) is a Souslin space. Then there exists a separable Banach space $(B, \|\cdot\|_B)$ and a weakly* compact convex subset Σ of the dual unit ball of B' such that $H^{1,p}(\mathbb{X})$ admits an isomorphic representation as $H^{1,p}(\Sigma, \tau_{w*}, \mathsf{d}_{B'}, \mathfrak{m}_B)$ where τ_{w*} is the weak* topology of B' ((Σ, τ_{w*}) is a compact geodesic metric space) and $\mathsf{d}_{B'}(v, w) := \|v - w\|_{B'}$. Moreover, we can choose the compatible algebra $\mathscr{A}$ of the smooth cylindrical functions generated by the elements of B (as linear functional on B').*

Proof Since (X, τ) is Souslin, we can find a metrizable and separable auxiliary topology τ' and a compatible algebra $\mathscr{A} \subset \mathrm{Lip}(X, \tau', \mathsf{d})$ which is countably generated. We can then apply the Gelfand compactification Theorem 2.1.34 with the construction described by Proposition 2.1.33. Since B is the closure of $\mathscr{A}$ in $C_b(X, \tau')$, B is a separable Banach space and Σ is a compact convex subset of the unit ball of B'. □

Appendix A

A.1 Nets

We recap here a few basic facts about nets (see e.g. [53, p.187-188]). Let I be a directed set, i.e. a set endowed with a partial order $\preceq$ satisfying

$$i \preceq i; \quad i \preceq j,\ j \preceq k \quad \Rightarrow \quad i \preceq k \quad \text{for every } i, j, k \in I, \tag{A.1}$$

$$\forall i, j \in I \quad \exists k \in I: \quad i \preceq k,\ j \preceq k. \tag{A.2}$$

As subset $J \subset I$ is called cofinal if for every $i \in I$ there exists $j \in J$ such that $i \preceq j$.

If (Y, τ_Y) is a Hausdorff topological space, a net in Y is a map $y : I \to Y$ defined in some directed set I; the notation $(y_i)_{i \in I}$ (or simply (y_i)) is often used to denote a net.

The net $(y_i)_{i \in I}$ converges to an element $y \in Y$ and we write $y_i \to y$ or $\lim_{i \in I} y_i = y$ if for every neighborhood U of y there exists $i_0 \in I$ such that $i_0 \preceq i \Rightarrow y_i \in U$.

y is an accumulation point of (y_i) if for every neighborhood U of y the set of indexes $\{i \in I : y_i \in U\}$ is cofinal.

A *subnet* $(y_{\mathrm{i}(j)})_{j\in J}$ of (y_i) is obtained by a composition $y \circ \mathrm{i}$ where $\mathrm{i} : J \to I$ is a map defined in a directed set J satisfying

$$j_1 \preceq j_2 \Rightarrow \mathrm{i}(j_1) \preceq \mathrm{i}(j_2), \quad \mathrm{i}(J) \text{ is cofinal in } I.$$

Nets are a useful substitution of the notion of sequences, when the topology τ_Y does not satisfy the first countable axiom. In particular we have the following properties:

(a) A point y belongs to the closure of a subset $A \subset Y$ if and only if there exists a net of points of A converging to y.
(b) A function $f : Y \to Z$ between Hausdorff topological spaces is continuous if and only if for every net $(y_i)_{i\in I}$ converging to y in Y we have $\lim_{i\in I} f(y_i) = f(y)$.
(c) y is an accumulation point of (y_i) if and only if there exists a subnet $(y_{\mathrm{i}(j)})_{j\in J}$ such that $\lim_{j\in J} y_{\mathrm{i}(j)} = y$.
(d) (Y, τ_Y) is compact if and only if every net in Y has a convergent subnet.

A.2 Initial Topologies

Let (Y, τ_Y) be a Hausdorff topological space and let $\mathcal{F} \subset \mathrm{C}(Y)$ be a collection of real continuous functions separating the points of Y. We say that τ_Y is generated by $\mathcal{F}$ if it is the coarsest topology for which all the functions of $\mathcal{F}$ are continuous (thus τ_Y coincides with the *initial* or *weak* topology induced by $\mathcal{F}$). A basis for the topology τ_Y is generated by the finite intersections of sets of the form $\{f^{-1}(U) : f \in \mathcal{F},\ U \text{ open in } \mathbb{R}\}$.

An important property of topologies generated by a separating family of functions is the characterization of convergence: for every net $(y_i)_{i\in I}$ in Y

$$\lim_{i\in I} y_i = y \text{ in } Y \quad \Leftrightarrow \quad \lim_{i\in I} f(y_i) = f(y) \quad \text{for every } f \in \mathcal{F}. \tag{A.3}$$

It is also easy to check that such topologies are completely regular: if F is a closed set and $y \in Y \setminus F$, we can find $f_1, \cdots, f_N \in \mathcal{F}$ and open sets $U_1, \cdots U_N \in \mathbb{R}$ such that $y \in \cap_{n=1}^N f_n^{-1}(U_n) \subset Y \setminus F$. Up to compositions with affine maps, it is not restrictive to assume that $f_n(y) = 1$ and $U_n \supset (0, 2)$ so that the function $f(x) := 0 \vee \min_{1\le n\le N} f_n(x)(2 - f_n(x))$ satisfies $f(y) = 1$ and $f_{|Y\setminus F} \equiv 0$.

A.3 Polish, Lusin, Souslin and Analytic Sets

Denote by $\mathbb{N}^\infty$ the collection of all infinite sequences of natural numbers and by $\mathbb{N}_0^\infty$ the collection of all finite sequences $(n_0, \ldots, n_i)$, with $i \geq 0$ and n_i natural numbers. Let $\mathcal{A} \subset \mathfrak{P}(Y)$ containing the empty set (typical examples are, in a topological space (Y, τ_Y), the classes $\mathscr{F}(Y)$, $\mathscr{K}(Y)$, $\mathscr{B}(Y)$ of closed, compact, and Borel sets respectively). We call *table of sets* (or *Souslin scheme*) in $\mathcal{A}$ [18, Definition 1.10.1] a map A associating to each finite sequence $(n_0, \ldots, n_i) \in \mathbb{N}_0^\infty$ a set $\mathrm{A}_{(n_0,\ldots,n_i)} \in \mathcal{A}$.

Definition A.1 ($\mathcal{A}$-Analytic Sets) $S \subset Y$ is said to be $\mathcal{A}$-analytic if there exists a table A of sets in $\mathcal{A}$ such that

$$S = \bigcup_{(n)\in\mathbb{N}^\infty} \bigcap_{i=0}^{\infty} \mathrm{A}_{(n_0,\ldots,n_i)}.$$

The collection of all the $\mathcal{A}$-analytic sets will be denoted by $\mathrm{S}(\mathcal{A})$.

Let us recall a list of useful properties (see [18, § 1.10])

(A1) Countable unions and countable intersections of elements of $\mathcal{A}$ belongs to $\mathrm{S}(\mathcal{A})$.
(A2) $\mathrm{S}(\mathrm{S}(\mathcal{A})) = \mathrm{S}(\mathcal{A})$
(A3) If the complement of every set of $\mathcal{A}$ belongs to $\mathrm{S}(\mathcal{A})$ then $\mathrm{S}(\mathcal{A})$ contains the σ-algebra generated by $\mathcal{A}$. In particular, in a metrizable space Y $\mathscr{B}(Y)$-analytic sets are $\mathscr{F}(Y)$-analytic.
(A4) In a topological space (E, τ), $\mathscr{B}(E)$-analytic sets are *universally measurable* [18, Theorem 1.10.5], i.e. they are μ-measurable for any finite Borel measure μ.

Definition A.2 ([56, Chap. II]) A Hausdorff topological space (Y, τ_Y) (in particular, a subset of a topological space (X, τ) with the relative topology) is *a Polish space* if it is separable and τ_Y is induced by a complete metric d_Y on Y.
(Y, τ_Y) is said to be *Souslin* (resp. *Lusin*) if it is the image of a Polish space under a continuous (resp. injective and continuous) map.

Differently from the Borel property, notice that the Souslin and Lusin properties for subsets of a topological space are intrinsic, i.e. they depend only on the induced topology.

We recall a few important properties of the class of Souslin and Lusin sets.

Proposition A.3 *The following properties hold:*

(a) *In a Hausdorff topological space (Y, τ_Y), Souslin sets are $\mathscr{F}(Y)$-analytic; if $\mathscr{S}(Y)$ denotes the class of Souslin sets, $\mathrm{S}(\mathscr{S}(Y)) = \mathscr{S}(Y)$.*

(b) *if (Y, τ_Y) is a Souslin space (in particular if it is a Polish or a Lusin space), the notions of Souslin and $\mathscr{F}(Y)$-analytic sets coincide and in this case Lusin sets are Borel and Borel sets are Souslin;*
(c) *if Y, Z are Souslin spaces and $f : Y \to Z$ is a Borel injective map, then f^{-1} is Borel;*
(d) *if Y, Z are Souslin spaces and $f : Y \to Z$ is a Borel map, then f maps Souslin sets to Souslin sets.*
(e) *If (Y, τ_Y) is Souslin then every finite nonnegative Borel measure in Y is Radon.*

Proof **(a)** is proved in [18, Theorems 6.6.6, 6.6.8]. In connection with **(b)**, the equivalence between Souslin and $\mathscr{F}(E)$-analytic sets is proved in [18, Theorem 6.7.2], the fact that Borel sets are Souslin in [18, Corollary 6.6.7] and the fact that Lusin sets are Borel in [18, Theorem 6.8.6]. **(c)** and **(d)** are proved in [18, Theorem 6.7.3]. For **(e)** we refer to [56, Thm. 9 & 10, p. 122]. □

Since in Souslin spaces (Y, τ_Y) we have at the same time tightness of finite Borel measures and coincidence of Souslin and $\mathscr{F}(E)$-analytic sets, the measurability of $\mathscr{B}(E)$-analytic sets yields in particular that for every $\mu \in \mathcal{M}_+(Y)$

$$\mu(B) = \sup\big\{\mu(K) \,:\, K \in \mathscr{K}(Y),\ K \subset B\big\} \quad \text{for every } B \in \mathscr{S}(Y). \tag{A.4}$$

We will also recall another useful property [56, Pages 103–105].

Lemma A.4 *Let us suppose that (Y, τ_Y) is a Souslin space.*

(a) *Y is strongly Lindeöf, i.e. every open cover of an open set has a countable subcover.*
(b) *Every family $\mathcal{F}$ of lower semicontinuous real functions defined in Y has a countable subfamily $(f_n)_{n\in\mathbb{N}} \subset \mathcal{F}$ such that $\sup_{f\in\mathcal{F}} f(x) = \sup_{n\in\mathbb{N}} f_n(x)$ for every $x \in Y$.*
(c) *If Y is regular, every open set is an F_σ (countable intersection of closed set), thus in particular is $\mathscr{F}(Y)$-analytic.*
(d) *If Y is completely regular, there exists a metrizable and separable topology τ' coarser than τ_Y.*

A.4 Choquet Capacities

Let us recall the definition of a *Choquet capacity* in related to a collection $\mathcal{A}$ of subsets of Y containing the empty set and closed under finite unions and countable intersections [24, Chap. III, § 2].

Definition A.5 A function $\mathcal{I} : \mathfrak{P}(Y) \to [0, +\infty]$ is a Choquet $\mathcal{A}$-capacity if it satisfies the properties

(C1) $\mathcal{I}$ is increasing: $A \subset B \quad \Rightarrow \quad \mathcal{I}(A) \le \mathcal{I}(B)$.
(C2) For every increasing sequence $A_n \subset Y$: $\mathcal{I}\big(\cup_n A_n\big) = \lim_{n\to\infty} \mathcal{I}(A_n)$.

(C3) For every decreasing sequence $K_n \in \mathcal{A}$: $\mathcal{I}\big(\bigcap_n K_n\big) = \lim_{n\to\infty} \mathcal{I}(K_n)$.

A subset $A \subset Y$ is called *capacitable* if $\mathcal{I}(A) = \sup\big\{\mathcal{I}(K) : K \subset A,\ K \in \mathcal{A}\big\}$.

Theorem A.6 (Choquet, [24, Chap. III, 28]) *If $\mathcal{I}$ is a $\mathcal{A}$-capacity then every $\mathcal{A}$-analytic set is capacitable.*

A.5 Measurable Maps with Values in Separable Banach Spaces

Let (Y, τ_Y) be a Hausdorff topological space endowed with a Radon measure $\mu \in \mathcal{M}_+(Y)$ and let $(V, \|\cdot\|_V)$ be a separable Banach space with dual V'. Since V is a Polish space, the classes of strong and weak Borel sets coincide.

A map $\boldsymbol{h} : Y \to V$ is Borel μ-measurable (recall the definition given in Sect. 2.1.1) then it is also Lusin μ-measurable, since V is metrizable; in particular, $\boldsymbol{h}$ admits a Borel representative $\tilde{\boldsymbol{h}}$ such that $\mathfrak{m}(\tilde{\boldsymbol{h}} \neq \boldsymbol{h}) = 0$. If $\int_Y \|\boldsymbol{h}\|\,\mathrm{d}\mathfrak{m} < \infty$ then $\boldsymbol{h}$ is also Bochner integrable, i.e. there exists a sequence $\boldsymbol{h}_n : Y \to V$ of simple Borel functions such that

$$\lim_{n\to\infty} \int_Y \|\boldsymbol{h}_n - \boldsymbol{h}\|\,\mathrm{d}\mathfrak{m} = 0.$$

We can then define its Bochner integral $\int_Y \boldsymbol{h}\,\mathrm{d}\mu$ as the limit $\lim_{n\to\infty}\int_Y \boldsymbol{h}_n\,\mathrm{d}\mathfrak{m}$ and the corresponding vector measure $\boldsymbol{\mu_h} := \boldsymbol{h}\mu$ defined by

$$\boldsymbol{\mu_h}(A) := \int_A \boldsymbol{h}\,\mathrm{d}\mu \quad \text{for every } \mu\text{-measurable set } A \subset Y.$$

A.6 Homogeneous Convex Functionals

Let us first recall a simple property of p-homogeneous convex functionals.

Lemma A.7 (Dual of p-Homogeneous Functionals) *Let C be a convex cone of some vector space V, $p > 1$, and $\phi, \psi : C \to [0, \infty]$ with $\psi = \phi^{1/p}$, $\phi = \psi^p$. We have the following properties:*

(a) *ϕ is convex and p-homogeneous (i.e. $\phi(\kappa v) = \kappa^p \phi(v)$ for every $\kappa \in \mathbb{R}$ and $v \in C$) in C if and only if ψ is convex and 1-homogeneous on C (a seminorm, if C is a vector space and ψ is finite).*

(b) *Under one of the above equivalent assumptions, setting for every linear functional $z : V \to \mathbb{R}$*

$$\frac{1}{q}\phi^*(z) := \sup_{v\in C}\, \langle z, v\rangle - \frac{1}{p}\phi(v), \quad \psi_*(z) := \sup\Big\{\langle z, v\rangle : v \in C,\ \psi(v) \le 1\Big\},$$

we have

$$\psi_*(z) = \inf\Big\{c \geq 0 : \langle z, v\rangle \leq c\,\psi(v) \quad \textit{for every } v \in C\Big\}, \qquad \phi^*(z) = (\psi_*(z))^q, \tag{A.5}$$

where in the first infimum we adopt the convention $\inf A = +\infty$ *if* A *is empty.*

Proof By setting $\phi(v) = \psi(v) = +\infty$ if $v \in V \setminus C$, it is not restrictive to assume that $C = V$.

1. Let us assume that ϕ is convex and p-homogeneous: we want to prove that ψ is a seminorm (this is the only nontrivial implication). Since ψ is 1-homogeneous, it is sufficient to prove that it is convex. Let $v_i \in V$, $i = 0, 1$, with $r_i := \psi(v_i) + \varepsilon$ for $\varepsilon > 0$, so that $\tilde{v}_i := v_i/r_i$ satisfies $\psi(\tilde{v}_i) < 1$. We fix $\alpha_i \geq 0$ with $\sum_i \alpha_i = 1$ and we set $r := \sum_i \alpha_i r_i$ and $\beta_i := \alpha_i r_i/r$ which still satisfy $\beta_i \geq 0$ and $\sum_i \beta_i = 1$. Since the set $K := \{\psi(v) \leq 1\} = \{\phi(v) \leq 1\}$ is convex we have $\sum \beta_i \tilde{v}_i \in K$. It follows that $\psi(\sum_i \beta_i \tilde{v}_i) \leq 1$; on the other hand $\sum_i \beta_i \tilde{v}_i = \frac{1}{r}\sum_i \alpha_i v_i$ and therefore $\psi(\sum_i \alpha_i v_i) = r\psi(\sum_i \beta_i \tilde{v}_i) \leq r = \varepsilon + \sum_i \alpha_i \psi(v_i)$. Since $\varepsilon > 0$ is arbitrary, we conclude.
2. We set $K_a := \{v \in V : \psi(v) = a\}$, $a \in \{0, 1\}$, and observe that

$$\psi_*(z) = \delta_{K_0}(z) + \sup_{v \in K_1} \langle z, v\rangle$$

where

$$\delta_{K_0}(z) = \sup_{v \in K_0} \langle z, v\rangle = \begin{cases} 0 & \text{if } \langle z, v\rangle \equiv 0 \;\forall\, v \in K_0 \\ +\infty & \text{otherwise.} \end{cases}$$

Similarly

$$\frac{1}{q}\phi^*(z) = \delta_{K_0}(z) + \sup_{v \in V\setminus K_0} \Big(\langle z, v\rangle - \frac{1}{p}\phi(v)\Big).$$

Since $V \setminus K_0 = \bigcup_{\kappa\in\mathbb{R}} \kappa K_1$ we have

$$\begin{aligned} \frac{1}{q}\phi^*(z) &= \delta_{K_0}(z) + \sup_{v \in K_1, \kappa\in\mathbb{R}} \kappa\langle z, v\rangle - \frac{\kappa^p}{p}\phi(v) \\ &= \delta_{K_0}(z) + \sup_{v \in K_1} \sup_{\kappa\in\mathbb{R}} \Big(\kappa\langle z, v\rangle - \frac{\kappa^p}{p}\Big) \\ &= \delta_{K_0}(z) + \frac{1}{p}\sup_{v \in K_1} \Big(\langle z, v\rangle\Big)^p = \big(\psi_*(z)\big)^p. \end{aligned}$$

□

A.7 Von Neumann Theorem

Let $\mathbb{A}, \mathbb{B}$ be convex sets of some vector spaces and let $\mathcal{L} : \mathbb{A} \times \mathbb{B} \to \mathbb{R}$ be a saddle function satisfying

$$a \mapsto \mathcal{L}(a, b) \quad \text{is concave in } \mathbb{A} \text{ for every } b \in \mathbb{B}, \tag{A.6}$$

$$b \mapsto \mathcal{L}(a, b) \quad \text{is convex in } \mathbb{B} \text{ for every } a \in \mathbb{A}. \tag{A.7}$$

It is always true that

$$\inf_{b \in \mathbb{B}} \sup_{a \in \mathbb{A}} \mathcal{L}(a, b) \geq \sup_{a \in \mathbb{A}} \inf_{b \in \mathbb{B}} \mathcal{L}(a, b). \tag{A.8}$$

The next result provides an important sufficient condition to guarantee the equality in (A.8): we use a formulation which is slightly more general than the statement of [58, Thm. 3.1], but it follows by the same argument.

Theorem A.8 (Von Neumann) *Let us suppose that* (A.6), (A.7) *hold, that* $\mathbb{B}$ *is endowed with some Hausdorff topology and that there exists* $a_\star \in \mathbb{A}$ *and* $C_\star > \sup_{a \in \mathbb{A}} \inf_{b \in \mathbb{B}} \mathcal{L}(a, b)$ *such that*

$$\mathbb{B}_\star := \Big\{ b \in \mathbb{B} : \mathcal{L}(a_\star, b) \leq C_\star \Big\} \quad \textit{is not empty and compact in } \mathbb{B}, \tag{A.9}$$

$$b \mapsto \mathcal{L}(a, b) \textit{ is lower semicontinuous in } \mathbb{B}_\star \textit{ for every } a \in \mathbb{A}. \tag{A.10}$$

Then

$$\min_{b \in \mathbb{B}} \sup_{a \in \mathbb{A}} \mathcal{L}(a, b) = \sup_{a \in \mathbb{A}} \inf_{b \in \mathbb{B}} \mathcal{L}(a, b). \tag{A.11}$$

Similarly, if $\mathbb{A}$ *is endowed with a Hausdorff topology and there exists* $b_\star \in \mathbb{B}$ *and* $D_\star < \inf_{b \in \mathbb{B}} \sup_{a \in \mathbb{A}} \mathcal{L}(a, b)$ *such that*

$$\mathbb{A}_\star := \Big\{ a \in \mathbb{A} : \mathcal{L}(a, b_\star) \geq D_\star \Big\} \quad \textit{is not empty and compact in } \mathbb{A}, \tag{A.12}$$

$$a \mapsto \mathcal{L}(a, b) \textit{ is upper semicontinuous in } \mathbb{A}_\star \textit{ for every } b \in \mathbb{B}. \tag{A.13}$$

Then

$$\inf_{b \in \mathbb{B}} \sup_{a \in \mathbb{A}} \mathcal{L}(a, b) = \max_{a \in \mathbb{A}} \inf_{b \in \mathbb{B}} \mathcal{L}(a, b). \tag{A.14}$$

We reproduce here the main part of the proof of (A.11); (A.14) follows simply by considering the Lagrangian $\tilde{\mathcal{L}}(b,a) := -\mathcal{L}(a,b)$ in $\mathbb{B}\times\mathbb{A}$ and inverting the role of $\mathbb{A}$ and $\mathbb{B}$.

Proof Let $s := \sup_{a\in\mathbb{A}}\inf_{b\in\mathbb{B}}\mathcal{L}(a,b)$ and let $\mathbb{B}_a := \{b\in\mathbb{B} : \mathcal{L}(a,b)\le s\}$, $\mathbb{B}_{a\star} := \{b\in\mathbb{B} : \mathcal{L}(a_\star,b)\le s\}$. We notice that $\mathbb{B}_{a\star}\subset\mathbb{B}_\star$ and that for every $a\in\mathbb{A}$ the set $\mathbb{B}_a\cap\mathbb{B}_\star = \{b\in\mathbb{B}_\star : \mathcal{L}(a,b)\le s\}$ is compact thanks to (A.9) and (A.10). If $A\subset\mathbb{A}$ is a collection containing $a_\star$ then

$$\mathbb{B}_A = \bigcap_{a\in A}\mathbb{B}_a = \bigcap_{a\in A}\big(\mathbb{B}_a\cap\mathbb{B}_{a\star}\big) = \bigcap_{a\in A}\big(\mathbb{B}_a\cap\mathbb{B}_\star\big)$$

so that $\mathbb{B}_A$ is a (possibly empty) compact set. The thesis follows if we check that $\mathbb{B}_\mathbb{A}$ contains a point $\bar{b}$, since in that case $\inf_{b\in\mathbb{B}}\sup_{a\in\mathbb{A}}\mathcal{L}(a,b)\le\sup_{a\in\mathbb{A}}\mathcal{L}(a,\bar{b})\le s$ by construction; on the other hand, (A.8) shows that $\sup_{a\in\mathbb{A}}\mathcal{L}(a,\bar{b}) = s\le\sup_{a\in\mathbb{A}}\mathcal{L}(a,b)$ for every $b\in\mathbb{B}$, so that the minimum in the left-hand side of (A.11) is attained at $\bar{b}$.

Since $\mathbb{B}_A$ are compact whenever $a_\star\in A$, it is sufficient to prove that for every finite collection $A = \{a_1,\cdots,a_n\}$ containing $a_\star$ the intersection $\mathbb{B}_A$ is not empty. To this aim, since $b\mapsto\mathcal{L}(a_k,b)$ are convex functions, [58, Lemma 2.1] yields

$$\inf_{b\in\mathbb{B}}\sup_{1\le k\le n}\mathcal{L}(a_k,b) = \inf_{b\in\mathbb{B}}\sum_{k=1}^{N}\chi_k\mathcal{L}(a_k,b)$$

for a suitable choice of nonnegative coefficients $\chi_k\in[0,1]$ with $\sum_{k=1}^n\chi_k = 1$. We thus get by concavity

$$\inf_{b\in\mathbb{B}}\sum_{k=1}^{N}\chi_k\mathcal{L}(a_k,b)\le\inf_{b\in\mathbb{B}}\mathcal{L}\Big(\sum_{k=1}^{N}\chi_k a_k,b\Big)\le s,$$

so that $\inf_{b\in\mathbb{B}}\sup_{1\le k\le n}\mathcal{L}(a_k,b)\le s$. On the other hand, since $\sup_{1\le k\le n}\mathcal{L}(a_k,b)\ge\mathcal{L}(a_\star,b)$, every $b\in\mathbb{B}$ such that $\sup_{1\le k\le n}\mathcal{L}(a_k,b)\le C_\star$ belongs to $\mathbb{B}_\star$ so that $C_\star > s$ yields

$$s = \inf_{b\in\mathbb{B}}\sup_{1\le k\le n}\mathcal{L}(a_k,b) = \inf_{b\in\mathbb{B}_\star}\sup_{1\le k\le n}\mathcal{L}(a_k,b) = \min_{b\in\mathbb{B}_\star}\sup_{1\le k\le n}\mathcal{L}(a_k,b),$$

where in the last identity we used the fact that $\mathbb{B}_\star$ is compact and that the restriction of the function $b\mapsto\sup_{1\le k\le n}\mathcal{L}(a_k,b)$ to $\mathbb{B}_\star$ is lower semicontinuous. We conclude that $\bigcap_{k=1}^N\{b\in\mathbb{B} : \mathcal{L}(a_k,b)\le s\}$ is not empty. □

References

1. L. Ambrosio, M. Colombo, S. Di Marino, Sobolev spaces in metric measure spaces: reflexivity and lower semicontinuity of slope, in *Variational Methods for Evolving Objects*, volume 67 of *Adv. Stud. Pure Math.* (Math. Soc. Japan, [Tokyo], 2015), pp. 1–58
2. L. Ambrosio, S. Di Marino, G. Savaré, On the duality between p-modulus and probability measures. J. Eur. Math. Soc. (JEMS) **17**(8), 1817–1853 (2015)
3. L. Ambrosio, M. Erbar, G. Savaré, Optimal transport, Cheeger energies and contractivity of dynamic transport distances in extended spaces. Nonlinear Anal. **137**, 77–134 (2016)
4. L. Ambrosio, R. Ghezzi, Sobolev and bounded variation functions on metric measure spaces, in *Geometry, Analysis and Dynamics on Sub-Riemannian Manifolds. vol. II*. EMS Ser. Lect. Math. (Eur. Math. Soc., Zürich, 2016), pp. 211–273
5. L. Ambrosio, N. Gigli, A. Mondino, T. Rajala, Riemannian Ricci curvature lower bounds in metric measure spaces with σ-finite measure. Trans. Am. Math. Soc. **367**, 4661–4701 (2015)
6. L. Ambrosio, N. Gigli, G. Savaré, *Gradient flows in metric spaces and in the space of probability measures*. Lectures in Mathematics ETH Zürich, 2nd edn. (Birkhäuser Verlag, Basel, 2008)
7. L. Ambrosio, N. Gigli, G. Savaré, Density of Lipschitz functions and equivalence of weak gradients in metric measure spaces. Rev. Mat. Iberoam. **29**(3), 969–996 (2013)
8. L. Ambrosio, N. Gigli, G. Savaré, Calculus and heat flow in metric measure spaces and applications to spaces with Ricci bounds from below. Invent. Math. **195**(2), 289–391 (2014)
9. L. Ambrosio, N. Gigli, G. Savaré, Calculus and heat flow in metric measure spaces and applications to spaces with Ricci bounds from below. Inventiones Mathematicae, 289–391 (2014)
10. L. Ambrosio, N. Gigli, G. Savaré, Metric measure spaces with Riemannian Ricci curvature bounded from below. Duke Math. J. **163**(7),, 1405–1490 (2014)
11. L. Ambrosio, N. Gigli, G. Savaré, Bakry-Émery curvature-dimension condition and Riemannian Ricci curvature bounds. Ann. Probab. **43**(1), 339–404 (2015)
12. L. Ambrosio, A. Mondino, G. Savaré, Nonlinear diffusion equations and curvature conditions in metric measure spaces. e-prints (Sept. 2015). ArXiv:1509.07273
13. L. Ambrosio, G. Savaré, L. Zambotti, Existence and stability for Fokker-Planck equations with log-concave reference measure. Probab. Theory Relat. Fields **145**(3-4), 517–564 (2009)
14. L. Ambrosio, P. Tilli, *Topics on Analysis in Metric Spaces*, vol. 25 of *Oxford Lecture Series in Mathematics and its Applications* (Oxford University Press, Oxford, 2004)
15. M. Biroli, U. Mosco, A Saint-Venant principle for Dirichlet forms on discontinuous media. Ann. Mat. Pura Appl. **169**, 125–181 (1995)
16. A. Björn, J. Björn, *Nonlinear Potential Theory on Metric Spaces*, vol. 17 of *EMS Tracts in Mathematics* (European Mathematical Society (EMS), Zürich, 2011)
17. V.I. Bogachev, *Gaussian Measures*, vol. 62 of *Mathematical Surveys and Monographs* (American Mathematical Society, Providence, RI, 1998)
18. V.I. Bogachev, *Measure Theory, Vols. I, II* (Springer, Berlin, 2007)
19. G. Bouchitte, G. Buttazzo, P. Seppecher, Energies with respect to a measure and applications to low-dimensional structures. Calc. Var. Partial Differ. Equ. **5**(1), 37–54 (1997)
20. H. Brezis, *Functional Analysis, Sobolev Spaces and Partial Differential Equations*. Universitext (Springer, New York, 2011)
21. D. Burago, Y. Burago, S. Ivanov, *A Course in Metric Geometry*, vol. 33 of *Graduate Studies in Mathematics* (American Mathematical Society, Providence, RI, 2001)
22. J. Cheeger, Differentiability of Lipschitz functions on metric measure spaces. Geom. Funct. Anal. **9**(3), 428–517 (1999)
23. G. Da Prato, J. Zabczyk, *Second Order Partial Differential Equations in Hilbert Spaces*, vol. 293 of *London Mathematical Society Lecture Note Series* (Cambridge University Press, Cambridge, 2002)

24. C. Dellacherie, P.-A. Meyer, *Probabilities and Potential*, vol. 29 of *North-Holland Mathematics Studies* (North-Holland Publishing, Amsterdam, 1978)
25. S. Di Marino, Sobolev and BV spaces on metric measure spaces via derivations and integration by parts. e-prints (Sep 2014). arXiv:1409.5620
26. J. Diestel, J.J. Uhl, Jr., *Vector Measures* (American Mathematical Society, Providence, RI, 1977). With a foreword by B. J. Pettis, Mathematical Surveys, No. 15
27. R.E. Edwards, *Functional Analysis* (Dover Publications, New York, 1995). Theory and applications, Corrected reprint of the 1965 original
28. R. Engelking, *General Topology*, vol. 6 of *Sigma Series in Pure Mathematics*, 2nd edn. (Heldermann Verlag, Berlin, 1989). Translated from the Polish by the author
29. M. Erbar, K. Kuwada, K.-T. Sturm, On the equivalence of the entropic curvature-dimension condition and Bochner's inequality on metric measure spaces. Invent. Math. **201**(3), 993–1071 (2015)
30. B. Fuglede, Extremal length and functional completion. Acta Math. **98**, 171–219 (1957)
31. M. Fukushima, Regular representations of Dirichlet spaces. Trans. Am. Math. Soc. **155**, 455–473 (1971)
32. M. Fukushima, Y. Oshima, M. Takeda, *Dirichlet Forms and Symmetric Markov Processes*, vol. 19 of *de Gruyter Studies in Mathematics*, extended edition (Walter de Gruyter & Co., Berlin, 2011)
33. E. Gagliardo, A unified structure in various families of function spaces. Compactness and closure theorems, in *Proc. Internat. Sympos. Linear Spaces (Jerusalem, 1960)* (Jerusalem Academic Press, Jerusalem, 1961), pp. 237–241
34. N. Gigli, On the differential structure of metric measure spaces and applications. Mem. Am. Math. Soc. **236**(1113), vi+91 (2015)
35. N. Gigli, Lecture notes on differential calculus on RCD spaces. Publ. Res. Inst. Math. Sci. **54**(4), 855–918 (2018)
36. N. Gigli, Nonsmooth differential geometry—an approach tailored for spaces with Ricci curvature bounded from below. Mem. Am. Math. Soc. **251**(1196), v+161 (2018)
37. N. Gigli, E. Pasqualetto, Differential structure associated to axiomatic Sobolev spaces. e-prints (Jul 2018). arXiv:1807.05417
38. N. Gigli, E. Pasqualetto, E. Soultanis, Differential of metric valued Sobolev maps. e-prints (Jul 2018). arXiv:1807.10063
39. N. Gozlan, C. Roberto, P.-M. Samson, From dimension free concentration to the Poincaré inequality. Calc. Var. Partial Differ. Equ. **52**(3-4), 899–925 (2015)
40. M. Gromov, *Metric Structures for Riemannian and Non-Riemannian Spaces*, vol. 152 of *Progress in Mathematics* (Birkhäuser Boston, Inc., Boston, MA, 1999). Based on the 1981 French original [MR0682063 (85e:53051)], With appendices by M. Katz, P. Pansu and S. Semmes, Translated from the French by Sean Michael Bates
41. P. Hajł asz, Sobolev spaces on an arbitrary metric space. Potential Anal. **5**(4), 403–415 (1996)
42. J. Heinonen, Nonsmooth calculus. Bull. Am. Math. Soc. (N.S.) **44**(2), 163–232 (2007)
43. J. Heinonen, T. Kilpeläinen, O. Martio, *Nonlinear Potential Theory of Degenerate Elliptic Equations*. Oxford Mathematical Monographs (The Clarendon Press, Oxford University Press, New York, 1993). Oxford Science Publications
44. J. Heinonen, P. Koskela, Quasiconformal maps in metric spaces with controlled geometry. Acta Math. **181**(1), 1–61 (1998)
45. J. Heinonen, P. Koskela, N. Shanmugalingam, J.T. Tyson, *Sobolev Spaces on Metric Measure Spaces*, vol. 27 of *New Mathematical Monographs* (Cambridge University Press, Cambridge, 2015). An approach based on upper gradients
46. P. Koskela, P. MacManus, Quasiconformal mappings and Sobolev spaces. Studia Math. **131**(1), 1–17 (1998)
47. K. Kuwada, Gradient estimate for Markov kernels, Wasserstein control and Hopf-Lax formula, in *Potential Theory and Its Related Fields*. RIMS Kôkyûroku Bessatsu, B43 (Res. Inst. Math. Sci. (RIMS), Kyoto, 2013), pp. 61–80

48. M. Liero, A. Mielke, G. Savaré, Optimal entropy-transport problems and a new Hellinger-Kantorovich distance between positive measures. Invent. Math. **211**(3), 969–1117 (2018)
49. G.G. Lorentz, *Bernstein Polynomials*, 2nd edn. (Chelsea Publishing Co., New York, 1986)
50. J. Lott, C. Villani, Ricci curvature for metric-measure spaces via optimal transport. Ann. Math. (2) **169**(3), 903–991 (2009)
51. G. Luise, G. Savaré, Contraction and regularizing properties of heat flows in metric measure spaces. eprints (2019). ArXiv:1904.09825
52. Z.-M. Ma, M. Röckner, *Introduction to the Theory of (Non-symmetric) Dirichlet Forms* (Springer, New York, 1992)
53. J.R. Munkres, *Topology* (Prentice Hall, Inc., Upper Saddle River, NJ, 2000). Second edition of [MR0464128]
54. E. Paolini, E. Stepanov, Decomposition of acyclic normal currents in a metric space. J. Funct. Anal. **263**(11), 3358–3390 (2012)
55. W. Rudin, *Functional Analysis*. International Series in Pure and Applied Mathematics, 2nd edn. (McGraw-Hill, Inc., New York, 1991)
56. L. Schwartz, *Radon Measures on Arbitrary Topological Spaces and Cylindrical Measures*. Published for the Tata Institute of Fundamental Research, Bombay by (Oxford University Press, London, 1973). Tata Institute of Fundamental Research Studies in Mathematics, No. 6
57. N. Shanmugalingam, Newtonian spaces: an extension of Sobolev spaces to metric measure spaces. Rev. Mat. Iberoamericana **16**(2), 243–279 (2000)
58. S. Simons, *Minimax and Monotonicity*, vol. 1693 of *Lecture Notes in Mathematics* (Springer, Berlin, 1998)
59. P. Stollmann, A dual characterization of length spaces with application to Dirichlet metric spaces. Polska Akademia Nauk. Instytut Matematyczny. Studia Mathematica **198**(3), 221–233 (2010)
60. K.-T. Sturm, Analysis on local Dirichlet spaces. II. Upper Gaussian estimates for the fundamental solutions of parabolic equations. Osaka J. Math. **32**(2), 275–312 (1995)
61. K.-T. Sturm, On the geometry of metric measure spaces. I. Acta Math. **196**(1), 65–131 (2006)
62. K.-T. Sturm, On the geometry of metric measure spaces. II. Acta Math. **196**(1), 133–177 (2006)
63. C. Villani, *Topics in Optimal Transportation*, vol. 58 of *Graduate Studies in Mathematics* (American Mathematical Society, Providence, RI, 2003)
64. C. Villani, *Optimal Transport. Old and New*, vol. 338 of *Grundlehren der Mathematischen Wissenschaften* (Springer, Berlin, 2009)

Brief Survey on Functions of Bounded Variation (BV) in Metric Setting

Nageswari Shanmugalingam

Abstract In this note we give an introductory description of the theory of functions of bounded variation (BV) in the nonsmooth setting of metric measure spaces based on the construction first proposed by Michele Miranda Jr. We will consider in particular the geometric properties of BV functions when the measure on the underlying metric measure space is doubling and supports a 1-Poincaré inequality with respect to function-upper gradient pairs.

1 Introduction

Functions of bounded variation (BV) in Euclidean and Riemannian domains form the backbone of geometric questions such as isoperimetric constants and regularity of minimal surfaces. They also are integral to the function of image analysis and image segmentation [8]. With the recent development of analysis in nonsmooth metric spaces, there was an interest in constructing analogs of BV functions in metric setting. This was first accomplished by Michele Miranda Jr. in the work [22], and further developed in [1, 4]. In this note we will give a brief description of the development and results associated with the notion of BV functions in metric spaces. The goal of this note is to introduce the notion of functions of bounded variation and sets of finite perimeter in the nonsmooth setting and their basic properties.

The rest of this section is devoted to a brief discussion on BV functions in Euclidean setting, followed in the next section by a description of Sobolev classes of functions in metric spaces based on upper gradients. Section 3 contains the construction of BV functions in this setting, together with the basic properties of

---

Prepared for the summer school at Levico Terme, Italy, June 2017.

N. Shanmugalingam (✉)
Department of Mathematical Sciences, University of Cincinnati, Cincinnati, OH, USA

Department of Mathematics (MAI), Linköping University, Linköping, Sweden
e-mail: shanmun@uc.edu

L. Ambrosio et al. (eds.), *New Trends on Analysis and Geometry in Metric Spaces*, C.I.M.E. Foundation Subseries 2296, https://doi.org/10.1007/978-3-030-84141-6_5

BV functions valid in all metric measure spaces. Following this, in Sect. 3.3 we discuss geometric properties associated with BV functions under the assumption that the metric measure space is doubling and supports a 1-Poincaré inequality.

In what follows, $1 \le p < \infty$ and n is an integer that is at least 2.

In considering existence of solutions to partial differential equations, for example of the form

$$-\mathrm{div}(|\nabla u|^{p-2}\nabla u) = 0 \text{ in } \Omega, \quad u = f \text{ on } \partial\Omega$$

for Euclidean domains, it is not always possible to find solutions that are of class C^2. Therefore trying to find functions that satisfy the above equation pointwise in Ω is difficult. In general it is only known that solutions are of class $C^{1,\alpha}_{loc}(\Omega)$, but if p is small enough, then there are solutions that are not twice differentiable, see for example [7, 16, 18]. Therefore, one needs to re-interpret the above equation in a weak sense, that is, $-\mathrm{div}(|\nabla u|^{p-2}\nabla u) = 0$ in Ω is interpreted to mean that for every compactly supported smooth function φ in Ω we have

$$\int_\Omega |\nabla u|^{p-2}\langle \nabla u, \nabla\varphi\rangle \, dx = 0.$$

To find solutions to this equation, one notes that the above is the Euler-Lagrange equation corresponding to an energy minimization problem, that is, finding a function u in a Sobolev class with the properties that $u = f$ (in a trace sense) on $\partial\Omega$ and

$$\int_\Omega |\nabla u|^p \, dx \le \int_\Omega |\nabla v|^p \, dx$$

whenever v is in the same Sobolev class with $v = f$ on $\partial\Omega$. Here we give up even more by asking only that $\nabla u : \Omega \to \mathbb{R}^n$ is a function in $L^p(\Omega : \mathbb{R}^n)$ such that whenever φ is a compactly supported smooth function in Ω, we have

$$\int_\Omega u \, \nabla\varphi \, dx + \int_\Omega \varphi \, \nabla u \, dx = 0.$$

We recognize in the above equation that ∇u acts like the derivative of u in satisfying an integration by parts. The function ∇u (if it exists) is uniquely determined (up to sets of Lebesgue measure zero) by the above equation, and is called the *weak derivative* of u. This naturally leads us to the notion of $W^{1,p}(\Omega)$ as the Sobolev class referred to above; it consists of functions $f \in L^p(\Omega)$ that have a weak derivative $\nabla u \in L^p(\Omega, \mathbb{R}^n)$.

Now, we also fix a smooth non-negative function ϕ, with compact support contained in the unit ball $B(\vec{0}, 1)$, such that $\int_{\mathbb{R}^n} \phi \, dx = 1$. For $u \in W^{1,p}(\mathbb{R}^n)$

and $\varepsilon > 0$ we set u_ε to be the convolution of u with $\phi_\varepsilon(\cdot) = \varepsilon^{-n}\phi(\cdot/\varepsilon)$, that is,

$$u_\varepsilon(x) = u * \phi_\varepsilon(x) := \varepsilon^{-n} \int_{\mathbb{R}^n} u(y)\phi([x-y]/\varepsilon)\, dy.$$

A direct computation shows that $\lim_{\varepsilon\to 0^+} u_\varepsilon = u$ in the $L^1(\mathbb{R}^n)$-sense, u_ε is smooth for each $\varepsilon > 0$, and that $\nabla u_\varepsilon = (\nabla u) * \phi_\varepsilon$ with $\nabla u_\varepsilon \to \nabla u$ in $L^1(\mathbb{R}^n, \mathbb{R}^n)$ as $\varepsilon \to 0^+$. In fact, we have the following result.

Theorem 1.1 *Let Ω be a domain in $\mathbb{R}^n$ and let $u \in L^p(\Omega)$. Then $u \in W^{1,p}(\Omega)$ if and only if there is a sequence of smooth functions u_k from $W^{1,p}(\Omega)$ such that $u_k \to u$ in $L^p(\Omega)$ and $\nabla u_k \to \vec{\psi}$ in $L^p(\Omega, \mathbb{R}^n)$ as $k \to \infty$. In the latter case, we also have that $\nabla u = \vec{\psi}$.*

For $p > 1$ we have a stronger version of the above theorem, thanks to the reflexivity of the space $L^p(\Omega, \mathbb{R}^n)$.

Theorem 1.2 *If Ω is a domain in $\mathbb{R}^n$ and $u \in L^p(\Omega)$, then $u \in W^{1,p}(\Omega)$ if and only if there is a sequence of functions u_k from $W^{1,p}(\Omega)$ such that $u_k \to u$ in $L^p(\Omega)$ and*

$$\liminf_{k\to\infty} \int_\Omega |\nabla u_k|^p\, dx < \infty.$$

For $p = 1$ the above fails, and this is leads us to the interesting space of functions of bounded variation (BV).

Example 1.3 Let u be the characteristic function of the ball $B(\vec{0}, r) \subset \mathbb{R}^n$ for some fixed $r > 0$. Then $u \in L^1(\mathbb{R}^n)$. For each positive integer k we set the function u_k on $\mathbb{R}^n$ by

$$u_k(x) = \Big[1 - k\,\text{dist}(x, \mathbb{R}^n \setminus B(\vec{0}, r + k^{-1}))\Big]_+.$$

It is easy to see that u_k is k-Lipschitz on $\mathbb{R}^n$ with

$$|\nabla u_k| = k\chi_{B(\vec{0},r+k^{-1})\setminus B(\vec{0},r)}.$$

Hence it follows that $u_k \to u$ in $L^1(\mathbb{R}^n)$ because

$$\int_{\mathbb{R}^n} |u - u_k|\, dx \le C_n[(r + k^{-1})^n - r^n] \to 0 \text{ as } k \to \infty.$$

Moreover, for large k,

$$\int_{\mathbb{R}^n} |\nabla u_k|\, dx \le C_n k\,[(r + k^{-1})^n - r^n] \le 2C_n r^{n-1}.$$

Thus the sequence u_k satisfies the hypotheses of the above theorem, but the L^1-limit function u fails to be in the class $W^{1,1}(\mathbb{R}^n)$.

2 Sobolev Classes in the Metric Setting

In this note, (X, d) is a separable metric space and μ a Radon measure supported on X such that whenever B is a ball in X, we have $0 < \mu(B) < \infty$. The triple (X, d, μ) will be called a metric measure space, or **mms** for short.

As one would expect, the notion of smooth functions does not exist in the non-smooth setting of general metric measure spaces, and so the concept of weak derivatives does not make sense there. So to define analogs of Sobolev functions in the metric setting, we need alternatives.

In Sect. 1 we discussed the problem $-\mathrm{div}|\nabla u|^{p-2}\nabla u = 0$ and we mentioned that it is the Euler-Lagrange equation corresponding to an energy minimization problem. In pursuing the problem of energy minimization, we only need an analog of $|\nabla u|$, not ∇u itself. There are many notions of Sobolev functions in the metric setting, but these all deal with constructing analogs of $|\nabla u|$. The analog of $|\nabla u|$ we will consider in this note is that of *upper gradients*. The notion of upper gradients is due to Heinonen and Koskela, see [15, 25], and takes its inspiration from the fundamental theorem of calculus, see also [23].

2.1 *Fundamental Theorem of Calculus and Upper Gradients*

If u is a smooth function in a Euclidean domain Ω, then whenever $\gamma : [a, b] \to \Omega$ is a smooth curve, we will have

$$u(\gamma(b)) - u(\gamma(a)) = \int_a^b \langle \nabla u(\gamma(t)), \gamma'(t)\rangle \, dt.$$

This is the much-celebrated fundamental theorem of calculus. From this it follows that

$$|u(\gamma(b)) - u(\gamma(a))| \le \int_a^b |\nabla u|(\gamma(t))\, |\gamma'(t)| \, dt = \int_\gamma |\nabla u| \, ds.$$

Taking this as inspiration, we give the following definition. For exposition on rectifiable curves and path integrals in the metric setting, see for example [5, 9, 10, 13, 17, 21, 23, 25].

Definition 2.1 Given an mms (X, d, μ), and a function $u : X \to \mathbb{R}$, we say that a non-negative Borel measurable function g on X is an *upper gradient* of u

if whenever γ is a non-constant compact rectifiable curve in X, we have

$$|u(x) - u(y)| \le \int_\gamma g\, ds. \tag{2.1.1}$$

In the above, x and y denote the two end points of γ. We take the above inequality to also mean that if at least one of $u(x)$ and $u(y)$ is not finite, then $\int_\gamma g\, ds = \infty$.

The fundamental theorem of calculus tells us that if u is a smooth function on a Euclidean domain, then $|\nabla u|$ is an upper gradient of u in Ω. Unfortunately (or fortunately, as this makes things more interesting), if $u \in W^{1,p}(\Omega)$ but we do not know that u is smooth, then we cannot conclude that $|\nabla u|$ is an upper gradient of u in Ω. So, what is $|\nabla u|$?

2.2 *p-Modulus*

We know from Sect. 1 that if $u \in W^{1,p}(\mathbb{R}^n)$ then the convolution approximations u_ε form good approximations of u in $W^{1,p}(\mathbb{R}^n)$. If γ is a smooth curve (or more generally, a rectifiable curve) in Ω, then for $\varepsilon > 0$ we have that

$$|u_\varepsilon(x) - u_\varepsilon(y)| \le \int_\gamma |\nabla u_\varepsilon|\, ds.$$

Here, x and y denote the end points of γ. We also know that $u_\varepsilon \to u$ both in $L^1(\mathbb{R}^n)$ and pointwise almost everywhere in $\mathbb{R}^n$; thus, if x and y are two points where this pointwise convergence happens, then

$$|u(x) - u(y)| \le \liminf_{\varepsilon \to 0^+} \int_\gamma |\nabla u_\varepsilon|\, ds.$$

thus if we also know that $\liminf_{\varepsilon \to 0^+} \int_\gamma |\nabla u_\varepsilon|\, ds \le \int_\gamma |\nabla u|\, ds$, then we have the inequality (2.1.1) for u and $g = |\nabla u|$ on γ. Unfortunately we do not know either of the above assumptions to hold on a given γ, so we hope that it holds for all except for a small class of rectifiable curves. This leads us to the idea of assigning (outer) measure to families of curves, so that we can talk about a class of rectifiable curves being "small".

From now on, let $\Gamma(X)$ denote the collection of all compact rectifiable curves in X, and $1 \le p < \infty$.

Remark 2.2 For $\Gamma \subset \Gamma(X)$, we set $\mathcal{A}(\Gamma)$ to be the collection of all non-negative Borel measurable functions ρ on X such that for all $\gamma \in \Gamma$ we have $\int_\gamma \rho\, ds \ge 1$. Observe that if Γ has even one constant curve, then $\mathcal{A}(\Gamma)$ is empty. On the other hand, if every curve γ in Γ has length at least $L > 0$ and the trajectory of γ lies in a Borel set $A \subset X$, then $L^{-1}\chi_A \in \mathcal{A}(\Gamma)$.

Definition 2.3 The p-modulus of Γ is the number

$$\text{Mod}_p(\Gamma) := \inf_{\rho \in \mathcal{A}(\Gamma)} \int_X \rho^p \, d\mu.$$

If $\mathcal{A}(\Gamma)$ is empty or if it contains no function in $L^p(X)$, then $\text{Mod}_p(\Gamma) = \infty$.

It turns out that Mod_p is an outer measure on $\Gamma(X)$, see for example [14, 17]. The book [14] has a good exposition on the history of development of Mod_p. It was shown by Fuglede that the only sets that are Mod_p-measurable are sets of Mod_p-measure zero and their complements. This is not a problem for us, as we are mostly interested in using (or discarding) sets of Mod_p zero.

Definition 2.4 For $E \subset X$ we set Γ_E to be the collection of all non-constant compact rectifiable curves in X that intersect the set E, and we denote by Γ_E^+ the collection of all non-constant compact rectifiable curves γ in X such that with γ_s its arc-length re-parametrization, we have $\mathcal{L}^1(\gamma_s^{-1}(E)) > 0$, that is, the arc-length of the part of γ inside E is positive.

Example 2.5 Suppose $A \subset \mathbb{R}^n$ be $\mathcal{L}^n$-measurable set with $\mathcal{L}(A) < \infty$, and let $L > 0$. Let Γ be the collection of all line segments γ_x in $\mathbb{R}^n \times \mathbb{R}$ that begin at a point $(x, 0) \in A \times \{0\}$ and end at the point (x, L). From Remark 2.2 above and Fubini's theorem, it follows that

$$\text{Mod}_p(\Gamma) = \frac{\mathcal{L}^n(A)}{L^{p-1}}.$$

Thus $\text{Mod}_p(\Gamma) = 0$ if and only if A has measure zero. Note that $\mathcal{L}^{n+1}(A \times \{0\}) = 0$. So just knowing that all the curves in a family γ intersects a set $E \subset X$ with $\mu(E) = 0$ we cannot conclude that Mod_p of that family is zero, and in particular, it may well be that $\text{Mod}_p(\Gamma_A) > 0$.

Remark 2.6 On the other hand (the above being on the one hand), if $E \subset X$ is a μ-null set, then with E_0 a Borel subset of X such that $E \subset E_0$ and $\mu(E_0) = 0$, we have that $\rho := \infty\chi_{E_0} \in \mathcal{A}(\Gamma_E^+)$, and hence $\text{Mod}_p(\Gamma_E^+) \le \int_X \rho^p \, d\mu = 0$. This fact will be of great use to us soon.

We now record some properties of Mod_p that would be useful in understanding Sobolev spaces. The following result is due to Koskela and MacManus [20].

Lemma 2.7 *If* $\Gamma \subset \Gamma(X)$, *then* $\text{Mod}_p(\Gamma) = 0$ *if and only if there is a non-negative Borel measurable function* $\rho \in L^p(X)$ *such that* $\int_\gamma \rho \, ds = \infty$ *for every* $\gamma \in \Gamma$.

Proof *(Sketch of Proof)* If there is such $\rho \in L^p(X)$, then for every $\varepsilon > 0$ we have that $\varepsilon\rho \in \mathcal{A}(\Gamma)$, and hence

$$\text{Mod}_p(\Gamma) \le \varepsilon^p \, \|\rho\|_{L^p(X)}^p \to 0 \text{ as } \varepsilon \to 0^+.$$

conversely, if $\mathrm{Mod}_p(\Gamma) = 0$, then for each positive integer k we can find $\rho_k \in \mathcal{A}(\Gamma)$ such that $\int_X \rho_k^p \, d\mu \le 2^{-kp}$. The choice of $\rho = \sum_k \rho_k$ gives the desired conclusion. □

The following result is due to Fuglede.

Lemma 2.8 *If $\rho_k \in L^p(X)$ is a non-negative Borel measurable function for each $k \in \mathbb{N}$ such that $\|\rho_k\|_{L^p(X)} \le 2^{-k/p}$. Set*

$$\Gamma = \{\gamma \in \Gamma(X) : \limsup_{k\to\infty} \rho_k \, ds > 0\}.$$

Then $\mathrm{Mod}_p(\Gamma) = 0$.

Proof For $n \in \mathbb{N}$ we set

$$\Gamma_n = \{\gamma \in \Gamma(X) : \limsup_{k\to\infty} \rho_k \, ds > 1/n\}.$$

If we can show that $\mathrm{Mod}_p(\Gamma_n) = 0$ for each positive integer n, then by the fact that Mod_p is an outer measure we are done. Fix a positive integer n. For positive integers m we set

$$\Gamma(n, m) = \{\gamma \in \Gamma(X) : \int_\gamma \rho_m \, ds > 1/n\}.$$

Then

$$\Gamma_n \subset \bigcap_{m_0 \in \mathbb{N}} \bigcup_{m \ge m_0} \Gamma(n, m).$$

Note that $n\rho_m \in \mathcal{A}(\Gamma(n, m))$, and so

$$\mathrm{Mod}_p(\Gamma(n, m)) \le n^p \int_X \rho_m^p \, d\mu \le n^p \, 2^{-m}.$$

Now by the countable sub-additivity of Mod_p, we see that for $m_0 \in \mathbb{N}$,

$$\mathrm{Mod}_p\Big(\bigcup_{m \ge m_0} \Gamma(n, m)\Big) \le n^p \sum_{m=m_0}^{\infty} 2^{-m} = n^p \, 2^{1-m_0}.$$

From this we can conclude that

$$\mathrm{Mod}_p(\Gamma_n) \le n^p \, 2^{1-m_0} \to 0 \text{ as } m_0 \to \infty.$$

□

As a corollary we obtain the following.

Corollary 2.9 *If ρ_k is a non-negative Borel measurable function in $L^p(X)$ for each $k \in \mathbb{N}$, and if ρ is a Borel function on X such that $\rho_k \to \rho$ in $L^p(X)$, then there is a subsequence, also denoted ρ_k, such that for* Mod_p*-almost every $\gamma \in \Gamma(X)$ we have*

$$\lim_{k\to\infty} \int_\gamma |\rho_k - \rho|\, ds = 0.$$

In particular, for such γ we have

$$\lim_{k\to\infty} \int_\gamma \rho_k\, ds = \int_\gamma \rho\, ds.$$

2.3 p-Weak Upper Gradients and Newton-Sobolev Classes $N^{1,p}(X)$

Back to the Euclidean domain setting- we know that $u_\varepsilon \to u$ in $L^p(\mathbb{R}^n)$ and $|\nabla u_\varepsilon| \to |\nabla u|$ in $L^p(\mathbb{R}^n)$. By modifying these functions on a set of measure zero if need be, we have both these functions to also be Borel on $\mathbb{R}^n$. We also have $u_\varepsilon \to u$ pointwise almost everywhere in $\mathbb{R}^n$. Let E be the set of all points where this convergence does not hold. Then $\mathcal{L}^n(E) = 0$, and so by the discussion in Example 2.5 we have that $\text{Mod}_p(\Gamma_E^+) = 0$. Let Γ_0 be the collection of curves $\gamma \in \Gamma(X)$ for which either $\int_\gamma |\nabla u|\, ds = \infty$ or the limit $\lim_{\varepsilon\to 0^+} \int_\gamma |\nabla u_\varepsilon|\, ds$ does not exist, or if it exists, is not equal to $\int_\gamma |\nabla u|\, ds$. Then $\text{Mod}_p(\Gamma_0) = 0$ by the properties of modulus discussed in the previous Sect. 2.2. Thus $\text{Mod}_p(\Gamma_E^+ \cup \Gamma_0) = 0$. Let $0 \le \rho_0 \in L^p(X)$ such that for each $\gamma \in \Gamma_E^+ \cup \Gamma_0$ we have $\int_\gamma \rho_0\, ds = \infty$, and set Γ to be the collection of all curves $\gamma \in \Gamma(X)$ for which $\int_\gamma \rho_0\, ds = \infty$. Then $\Gamma_E^+ \cup \Gamma_0 \subset \Gamma$ and $\text{Mod}_p(\Gamma) = 0$. Now, for non-constant $\gamma \in \Gamma(X) \setminus \Gamma$, we have that with $\gamma : [a, b] \to X$, for $\mathcal{L}^1$-almost every $t, s \in [a, b]$ with $s < t$ we have

$$|u(\gamma(t)) - u(\gamma(s))| \le \int_{\gamma|_{[s,t]}} |\nabla u|\, ds.$$

Thus, as $\int_\gamma |\nabla u|\, ds$ is finite, by Lusin's condition for integrable functions we have that $u \circ \gamma$ can be modified on a set of $\mathcal{L}^1$-measure zero in a unique manner so that it becomes absolutely continuous on $[a, b]$ with

$$|u(\gamma(t)) - u(\gamma(s))| \le \int_{\gamma|_{[s,t]}} |\nabla u|\, ds$$

holding for *all* $s, t \in [a, b]$ with $s < t$. A simple concatenation argument (if γ, β are two curves in $\Gamma(X) \setminus \Gamma$, with the end point of γ the starting point of β, then the concatenation of γ and β also lies in $\Gamma(X) \setminus \Gamma$), we see that there is a way of modifying u on a set of $\mathcal{L}^n$-measure zero of $\mathbb{R}^n$ such that $u, g = |\nabla u|$ satisfies (2.1.1) for every $\gamma \in \Gamma(X)\setminus\Gamma$. This leads us to the following definition in the metric setting [25].

Definition 2.10 We say that a non-negative Borel measurable function g on X is a p-weak upper gradient of a function u on X if u, g fails (2.1.1) only for a Mod_p-family of curves. The space $N^{1,p}(X)$ consists of functions f on X such that $f \in L^p(X)$ and $D_p(X)$ is non-empty. We set

$$\|f\|_{N^{1,p}(X)} = \|f\|_{L^p(X)} + \|g_f\|_{L^p(X)}.$$

Lemma 2.11 *If $\gamma \in \Gamma$ such that $\int_\gamma g\,ds$ is finite and* (2.1.1) *holds for u, g for every sub-curve of γ, then $u \circ \gamma$ is absolutely continuous. If $g \in L^p(X)$ is a p-weak upper gradient of u, then* Mod_p*-almost every non-constant $\gamma \in \Gamma(X)$ is such a curve.*

Proof To see this, let Γ be the collection of all curves γ on which the upper gradient inequality (2.1.1) fails or $\int_\gamma g\,ds = \infty$. Then $\mathrm{Mod}_p(\Gamma) = 0$, and so there is some $\rho \in L^p(X)$ such that for all $\gamma \in \Gamma$ we have $\int_\gamma \rho\,ds = \infty$. Let Γ_1 be the collection of all non-constant compact rectifiable curves β in X (that is, all $\beta \in \Gamma(X)$) for which $\int_\beta \rho\,ds = \infty$. Then

1. $\mathrm{Mod}_p(\Gamma_1) = 0$,
2. Whenever $\gamma_1, \gamma_2 \in \Gamma(X) \setminus \Gamma_1$ are non-constant compact rectifiable curves, then every subcurve of γ_1 is in $\Gamma(X) \setminus \Gamma_1$ and if an end-point of γ_1 is in common with an end point of γ_2, then the concatenation of γ_1 and γ_2 is also in $\Gamma(X) \setminus \Gamma_1$.

In particular, if γ is a non-constant compact rectifiable curve in X such that $\gamma \notin \Gamma_1$, then for each subcurve β of γ we have that (2.1.1) holds for u, g on β *and* $\int_\beta g\,ds < \infty$. Now an application to the absolute continuity of the integral $\int_\gamma g\,ds$ implies that $u \circ \gamma$ is absolutely continuous. □

2.4 Some Properties of *p*-Weak Upper Gradient

Given a function u on X, let $D_p(u)$ denote the collection of all p-weak upper gradients of u that lie in $L^p(X)$.

Lemma 2.12 *If $D_p(u)$ is non-empty, then it is a closed convex subset of $L^p(X)$. If $g, h \in D_p(u)$ and A is a Borel subset of X, then $g\chi_A + h\chi_{X\setminus A} \in D_p(u)$. Consequently, there is a unique function $g_u \in D_p(u)$ such that whenever $g \in D_p(u)$, we must have $g_u \leq g$ μ-a.e. in X.*

For Lipschitz functions η on X, and $x \in X$, we set

$$\mathrm{Lip}\,\eta(x) := \limsup_{x\neq y\to x} \frac{|\eta(y)-\eta(x)|}{d(y,x)}.$$

Proof *(Sketch of Proof)* An appeal to Corollary 2.9 tells us that if $g_k \in D_p(u)$ and $g \in L^p(X)$ such that $g_k \to g$ in $L^p(X)$, then for p-modulus almost every non-constant compact rectifiable curve γ in X, after passing to a subsequence of g_k if need be, we have $\lim_k \int_\gamma g_k\, ds = \int_\gamma g\, ds$. Let Γ_0 denote the collection of all non-constant compact rectifiable curves in X for which this does not happen, and for each positive integer k set Γ_k to be the collection of all non-constant compact rectifiable curves in X for which u, g_k fail the upper gradient inequality (2.1.1). Then $\mathrm{Mod}_p(\Gamma_0 \cup \bigcup_k \Gamma_k) = 0$. We can then see that if γ is not in $\Gamma_0 \cup \bigcup_k \Gamma_k$, then u, g satisfies (2.1.1) on γ. Thus $g \in D_p(u)$, that is, $D_p(u)$ is a closed set in the topology of $L^p(X)$.

It can be directly seen that convex combinations of functions in $D_p(u)$ lie in $D_p(u)$; this is a short exercise.

Finally, if $g, h \in D_p(u)$ and A is a Borel subset of X, then to show that $g\chi_A + h\chi_{X\setminus A} \in D_p(u)$, we can first do this by proving this claim for the case that A is a closed set (and hence $\gamma^{-1}(X \setminus A)$ is a pairwise disjoint countable collection of relatively open intervals). A limiting argument then completes the proof. □

Lemma 2.13 *Let u be a function on X and suppose that g is a non-negative Borel function on X such that for* Mod_p*-almost every non-constant $\gamma \in \Gamma(X)$ we have that $u \circ \gamma$ is absolutely continuous and (denoting the domain of γ to be $[a, b] \subset \mathbb{R}$),*

$$|(u \circ \gamma)'| \leq g \circ \gamma \qquad \mathcal{L}^1 - a.e.\ in\ [a, b].$$

Then g is a p-weak upper gradient of u.

Consequently, if η is Lipschitz on X, then $\mathrm{Lip}\,\eta$ is a p-weak upper gradient on X for all $1 \leq p < \infty$ (indeed, it is an upper gradient of η, but that is not needed here).

Proof *(Sketch of Proof)* This is done by applying Lemma 2.11 together with the properties of absolutely continuous functions on an interval. □

Remark 2.14 In the Euclidean setting ($\mathbb{R}^n$), if we know that $u \circ \gamma$ is absolutely continuous for Mod_p-almost every line segment parallel to the (pre-chosen) coordinate axes, then $\partial_i u$, $i = 1, \cdots, n$ exists on the space and is measurable. Hence to ensure that such u is in the Sobolev class $W^{1,p}(\mathbb{R}^n)$ we only need to ensure that $\partial_i u \in L^p(\mathbb{R}^n)$ for $i = 1, \cdots, n$. In the metric setting, no such canonical directions exist, and if we set $\partial u(x)$ to be the supremum of $(u \circ \gamma)'(t)$ for Mod_p-almost every γ (with $\gamma(t) = x$), there is no reason why $x \mapsto \partial u(x)$ should be measurable. Thus the function g in the above lemma acts as a "measurable blanket" draped over $(u \circ \gamma)'$ and ensures measurability and integrability.

Another use of the first part of the above lemma gives the following sublinearity and Leibniz inequality properties of p-weak upper gradients.

Lemma 2.15 *Let u, v be two Borel functions in $L^p(X)$. If $g \in D_p(u)$, $h \in D_p(v)$, $\alpha, \beta \in \mathbb{R}$, and η is a Lipschitz continuous function on X, then*

$$|\alpha| g + |\beta| h \in D_p(\alpha u + \beta v)$$

and

$$g + h + |u - v| Lip\, \eta \in D_p(\eta u + (1-\eta)v).$$

A good resource for the above two results is [9]. It is known that $N^{1,p}(X)$ is a Banach space, see for example [9, 17, 25]. A slightly weaker property can be seen with a more elementary proof.

Lemma 2.16 *Let $1 < p < \infty$, and $u_k \in N^{1,p}(X)$ have upper gradient $g_k \in L^p(X)$. Suppose that $u_k \to u$ in $L^p(X)$ and $g_k \to g$ in $L^p(X)$ (and hence, $u, g \in L^p(X)$ automatically). Then a modification of u on a set of measure zero yields a function in $N^{1,p}(X)$ with g as a p-weak upper gradient.*

Proof *(Sketch of Proof)* By passing to a subsequence if necessary, we can also assume that

1. $u_k \to u$ pointwise almost everywhere in X as well,
2. for Mod_p-almost every non-constant compact rectifiable curve γ in X we have $\int_\gamma |g_k - g|\, ds \to 0$ as $k \to \infty$ and $\int_\gamma g\, ds < \infty$.

Find a family $\Gamma \subset \Gamma(X)$ with $\mathrm{Mod}_p(\Gamma) = 0$ such that for each non-constant compact rectifiable curve γ in X that does not belong to Γ, (we can assume that γ is arc-length re-parametrized) we have

1. At $\mathcal{L}^1$-almost every point t in the domain of γ satisfies $\lim_k u_k \circ \gamma(t) = u \circ \gamma(t) \in \mathbb{R}$,
2. $\int_\gamma |g_k - g|\, ds \to 0$ as $k \to \infty$ and $\int_\gamma g\, ds < \infty$,
3. u_k, g_k satisfies the upper gradient inequality (2.1.1) on γ and all of its sub-curves.

Using the above, it is now straight-forward to check that at $\mathcal{L}^1$-almost every pair of points is in the domain of such γ, the restriction β of γ to the corresponding sub-interval satisfies the upper gradient inequality (2.1.1) for u, g. Using this and the fact that $\int_\gamma g\, ds$ is finite (and absolute continuity of integrals), we can see that u has a unique extension to the entire domain of γ so that the extended function and g together satisfy (2.1.1).

Next, noting that concatenation of two curve that do not lie in Γ yields a curve that does not lie in Γ, and this can be used to show that the modification of u on the respective curves also gives the desired modification of u on the concatenated curve.

Finally, show that the set where u needs to be modified is a subset of the set of points in X at which u_k does not converge to u; as this is a measure zero subset of X, the proof can be now easily completed. □

Given a function $u \in L^1_{loc}(X)$, and a measurable bounded set $A \subset X$ with $0 < \mu(A) < \infty$, we set

$$u_A := \frac{1}{\mu(A)} \int_A u\, d\mu =: \fint_A u\, d\mu.$$

Definition 2.17 We say that μ supports a p-Poincaré inequality on X if there are constants $C > 0$ and $\lambda \geq 1$ such that whenever $u \in L^1_{loc}(X)$ with p-weak upper gradient g, and whenever B is a ball in X, we have

$$\fint_B |u - u_B|\, d\mu \leq C\,\text{rad}(B) \left(\fint_{\lambda B} g^p\, d\mu\right)^{1/p}.$$

If μ is doubling and supports a p-Poincaré inequality for some $1 \leq p < \infty$, then

1. if X is also locally compact, then X is quasiconvex, that is, there is some $C_q \geq 1$ such that whenever $x, y \in X$ there is a rectifiable curve γ in X with end points x, y such that $\ell(\gamma) \leq C_q\, d(x, y)$. This can be seen by using an ε-chaining argument; given $x, y \in X$ and $\varepsilon > 0$ we consider all finite sequences $x = x_0, x_1, \cdots, x_k = y$ such that for $j = 0, \cdots, k-1$ we have $d(x_j, x_{j+1}) < \varepsilon$. Use the Poincaré inequality first to see that X is a connected set, and then that given $x \in X$ every $y \in X \setminus \{x\}$ can be connected to x by such an ε-chain. We set

$$\varphi_{\varepsilon,x}(y) = \inf_{x=x_0,\cdots,x_k=y \text{ an } \varepsilon-\text{chain}} \sum_{j=0}^{k-1} d(x_j, x_{j+1}),$$

   and note that the constant function 1 is an upper gradient of $\varphi_{\varepsilon,x}$ and that this function is locally 1-Lipschitz continuous. A telescoping series argument together with the Poincaré inequality and a completion argument yields the existence of a desired quasiconvex curve [10].
2. Lipschitz functions form a dense subclass of $N^{1,p}(X)$ and locally Lipschitz continuous functions form a dense subclass of $N^{1,p}(U)$ for every open set $U \subset X$. This is proved using a discrete convolution argument [25].

3 BV Functions in Metric Setting

Recall that the conclusion of Theorem 1.2 fails when $p = 1$. Keeping this in mind, we give the following definition of BV functions in an mms (X, d, μ). This definition is inspired by the one found in [22].

Definition 3.1 We say that $u \in BV(X)$ if $u \in L^1(X)$ and there is a sequence u_k from $N^{1,1}(X)$ such that $u_k \to u$ in $L^1(X)$ such that

$$\liminf_{k\to\infty} \int_X g_{u_k}\, d\mu < \infty.$$

For $u \in L^1(X)$ we set

$$\|Du\|(X) = \inf_{(u_k)} \liminf_{k\to\infty} \int_X g_{u_k}\, d\mu.$$

A Cantor-type diagonalization argument leads to the following "lowe semicontinuity of BV energy" lemma.

Lemma 3.2 *Let $u_k \in BV(X)$ be a sequence of functions such that $u_k \to u \in L^1(X)$. Then*

$$\|Du\|(X) \le \liminf_{k\to\infty} \|Du_k\|(X).$$

In the Euclidean setting, another definition of $BV(\mathbb{R}^n)$ (or $BV(\Omega)$ for some domain in $\mathbb{R}^n$) is as follows. We say that $u \in BV(\mathbb{R}^n)$ if $u \in L^1(\mathbb{R}^n)$ (or locally so, but we will not be picky here for simplicity) and the linear operator $T_u : C_c^\infty(\mathbb{R}^n, \mathbb{R}^n) \to \mathbb{R}$ given by

$$T_u(\vec{\phi}) = \int_{\mathbb{R}^n} u \, \mathrm{div}\vec{\phi}\, dx$$

is a bounded operator. Should this be a bounded operator, we set

$$\|Du\|(\mathbb{R}^n) = \sup\{T_u(\vec{\phi}) \,:\, \vec{\phi} \in C_c^\infty(\mathbb{R}^n, \mathbb{R}^n),\, \|\vec{\phi}\|_{L^\infty} \le 1\}.$$

Riesz representation theorem then gives us a vector-valued signed Radon measure

$$Du = (D_1 u, \cdots, D_n u)$$

on $\mathbb{R}^n$ such that

$$T_u(\vec{\phi}) = -\sum_{i=1}^n \int_{\mathbb{R}^n} \phi_i \, dD_i u.$$

The above two notions of BV functions coincide; this can be seen with the aid of convolutions $u_\varepsilon = u * \varphi_\varepsilon$ and $\nabla(u_\varepsilon) = \varphi_\varepsilon * Du$, see for example [12]. The above use of Riesz representation theorem gives us additional information about $\|Du\|$ in the Euclidean setting, namely, that it is a Radon measure on $\mathbb{R}^n$. We will see in the next subsection that this property holds also in the metric setting.

The notion of BV found in [22] is very similar to the one above, except that in [22] the sequence u_k is assumed to be locally Lipschitz continuous and g_{u_k} is replaced by $\mathrm{Lip}\, u_k$. It turns out that if locally Lipschitz continuous functions are dense in $N^{1,1}(X)$ or if X is complete, then the two definitions coincide, see [2, Theorem 1.1]. The paper [2] also studies another equivalent notion of BV using optimal mass transport theory. We point out here that if X is complete, then a deep result of Ambrosio, Gigli and Savaré shows that locally Lipschitz functions are dense in $N^{1,1}(X)$, see [3].

3.1 Outer Measure Property

Suppose $u \in BV(X)$; then for open sets $U \subset X$ we set

$$\|Du\|(U) = \inf_{(u_k)} \liminf_{k\to\infty} \int_U g_{u_k}\, d\mu,$$

where the infimum is over all sequences (u_k) from $N^{1,1}(U)$ for which $u_k \to u$ in $L^1(U)$. We first list some elementary properties of $\|Du\|$.

Lemma 3.3 *Let U and V be two open subsets of X. Then*

1. $\|Du\|(\emptyset) = 0$;
2. $\|Du\|(U) \le \|Du\|(V)$ *if* $U \subset V$,
3. $\|Du\|(\bigcup_i U_i) = \sum_i \|Du\|(U_i)$ *if* $\{U_i\}_i$ *is a pairwise disjoint subfamily of open subsets of* X.

Proof We will only prove the third property here, as other two are quite direct consequences of the definition of $\|Du\|$. Since any function $u \in N^{1,1}(\bigcup_i U_i)$ has restrictions $u_i = f|_{U_i} \in N^{1,1}(U_i)$ with $\int_{\bigcup_i U_i} g_u\, d\mu = \sum_i \int_{U_i} g_{u_i}\, d\mu$, it follows that

$$\|Df\|(\bigcup_i U_i) \ge \sum_i \|Df\|(U_i).$$

In the above we also used the fact that as u gets closer to f in the $L^1(\bigcup_u U_i)$ sense, u_i gets closer to f in the $L^1(U_i)$ sense.

To prove the reverse inequality, for $\varepsilon > 0$ we can choose $u_i \in N^{1,1}(U_i)$ for each i such that

$$\int_{U_i} |f - u_i|\, d\mu < 2^{-i-2}\varepsilon$$

and

$$\int_{U_i} g_{u_i}\, d\mu < \|Df\|(U_i) + 2^{-i-2}\varepsilon.$$

Now the function $u_\varepsilon = \sum_i u_i \chi_{U_i}$ is in $N^{1,1}(\bigcup_i U_i)$ because there are no rectifiable curves in $\bigcup_i U_i$ with end points in two different U_i's. Therefore

$$\int_{\bigcup_i U_i} |f - u_\varepsilon|\, d\mu \le \sum_i \int_{U_i} |f - u_i|\, d\mu \le \frac{\varepsilon}{2}$$

and

$$\int_{\bigcup_i U_i} g_{u_\varepsilon}\, d\mu = \sum_i \int_{U_i} g_{u_i}\, d\mu \le \frac{\varepsilon}{2} + \sum_i \|Df\|(U_i).$$

From the first of the above two inequalities it follows that $\lim_{\varepsilon\to 0^+} u_\varepsilon = f$ in $L^1(\bigcup_i U_i)$, and therefore

$$\|Df\|(\bigcup_i U_i) \le \liminf_{\varepsilon\to 0^+}\left(\frac{\varepsilon}{2} + \sum_i \|Df\|(U_i)\right) = \sum_i \|Df\|(U_i).$$

□

We use the above definition of $\|Du\|$ on open sets to consider the following Caratheodory construction.

Definition 3.4 For $A \subset X$, we set

$$\|Du\|^*(A) := \inf\{\|Du\|(O) \,:\, O \text{ is open subset of } X,\, A \subset O\}.$$

By the second property listed in the above lemma, we note that if A is an open subset of X, then $\|Du\|^*(A) = \|Du\|(A)$. With this observation, we re-name $\|Du\|^*(A)$ as $\|Du\|(A)$ even when A is not open.

The following lemma is due to De Giorgi and Letta [11, Theorem 5.1], see also [6, Theorem 1.53].

Lemma 3.5 *If ν is a non-negative function on the class of all open subsets of X such that*

1. $\nu(\emptyset) = 0$,
2. *if $U_1 \subset U_2$ and they are both open sets, then $\nu(U_1) \le \nu(U_2)$,*
3. $\nu(U_1 \cup U_2) \le \nu(U_1) + \nu(U_2)$ *for open sets U_1, U_2 in X,*
4. *if $U_1 \cap U_2$ is empty and U_1, U_2 are open, then $\nu(U_1 \cup U_2) = \nu(U_1) + \nu(U_2)$,*
5. *for open sets U*

$$\nu(U) = \sup\{\nu(V) \,:\, V \text{ is open in } X, \overline{V} \text{ is compact subset of } U\}.$$

Then the Caratheodory extension of ν to all subsets of X gives a Borel regular outer measure on X.

Theorem 3.6 *If X is proper (that is, closed and bounded subsets of X are compact) and $f \in BV(X)$, then $\|Df\|$ is an Radon outer measure on X.*

The properness of X is not essential in the above theorem; local compactness suffices. For simplicity we assume that X itself is proper.

Proof Thanks to the lemma of De Giorgi and Letta, it suffices to verify that $\|Du\|$ satisfies the five conditions set forth in Lemma 3.5. By Lemma 3.3, we know that

$\|Du\|$ satisfies Conditions 1, 2, and 4. Thus it suffices for us to verify Conditions 3 and 5. We will first show the validity of Condition 5, and use it (or rather, it's proof) to show that Condition 3 holds. We will do so for bounded open subsets of X; a simple modification (by truncating U_δ by balls) would complete the proof for unbounded sets; we leave this part of the extension as an exercise.

Proof of Condition 5 From the monotonicity Condition 2, it suffices to prove that

$$\|Df\|(U) \le \sup\{\|Df\|(V) \,:\, V \text{ is open in } X, \overline{V} \text{ is a compact subset of } U\}.$$

For $\delta > 0$ we set

$$U_\delta = \{x \in U \,:\, \operatorname{dist}(x, X \setminus U) > \delta\}.$$

For $0 < \delta_1 < \delta_2 < \operatorname{diam}(U)/2$, let $V = U_{\delta_1}$ and $W = U \setminus \overline{U_{\delta_2}}$. Then V and W are open subsets of U, and the closure of V is a compact subset of U. Note also that $U = V \cup W$ and that $\partial V \cap \partial W$ is empty. Thus we can find a Lipschitz function η on U that can be used as a "needle+thread" to stitch Sobolev functions on V to Sobolev functions on W as follows to obtain a Sobolev function on U: $0 \le \eta \le 1$ on U, $\eta = 1$ on $V \setminus W = \overline{U_{\delta_2}}$, $\eta = 0$ on $W \setminus V = U \setminus U_{\delta_1}$, and

$$\operatorname{Lip} \eta \le \frac{2}{\delta_2 - \delta_1} \chi_{V \cap W}.$$

Now, for $v \in N^{1,1}(V)$ and $w \in N^{1,1}(W)$ we set $u = \eta v + (1 - \eta) w$. With the aid of Lemma 2.15 we can see that $u \in N^{1,1}(U)$ and

$$\int_U g_u \, d\mu \le \int_V g_v \, d\mu + \int_W g_w \, d\mu + \frac{2}{\delta_2 - \delta_1} \int_{V \cap W} |v - w| \, d\mu. \tag{3.1.1}$$

Furthermore, whenever $h \in L^1(U)$, we can write $h = \eta h + (1 - \eta) h$ to see that

$$\int_U |u - h| \, d\mu \le \int_V |v - h| \, d\mu + \int_W |w - h| \, d\mu. \tag{3.1.2}$$

Now, we take v_k from $N^{1,1}(V)$ such that $v_k \to f$ in $L^1(V)$ and $\lim_{k\to\infty} \int_V g_{v_k} \, d\mu = \|Df\|(V)$, and take $w_k \in N^{1,1}(W)$ analogously. We then follow through by stitching together v_k and w_k into the function u_k as prescribed above. By (3.1.2) with $h = f$, we have that

$$\int_U |f - u_k| \, d\mu \le \int_V |v_k - f| \, d\mu + \int_W |w_k - f| \, d\mu \to 0 \text{ as } k \to \infty.$$

It follows from (3.1.1) and the fact $\int_{V\cap W}|v_k - w_k|\,dmu \to 0$ as $k \to \infty$ that

$$\|Df\|(U) \le \liminf_{k\to\infty} \int_U g_{u_k}\,d\mu \le \|Df\|(V) + \|Df\|(W).$$

Remembering again that the closure of V is a compact subset of U, we see that

$$\|Df\|(U) \le \sup\{\|Df\|(V) : V \text{ is open in } X, \overline{V} \text{ is compact subset of } U\} + \|Df\|(U \setminus \overline{U_{\delta_2}}).$$

So now it suffices to prove that

$$\lim_{\delta\to 0^+} \|Df\|(U \setminus \overline{U_\delta}) = 0. \tag{3.1.3}$$

To prove this, we note first that the above limit exists as $\|Df\|(U \setminus \overline{U_\delta})$ decreases as δ decreases. We fix a strictly monotone decreasing sequence of real numbers δ_k with $\lim_{k\to\infty}\delta_k = 0$, and for $k \ge 2$ we set $V_k := U_{\delta_{2k-3}} \setminus \overline{U_{\delta_{2k}}}$. (*It helps to sketch a picture here!*) Observe that the family $\{V_{2k}\}_k$ is a pairwise disjoint family of open subsets of X and that the family $\{V_{2k+1}\}_k$ is also a pairwise disjoint family of open subsets of X.

By Lemma 3.3, we know that

$$\infty > \|Df\|(U) \ge \|Df\|(\bigcup_{k\ge 1} V_{2k}) = \sum_{k=1}^{\infty} \|Df\|(V_{2k}),$$

and

$$\infty > \|Df\|(U) \ge \|Df\|(\bigcup_{k\ge 1} V_{2k+1}) = \sum_{k=1}^{\infty} \|Df\|(V_{2k+1}).$$

It follows that for $\varepsilon > 0$ there is some positive integer $k_\varepsilon \ge 2$ such that

$$\sum_{k=k_\varepsilon}^{\infty} \|Df\|(V_{2k}) + \sum_{k=k_\varepsilon}^{\infty} \|Df\|(V_{2k+1}) < \varepsilon.$$

Now we stitch together approximations on V_{2k} to approximations on V_{2k+1}, and from there to V_{2k+2} and so on. For each k we choose a "stitching function" η_k as a Lipschitz function on $\bigcup_{j=k_\varepsilon}^{k+1} V_j$ such that $0 \le \eta_k \le 1$, with $\eta_k = 1$ on $V_k \setminus V_{k-1}$, $\eta_k = 0$ on $\bigcup_{j=k_\varepsilon}^{k-1} V_j \setminus V_k$, and $g_{\eta_k} \le C_k \chi_{V_k \cap V_{k-1}}$.

Next, for each k we can find $v_{k,j} \in N^{1,1}(V_k)$ such that

$$\int_{V_k} |v_{k,j} - f|\,d\mu \le \frac{2^{-k-j}}{3(1+C_k)}$$

and

$$\int_{V_k} g_{v_{k,j}}\,d\mu \le \|Df\|(V_k) + 2^{-j-k}.$$

We now inductively stitch the functions together. To do so, we first fix $i \in \mathbb{N}$.

Starting with $k = k_\varepsilon$, we stitch $u_{k,i}$ to $u_{k+1,i}$ using $\eta_{k+1} = \eta_{k_\varepsilon+1}$ to obtain $w_{i,k} \in N^{1,1}(V_{k_\varepsilon} \cup V_{k_\varepsilon+1})$ so that we have

$$\int_{V_{k_\varepsilon} \cup V_{k_\varepsilon+1}} |w_{i,k} - f|\,d\mu \le \frac{2^{-i-k_\varepsilon}}{1 + C_{k_\varepsilon+1}}$$

and

$$\int_{V_{k_\varepsilon} \cup V_{k_\varepsilon+1}} g_{w_{i,k}}\,d\mu \le \sum_{j=k_\varepsilon}^{k_\varepsilon+1} \|Df\|(V_j) + 2^{1-i-k_\varepsilon}.$$

Suppose now that for some $k \in \mathbb{N}$ with $k \ge k_\varepsilon + 1$ we have constructed $w_{i,k} \in N^{1,1}(\bigcup_{j=k_\varepsilon}^{k} V_j)$ such that

$$\int_{\bigcup_{j=k_\varepsilon}^{k} V_j} |w_{i,k} - f|\,d\mu \le \sum_{k=k_\varepsilon}^{k} \frac{2^{-i-j}}{1 + C_j}$$

and

$$\int_{\bigcup_{j=k_\varepsilon}^{k} V_j} g_{w_{i,k}}\,d\mu \le \sum_{j=k_\varepsilon}^{k} [\|Df\|(V_j) + 2^{1-i-j}].$$

Then we stitch $u_{k+1,i}$ to $w_{i,k}$ using η_{k+1} to obtain $w_{i,k+1}$ satisfying inequalities analogous to the above two. Note that $w_{i,k+1} = w_{i,k-1}$ on V_{k-1} for $k \ge k_\varepsilon + 2$. Thus, in the limit, we obtain a function $w_i = \lim_k w_{i,k} \in N^{1,1}(\bigcup_{k=k_\varepsilon}^{\infty} V_k)$ satisfying

$$\int_{\bigcup_{j=k_\varepsilon}^{\infty} V_j} |w_i - f|\,d\mu \le \sum_{k=k_\varepsilon}^{k} \frac{2^{-i-j}}{1 + C_j} < 2^{1-i},$$

$$\int_{\bigcup_{j=k_\varepsilon}^{\infty} V_j} g_{w_i}\,d\mu \le \sum_{j=k_\varepsilon}^{\infty} \|Df\|(V_j) + 2^{2-i} < \varepsilon + 2^{2-i}.$$

From the first of the above two inequalities, we see that $w_i \to f$ in $L^1(\bigcup_{j=k_\varepsilon}^\infty V_j)$ as $i \to \infty$, and so from the second of the above two inequalities we obtain

$$\|Df\|(\bigcup_{j=k_\varepsilon}^\infty V_j) = \|Df\|(U \setminus \overline{U_{\delta_{k_\varepsilon}}}) \le \liminf_{i\to\infty} \int_{\bigcup_{j=k_\varepsilon}^\infty V_j} g_{w_i}\, d\mu \le \varepsilon.$$

This last inequality above tells us that the claim we set out to prove, namely

$$\lim_{\delta\to 0^+} \|Df\|(U \setminus \overline{U_\delta}) = 0.$$

this completes the proof of Condition 5.

Proof of Condition 3 By Condition 5 (which we have now proved above, so no circular argument here!), for each $\varepsilon > 0$ we can find relatively compact open subsets $U_1' \Subset U_1$ and $U_2' \Subset U_2$ such that $\|Df\|(U_1 \cup U_2) \le \|Df\|(U_1' \cup U_2') + \varepsilon$. We then choose a Lipschitz "stitching function" η on X such that $0 \le \eta \le 1$ on X, $\eta = 1$ on U_1', $\eta = 0$ on $X \setminus U_1$, and

$$g_\eta \le \frac{1}{C_{U_1,U_1'}} \chi_{U_1\setminus U_1'}.$$

For $u_1 \in N^{1,1}(U_1)$ and $u_2 \in N^{1,1}(U_2)$, we obtain the stitched function $w = \eta u_1 + (1-\eta)u_2$ and note that $w \in N^{1,1}(U_1' \cup U_2')$. Observe that we cannot in general have $w \in N^{1,1}(U_1 \cup U_2)$ as w is not defined in $U_1 \setminus (U_1' \cup U_2)$ because $1-\eta$ is non-vanishing there and u_2 is not defined there, for example. Then we have

$$\int_{U_1'\cup U_2'} g_w\, d\mu \le \int_{U_1} g_{u_1}\, d\mu + \int_{U_2} g_2\, d\mu + \frac{1}{C_{U_1,U_1'}} \int_{U_1\cap U_2} |u_1 - u_2|\, d\mu$$

and

$$\int_{U_1'\cup U_2'} |w - f|\, d\mu \le \int_{U_1} |u_1 - f|\, d\mu + \int_{U_2} |u_2 - f|\, d\mu.$$

As before, choosing u_{1k} to be the optimal approximating sequence for f on U_1 and $u_{2,k}$ correspondingly for f on U_2, we see from the first of the above two inequalities that the stitched sequence w_k approximates f on $U_1' \cup U_2'$. Therefore we obtain

$$\begin{aligned}\|Df\|(U_1 \cup U_2) &\le \varepsilon + \|Df\|(U_1' \cup U_2') \le \varepsilon + \liminf_{k\to\infty} \int_{U_1\cup U_2} g_{w_k}\, d\mu \\ &\le \|Df\|(U_1) + \|Df\|(U_2) + \varepsilon.\end{aligned}$$

Letting $\varepsilon \to 0$ now gives the desired inequality (3).

□

3.2 Sets of Finite Perimeter

We now consider BV functions that arise as characteristic functions of subsets of X.

Definition 3.7 Let $E \subset X$ be a measurable set. We say that E is of finite perimeter in X if $\chi_E \in BV(X)$. If E is of finite perimeter, then for $F \subset X$ we set $P(E, F) := \|D\chi_E\|(F)$.

Note that if E is of finite perimeter, then $P(E, \text{int}(E)) = 0$ and $P(E, \text{int}(X \setminus E)) = 0$, that is, the measure $P(E, \cdot)$ lives inside ∂E. However, in general $P(E, \cdot)$ lives in a much smaller subset of ∂E, as the following example of enlarged rationals shows.

Example 3.8 Let the countable dense set $\mathbb{Q} \times \mathbb{Q} \subset \mathbb{R}^2$ be enumerated as $\{q_i\}_{i\in\mathbb{N}}$. Fixing $\varepsilon > 0$, we set

$$E = \bigcup_{i\in\mathbb{N}} B(q_i, 2^{-i}\varepsilon).$$

Then subadditivity of Lebesgue 2-dimensional measure $\mathcal{L}^2$, together with our knowledge of measures of disks of radius R in the plane, shows that $\mathcal{L}^2(E) \le 4\pi\varepsilon^2$. For $u_k = \chi_{\bigcup_{j=1}^k B(q_j, 2^{-j}\varepsilon)}$, we see that $u_k \to \chi_E$ in $L^1(\mathbb{R}^2)$ and that

$$\|Du_k\|(\mathbb{R}^2) \le \sum_{j=1}^{k} P(B(q_j, 2^{-j}\varepsilon) = 2\pi\varepsilon \sum_{j=1}^{k} 2^{-j} \le 4\pi\varepsilon < \infty.$$

It follows that E is of finite perimeter, with $P(E, \mathbb{R}^2) \le 4\pi\varepsilon$.

Note that $\partial E = \mathbb{R}^2 \setminus E$, and this set has $\mathcal{L}^2$-measure infinite, but $P(E, \cdot)$ lives on a set of σ-finite 1-dimensional Hausdorff measure subset of ∂E, which in turn has $\mathcal{L}^2$-measure zero.

The following *coarea formula* tells us that to understand BV functions one should understand sets of finite perimeter.

Lemma 3.9 (Co-Area Formula) *If $u \in BV(X)$ and $A \subset X$ is a Borel set, then*

$$\|Du\|(A) = \int_{\mathbb{R}} P(E_t, A)\, dt,$$

where

$$E_t = \{x \in X : u(x) > t\} \text{ for } t \in \mathbb{R}.$$

Proof We first prove the formula for open sets A.

Suppose first that $u \in BV(X)$ with $\|Du\|(A) < \infty$. For $s \in \mathbb{R}$ we set $E_s := \{x \in X : u(x) > s\}$. Consider the function $m : \mathbb{R} \to \mathbb{R}$ given by

$$m(t) = \|Du\|(A \cap E_t).$$

Then m is a monotone decreasing function, and hence is differentiable almost everywhere. Let $t \in \mathbb{R}$ such that $m'(t)$ exists. Then

$$|m'(t)| = \lim_{h \to 0^+} \frac{\|Du\|(A \cap E_t \setminus E_{t+h})}{h}.$$

Note that the functions

$$u_{t,h} := \frac{\max\{t, \min\{t+h, u\}\} - t}{h}$$

converge in $L^1(X)$ to χ_{E_t} as $h \to 0^+$. It follows that as A is open,

$$P(E_t, A) \le \liminf_{h \to 0^+} \|Du_{t,h}\|(A) = \liminf_{h \to 0^+} \frac{\|Du\|(A \cap E_t \setminus E_{t+h})}{h} = |m'(t)|.$$

Note also that by this lower semicontinuity of BV energy, $t \mapsto P(E_t, A)$ is a lower semicontinuous function, and hence is measurable; and as it is non-negative, we can talk about its integral, whether that integral is finite or not. Therefore, by the fundamental theorem of calculus for monotone functions,

$$\int_{\mathbb{R}} P(E_t, A)\, dt \le \int_{\mathbb{R}} |m'(t)|\, dt \le \lim_{s,\tau \to \infty} m(s) - m(-\tau) = \|Du\|(A).$$

The above in particular tells us that if $u \in BV(X)$ then almost all of its superlevel sets E_t have finite perimeter. If u is not a BV function on A, then $\|Du\|(A) = \infty$, and hence we also have

$$\int_{\mathbb{R}} P(E_t, A)\, dt \le \|Du\|(A). \tag{3.2.1}$$

In particular, it also follows that $\int_{\mathbb{R}} P(E_t, A)\, dt < \infty$ if $u \in BV(X)$.

We still continue to assume that A is open, and prove the reverse of the above inequality. If $\int_{\mathbb{R}} P(E_t, A)\, dt = \infty$, then trivially

$$\|Du\|(A) \le \int_{\mathbb{R}} P(E_t, A)\, dt.$$

So we may assume without loss of generality that $\int_{\mathbb{R}} P(E_t, A)\, dt$ is finite. Note also by the truncation property of upper gradients (that is, $g\chi_{\{a \le u \le b\}}$ is a p-weak upper gradient of the truncated function $\max\{a, \min\{u, b\}\}$), filtered down to the level of the measure $|\nabla u|$, we have that

$$\|Du\|(A) = \lim_{s,\tau\to\infty} \|Du_{s,\tau}\|(A),$$

where $u_{s,\tau} = \max\{-\tau, \min\{u, s\}\}$. So without loss of generality we may assume that $0 \le u \le 1$. For positive integers k we can divide $[0, 1]$ into k equal sub-intervals $[t_i, t_{i+1}]$, $i = 0, \cdots, k$ with $t_{i+1} - t_i = 1/k$. Then we can find $\rho_{k,i} \in (t_i, t_{i+1})$ such that

$$\frac{1}{k} P(E_{\rho_{k,i}}, A) \le \int_{t_i}^{t_{i+1}} P(E_s, A)\, ds.$$

We set

$$u_k = \sum_{j=1}^{k} \frac{1}{k} \chi_{E_{\rho_{k,i}}}.$$

Then it is an exercise to verify that $u_k \to u$ in $L^1(A)$ as $k \to \infty$, and so

$$\|Du\|(A) \le \liminf_{k\to\infty} \|Du_k\|(A) = \liminf_{k\to\infty} \sum_{j=1}^{k} \frac{1}{k} P(E_{\rho_{k,i}}, A) \le \int_0^1 P(E_s, A)\, ds. \tag{3.2.2}$$

Note now that by the proofs of inequalities (3.2.1) and (3.2.2), if A is an open set then $u \in BV(A)$ if and only if $\int_{\mathbb{R}} P(E_t, A)\, dt$ is finite.

Finally, we remove the requirement that A be open. By the above comment, it suffices to prove this for the case that $u \in BV(X)$. In this case, the maps $A \mapsto \|Du\|(A)$ and $A \mapsto \int_{\mathbb{R}} P(E_t, A)\, dt$ are both Radon measures on X that agree on open subsets of X (that is, they are equal for open A). Hence it follows that they agree on Borel subsets of X. This completes the proof of the coarea formula. □

Definition 3.10 The co-dimension 1 Hausdorff measure $\mathcal{H}(F)$ of a set $F \subset X$ is the number

$$\mathcal{H}(F) = \lim_{\varepsilon\to 0^+} \inf\{\sum_i \frac{\mu(B_i)}{\text{rad}(B_i)} : F \subset \bigcup_{i\in I\subset\mathbb{N}} B_i,\ B_i \text{ are balls in } X \text{ with } \text{rad}(B_i) < \varepsilon\}.$$

Lemma 3.11 *Let μ be doubling on X. That is, there is a constant $C_D \ge 1$ such that whenever B is a ball in X, we have $\mu(2B) \le C_D\, \mu(B)$. If $E \subset X$ is a measurable set such that $\mathcal{H}(\partial E) < \infty$, then E is of finite perimeter.*

Proof For $1 \geq \varepsilon > 0$ we can find a countable cover of ∂E by balls B_i such that $\mathrm{rad}(B_i) < \varepsilon$ and

$$\sum_i \frac{\mu(B_i)}{\mathrm{rad}(B_i)} \leq \mathcal{H}(\partial E) + \varepsilon.$$

For each i we fix a Lipschitz function φ_i on X such that $0 \leq \varphi_i \leq 1$ on X, $\varphi_i = 1$ on B_i, $\varphi_1 = 0$ on $X \setminus 2B_i$, and $g_{\varphi_i} \leq \frac{2}{\mathrm{rad}(B_i)} \chi_{2B_i \setminus B_i}$.

Now we set $u_\varepsilon : X \to \mathbb{R}$ by

$$u_\varepsilon(x) = \sup\{\chi_E(x), \varphi_i(x) : i \in I \subset \mathbb{N}\}.$$

Then $0 \leq u_\varepsilon \leq 1$ on X, $u_\varepsilon = 1$ on E, and

$$g_{u_\varepsilon} \leq \sum_{i \in I} \frac{2}{\mathrm{rad}(B_i)} \chi_{2B_i \setminus B_i}.$$

It follows that

$$\begin{aligned}
\int_X g_{u_\varepsilon} \, d\mu &\leq \sum_{i \in I} \frac{2}{\mathrm{rad}(B_i)} \mu(2B_i \setminus B_i) \\
&\leq \sum_{i \in I} \frac{2}{\mathrm{rad}(B_i)} \mu(2B_i) \\
&\leq C_D \sum_{i \in I} \frac{2}{\mathrm{rad}(B_i)} \mu(B_i) \leq 2C_D[\mathcal{H}(\partial E) + \varepsilon] \leq 2C_D[\mathcal{H}(\partial E) + 1] < \infty.
\end{aligned}$$

Finally, note that u_ε differs from χ_E only on $\bigcup_{i \in I} 2B_i$. Therefore, as each $\mathrm{rad}(B_i) < \varepsilon$,

$$\begin{aligned}
\int_X |u_\varepsilon - \chi_E| \, d\mu \leq \sum_{i \in I} \mu(2B_i) \leq C_D \sum_{i \in I} \frac{\mu(B_i)}{\mathrm{rad}(B_i)} \varepsilon \\
\leq C_D \, \varepsilon \, [\mathcal{H}(\partial E) + \varepsilon] \to 0 \text{ as } \varepsilon \to 0^+.
\end{aligned}$$

Thus we have a sequence of functions $u_\varepsilon \in N^{1,1}(X)$ converging to χ_E in $L^1(X)$, with uniformly bounded $N^{1,1}$-energy $\int_X g_{u_\varepsilon} \, d\mu$. It follows that E is of finite perimeter. □

Asking that $\mathcal{H}(\partial E)$ be finite is overly restrictive, as there are many sets of finite perimeter that do not satisfy this condition. The more natural boundary to consider is the so-called measure-theoretic boundary.

Definition 3.12 Let $E \subset X$ be a measurable set. We say that $x \in \partial E$ is in the *measure-theoretic boundary* $\partial_m E$ of E if

$$\limsup_{r\to 0^+} \frac{\mu(B(x,r)\cap E)}{\mu(B(x,r))} > 0 \qquad \text{and} \qquad \limsup_{r\to 0^+} \frac{\mu(B(x,r)\setminus E)}{\mu(B(x,r))} > 0.$$

Remark 3.13 It turns out that if E is of finite perimeter and μ is doubling and supports a 1-Poincaré inequality (see the next section), then whenever $A \subset X$ is a Borel set,

$$\frac{1}{C}\, P(E, A) \le \mathcal{H}(A \cap \partial_m E) \le C\, P(E, A) < \infty.$$

This result is due to Ambrosio, see [1, 4].

It is a deep theorem of Federer that a measurable subset of a Euclidean space is of finite perimeter if and only if its measure-theoretic boundary has finite $\mathcal{H}$-measure. In the setting of metric measure spaces we lack the knowledge of whether E is of finite perimeter if $\mathcal{H}(\partial_m E)$ is finite. For special classes of doubling metric measure spaces supporting a geometric Semmes pencil of curves [23, 24], this problem has been solved, see [19], but in the wider class of doubling metric measure spaces supporting a 1-Poincaré inequality this problem is still wide open.

3.3 1-Poincaré Inequality and BV

By using the definition of $BV(X)$, we see that if μ supports a 1-Poincaré inequality, then for every $u \in BV(X)$ and every ball $B \subset X$ we have that

$$\fint_B |u - u_B|\, d\mu \le C\, \text{rad}(B)\, \frac{\|Du\|(\lambda B)}{\mu(\lambda B)}.$$

To verify the above inequality, by the lower semicontinuity of the BV energy (see Sect. 3), it suffices to verify the above for $u \in N^{1,1}(X)$ and $\|Du\|(\lambda B)$ replaced with $\fint_{\lambda B} g\, d\mu$ whenever g is a 1-weak upper gradient of u.

Let $u \in BV(X)$. Now that we know that $\|Du\|$ is an outer measure on X, we can obtain a decomposition of this measure into the sum of two measures, $\|Du\|_a$ which is absolutely continuous with respect to μ, and $\|Du\|_s$ which is singular with respect to μ.

Theorem 3.14 *Let $u \in BV(X)$. Then $\|Du\|_s = 0$ if and only if a modification of u on a set of measure zero, also denoted u, satisfies $u \in N^{1,1}(X)$. Furthermore, if μ is doubling on X and supports a 1-Poincaré inequality, then for such u we have that*

$$\|Du\|(X) \le \int_X g_u\, d\mu \le C\, \|Du\|(X).$$

For general X the proof of the first part of the above theorem is more complicated, so we will assume in the proof of the theorem that X is compact.

Proof *(Sketch of Proof)* First, note that we can choose a sequence of functions $u_k \in N^{1,1}(X)$ such that $u_k \to u$ in $L^1(X)$ and

$$\lim_{k\to\infty} \int_X g_{u_k}\, d\mu = \|Df\|(X) < \infty.$$

Such a sequence can be obtained through a Cantor-type diagonalization argument. From the above, it is clear that the Radon measures μ_k given by $d\mu_k = g_{u_k}\, d\mu$ is uniformly bounded in total mass, and so has a weak limit measure (after passing to a subsequence if necessary– we will denote the subsequence also as u_k). Denoting this limit measure as ν, we see that as X is compact, $\nu(X) = \|Df\|(X)$.

If $U \subset X$ is an open set such that $\nu(\partial U) = 0$, then we know from the weak limits of measures that

$$\nu(\overline{U}) = \nu(U) = \lim_{k\to\infty} \int_U g_{u_k}\, d\mu, \quad \nu(X \setminus \overline{U}) = \nu(X \setminus U) = \lim_{k\to\infty} \int_U g_{u_k}\, d\mu.$$

Hence we must have $\|Df\|(U) = \nu(U)$ and $\|Df\|(X \setminus \overline{U}) = \nu(X \setminus \overline{U})$. Finally, we can use the fact that for $\mathcal{L}^1$-almost every $\delta > 0$ we must have $\nu(\partial U_\delta) = 0$, and therefore as $\nu(U)$ and $\|Df\|(U)$ are approximated by $\nu(U_\delta)$ and $\|Df\|(U_\delta)$ respectively, it follows that $\nu(U) = \|Df\|(U)$ for all open sets $U \subset X$. So we conclude that $\nu = \|Df\|$.

As $\|Du\|_s = 0$, it follows that there is some non-negative $g \in L^1(X)$ such that $d\nu = g\, d\mu$, that is g_{u_k} converges weak-$*$ in $L^1(X)$. An appeal to Mazur's lemma completes the proof. □

We also have the following relative isoperimetric inequality.

Lemma 3.15 (Relative Isoperimetric Inequality) *Suppose that μ is doubling and supports a 1-Poincaré inequality. Then whenever $E \subset X$ is a measurable set and B is a ball in X, we have*

$$\frac{\mu(B \cap E)}{\mu(B)} \times \frac{\mu(B \setminus E)}{\mu(B)} \le C\, rad(B)\, \frac{P(E, \lambda B)}{\mu(\lambda B)}.$$

Proof With $u = \chi_E$, we have

$$u_B = \frac{\mu(B \cap E)}{\mu(B)}.$$

Moreover,

$$1 - u_B = \frac{\mu(B \setminus E)}{\mu(B)}.$$

Therefore

$$\fint_B |u-u_B|\,d\mu = \frac{1}{\mu(B)}\,[\mu(E\cap B)(1-u_B)+\mu(B\setminus E)u_B] = \frac{\mu(B\cap E)}{\mu(B)}\times\frac{\mu(B\setminus E)}{\mu(B)}.$$

Now an application of the 1-Poincaré inequality to u yields the desired claim. □

References

1. L. Ambrosio, Fine properties of sets of finite perimeter in doubling metric measure spaces, calculus of variations, nonsmooth analysis and related topics. Set Valued Anal. **10**(2–3), 111–128 (2002)
2. L. Ambrosio, S. Di Marino, Equivalent definitions of BV space and of total variation on metric measure spaces. J. Funct. Anal. **266**, 4150–4188 (2014)
3. L. Ambrosio, N. Gigli, G. Savaré, Density of Lipschitz functions and equivalence of weak gradients in metric measure spaces. Rev. Mat. Iberoamericana **29**(3), 969–996 (2013)
4. L. Ambrosio, M. Miranda Jr., D. Pallara, Special functions of bounded variation in doubling metric measure spaces. Calculus of variations: topics from the mathematical heritage of E. De Giorgi, 1–45. Quad. Mat., vol. 14, Dept. Math., Seconda Univ. Napoli, Caserta, 2004
5. L. Ambrosio, P. Tilli, Topics on analysis in metric spaces. *Oxford Lecture Series in Mathematics and Its Applications,* vol. 25, viii+133 (2003)
6. L. Ambrosio, N. Fusco, D. Pallara, Functions of bounded variation and free discontinuity problems. *Oxford Mathematical Monographs* (The Clarendon Press, Oxford University Press, New York 2000)
7. G. Aronsson, Representation of a p-harmonic function near a critical point in the plane. Manuscripta Math. **66**(1), 73–95 (1989)
8. M. Bertalmo, V. Caselles, G. Haro, G. Sapiro, PDE-based image and surface inpainting. *Handbook of Mathematical Models in Computer Vision* (Springer, New York, 2006), pp. 33–61
9. A. Björn, J. Björn, Nonlinear potential theory on metric spaces. *EMS Tracts Mathematics*, vol. 17 (European Math. Soc., Zurich, 2011)
10. J. Cheeger, Differentiability of Lipschitz Functions on metric measure spaces. Geom. Funct. Anal. **9**, 428–517 (1999)
11. E. De Giorgi, G. Letta, Une notion générale de convergence faible pour des fonctions croissantes d'ensemble, Annali della Scuola Normale Superiore di Pisa - Classe di Scienze **4**(1), 61–99 (1977)
12. L.C. Evans, R. Gariepy, Measure theory and fine properties of functions. *Studies in Advanced Mathematics* (CRC Press, 1991), 288pp.
13. P. Hajłasz, P. Koskela, Sobolev met Poincaré. Sobolev met Poincaré. Mem. Am. Math. Soc. **145**(688), x+101 pp. (2000)
14. J. Heinonen, *Lectures on Analysis on Metric Spaces* (Springer, 2001)
15. J. Heinonen, P. Koskela, Quasiconformal maps in metric spaces with controlled geometry. Acta Math. **181**, 1–61 (1998)
16. J. Heinonen, T. Kilpeläinen, O. Martio, *Nonlinear Potential Theory of Degenerate Elliptic Equations*. Unabridged republication of the 1993 original (Dover Publications, Mineola, NY, 2006), xii+404 pp.
17. J. Heinonen, P. Koskela, N. Shanmugalingam, J. Tyson, *Sobolev Spaces on Metric Measure Spaces: An Approach Based on Upper Gradients*. New Mathematical Monographs series (Cambridge University Press, 2015)
18. T. Iwaniec, J. Manfredi, Regularity of p-harmonic functions on the plane. Rev. Mat. Iberoamericana **5**(1-2), 1–19 (1989)

19. R. Korte, P. Lahti, N. Shanmugalingam, Semmes family of curves and a characterization of functions of bounded variation in terms of curves. Calc. Var. Partial Differ. Equ. **54**(2), 1393–1424 (2015)
20. P. Koskela, P. McManus, Quasiconformal mappings and Sobolev spaces. Studia Math. **131**(1), 1–17 (1998)
21. S. Keith, X. Zhong, The Poincaré inequality is an open ended condition. Ann. Math. **167**(2), 575–599 (2008)
22. M. Miranda jr., Functions of bounded variation on "good? metric spaces. J. Math. Pures Appl. **82**, 975–1004 (2003)
23. S. Semmes, Finding curves on general spaces through quantitative topology, with applications to Sobolev and Poincaré inequalities. Selecta Math. New Series **2**(2), 155–295 (1996)
24. S. Semmes, *Some Novel Types of Fractal Geometry* (Oxford Science Publications, 2001)
25. N. Shanmugalingam, Newtonian spaces: An extension of Sobolev spaces to metric measure spaces. Rev. Mat. Iberoamericana **16**, 243–279 (2000)

Index

L. Ambrosio et al. (eds.), *New Trends on Analysis and Geometry in Metric Spaces*,
C.I.M.E. Foundation Subseries 2296, https://doi.org/10.1007/978-3-030-84141-6

Editorial Policy

1. Lecture Notes aim to report new developments in all areas of mathematics and their applications – quickly, informally and at a high level. Mathematical texts analysing new developments in modelling and numerical simulation are welcome.

   Manuscripts should be reasonably self-contained and rounded off. Thus they may, and often will, present not only results of the author but also related work by other people. They may be based on specialised lecture courses. Furthermore, the manuscripts should provide sufficient motivation, examples and applications. This clearly distinguishes Lecture Notes from journal articles or technical reports which normally are very concise. Articles intended for a journal but too long to be accepted by most journals, usually do not have this "lecture notes" character. For similar reasons it is unusual for doctoral theses to be accepted for the Lecture Notes series, though habilitation theses may be appropriate.

2. Besides monographs, multi-author manuscripts resulting from SUMMER SCHOOLS or similar INTENSIVE COURSES are welcome, provided their objective was held to present an active mathematical topic to an audience at the beginning or intermediate graduate level (a list of participants should be provided).

   The resulting manuscript should not be just a collection of course notes, but should require advance planning and coordination among the main lecturers. The subject matter should dictate the structure of the book. This structure should be motivated and explained in a scientific introduction, and the notation, references, index and formulation of results should be, if possible, unified by the editors. Each contribution should have an abstract and an introduction referring to the other contributions. In other words, more preparatory work must go into a multi-authored volume than simply assembling a disparate collection of papers, communicated at the event.

3. Manuscripts should be submitted either online at www.editorialmanager.com/lnm to Springer's mathematics editorial in Heidelberg, or electronically to one of the series editors. Authors should be aware that incomplete or insufficiently close-to-final manuscripts almost always result in longer refereeing times and nevertheless unclear referees' recommendations, making further refereeing of a final draft necessary. The strict minimum amount of material that will be considered should include a detailed outline describing the planned contents of each chapter, a bibliography and several sample chapters. Parallel submission of a manuscript to another publisher while under consideration for LNM is not acceptable and can lead to rejection.

4. In general, **monographs** will be sent out to at least 2 external referees for evaluation.

   A final decision to publish can be made only on the basis of the complete manuscript, however a refereeing process leading to a preliminary decision can be based on a pre-final or incomplete manuscript.

   Volume Editors of **multi-author works** are expected to arrange for the refereeing, to the usual scientific standards, of the individual contributions. If the resulting reports can be forwarded to the LNM Editorial Board, this is very helpful. If no reports are forwarded or if other questions remain unclear in respect of homogeneity etc, the series editors may wish to consult external referees for an overall evaluation of the volume.

5. Manuscripts should in general be submitted in English. Final manuscripts should contain at least 100 pages of mathematical text and should always include

   - a table of contents;
   - an informative introduction, with adequate motivation and perhaps some historical remarks: it should be accessible to a reader not intimately familiar with the topic treated;
   - a subject index: as a rule this is genuinely helpful for the reader.
   - For evaluation purposes, manuscripts should be submitted as pdf files.

6. Careful preparation of the manuscripts will help keep production time short besides ensuring satisfactory appearance of the finished book in print and online. After acceptance of the manuscript authors will be asked to prepare the final LaTeX source files (see LaTeX templates online: https://www.springer.com/gb/authors-editors/book-authors-editors/manuscriptpreparation/5636) plus the corresponding pdf- or zipped ps-file. The LaTeX source files are essential for producing the full-text online version of the book, see http://link.springer.com/bookseries/304 for the existing online volumes of LNM). The technical production of a Lecture Notes volume takes approximately 12 weeks. Additional instructions, if necessary, are available on request from lnm@springer.com.

7. Authors receive a total of 30 free copies of their volume and free access to their book on SpringerLink, but no royalties. They are entitled to a discount of 33.3 % on the price of Springer books purchased for their personal use, if ordering directly from Springer.

8. Commitment to publish is made by a *Publishing Agreement*; contributing authors of multiauthor books are requested to sign a *Consent to Publish form*. Springer-Verlag registers the copyright for each volume. Authors are free to reuse material contained in their LNM volumes in later publications: a brief written (or e-mail) request for formal permission is sufficient.

Addresses:

Professor Jean-Michel Morel, CMLA, École Normale Supérieure de Cachan, France
E-mail: moreljeanmichel@gmail.com

Professor Bernard Teissier, Equipe Géométrie et Dynamique,
Institut de Mathématiques de Jussieu – Paris Rive Gauche, Paris, France
E-mail: bernard.teissier@imj-prg.fr

Springer: Ute McCrory, Mathematics, Heidelberg, Germany,
E-mail: lnm@springer.com

Printed in the United States
by Baker & Taylor Publisher Services